A Land Imperiled

A Land Imperiled

The Declining Health of the Southern Appalachian Bioregion

John Nolt

Outdoor Tennessee Series
Jim Casada, Series Editor

The University of Tennessee Press
Knoxville

The Outdoor Tennessee Series covers a wide range of topics of interest to the general reader, including titles on the flora and fauna, the varied recreational activities, and the rich history of outdoor Tennessee. With a keen appreciation of the importance of protecting our state's natural resources and beauty, the University of Tennessee Press intends the series to emphasize environmental awareness and conservation.

Library of Congress Cataloging-in-Publication Data

Nolt, John Eric.
A land imperiled: the declining health of the southern appalachian bioregion / John Nolt.— 1st ed.
    p. cm.—(Outdoor Tennessee series)
Includes bibliographical references and index.

ISBN 1-57233-326-X (pbk.: alk. paper)

1. Appalachian Region, Southern—Environmental conditions.
2. Biogeography—Appalachian Region, Southern.
I. Title. II. Series.

GE155.A58N65 2005
333.95'137'0975—dc22   2004028991

Contents

Illustrations

Figures

Maps

Tables

Preface

THE SOUTHErn APPALACHIAN BIOregion

This book is an extensive rewriting and updating of *What Have We Done? The Foundation for Global Sustainability's State of the Bioregion Report for the Upper Tennessee Valley and Southern Appalachian Mountains* (Earth Knows Publications, 1997). That book saw use as a reference work, a high school and college textbook, and a general assessment of bioregional health for policymakers and general reader. But the region is changing rapidly, and much of the information it contains is already out of date—hence the need for this new volume.

Like the original, this book's aim is to assess the health (in a very broad sense) of the Southern Appalachian bioregion. The relevant concept of health is explained in the introduction. A bioregion is a place with a unique ecology and (at least historically) a unique culture. Bioregions are defined, not by current political boundaries, but by boundaries of ecological and cultural significance. Since, both here and elsewhere, watersheds play a central role in ecology and in human settlement, they are typically the most appropriate bioregional units.

The Southern Appalachian bioregion, as understood in this book, is the watershed of the upper Tennessee valley. With regard to political boundaries, it occupies portions of four states: Tennessee, North Carolina, Virginia, and Georgia (see map 1). Though watershed boundaries do not correspond exactly to county lines, large portions of the following counties are included within the region covered by this study:

> **TENNESSEE.** Anderson, Bledsoe, Blount, Bradley, Campbell, Carter, Claiborne, Cocke, Cumberland, Grainger, Greene, Hamblen, Hamilton, Hancock, Hawkins, Jefferson, Johnson, Knox, Loudon, Marion, McMinn, Meigs, Monroe, Morgan, Polk, Rhea, Roane, Sequatchie, Sevier, Sullivan, Unicoi, Union, Washington.

NORTH CAROLINA. Avery, Buncombe, Cherokee, Clay, Graham, Haywood, Henderson, Jackson, Macon, Madison, Mitchell, Swain, Transylvania, Watauga, Yancey.

VIRGINIA. Lee, Russell, Scott, Smyth, Tazewell, Washington, Wise.

GEORGIA. Catoosa, Dade, Fannin, Towns, Union, Walker.

The Great Smoky Mountains National Park is wholly included within the bioregion.

Since our definition of the bioregion does not precisely correspond to anyone else's, we have often had to make use of data for regions that only partially overlap our bioregion. For example, we make frequent use of data for the whole state of Tennessee, the state most central to the bioregion, or for the whole Tennessee valley, or for the whole Southern Appalachian Assessment region. In such cases, we always specify the actual area covered; and, where we have further information, we note it, in an attempt to correct for inaccuracies that may be introduced by geographic disparity.

We regret that what we have to report is often disheartening. But to be conscious of dysfunction is a necessary step toward cure. Some people still insist that there are no serious environmental problems. Others, while acknowledging some general problems, seek in particular instances to reassure us that "the impact is so small it's not worth bothering about" or "it's still not good, but we're doing much better than we were twenty years ago," or "don't worry, we're only cutting 20 percent of the forest here." Taken in isolation, each of the problems represented by such remarks may in fact be insignificant. But the mistake is to take them in isolation. This book aims to view them synoptically in order to gauge their cumulative significance.

Many dedicated people provided valuable assistance in creating this volume. The research of many of those acknowledged in the preface to *What Have We Done?* served as background for much of the work done for this updated version. Where old sources have not been superceded, they have sometimes been reused here.

We are also indebted to the following reviewers for careful comments and corrections: Dr. Jonathan Scurlock, adjunct professor, Department of Ecology and Evolutionary Biology, University of Tennessee, Knoxville, for

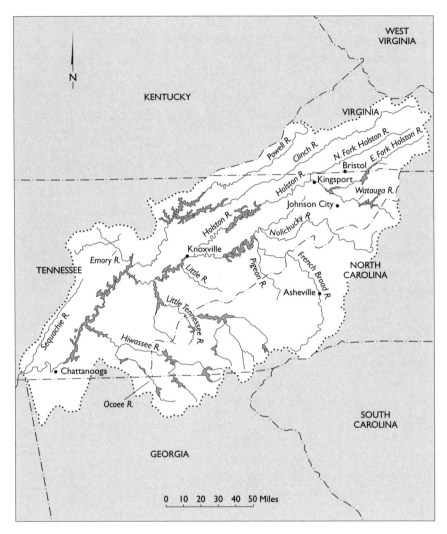

Map. 1. *The Southern Appalachian bioregion.*

extensive corrections on chapter 7; Thomas Fraser, features and news editor of *The Hellbender Press* for a very thorough review of chapter 4; Professor Scott Schlarbaum, Department of Forestry, Wildlife and Fisheries, University of Tennessee, Knoxville, for a very helpful review of the final section of chapter 4; Cielo Sand of the Dogwood Alliance and Doug Murray of ForestWatch for help with the material on forests.

Thanks are due also to the Foundation for Global Sustainability for supporting the research of Keith Bustos and Sarah Kenehan.

Mignon and Wolf Naegeli collaborated on producing many superb photographs.

All proceeds from this book support the good work of the Foundation for Global Sustainability and Narrow Ridge Earth Literacy Center.

John Nolt
University of Tennessee, Knoxville
Summer 2003

Introduction

THE VALUE OF HEALTH

John Nolt

Only by restoring the broken connections can we be healed. Connection *is* health.

—Wendell Berry, *The Unsettling of America,* 138

The Concept of Health

Our concern is with the health of Southern Appalachia. The term "health" is intended in its broadest sense—a sense derived from its Old English root "hāl," which is likewise the source of the modern terms "heal," "hale," "whole," "wholesome"—and, significantly, also "hallow," and "holy."[1] Together, this family of words connotes the design-specific integrity of the whole—of Creation itself. Our concern, therefore, is with the integral organization not just of the human body, but also of the human mind and spirit—and not just of what is human, but of the entire system of life in the Southern Appalachian bioregion, both now and in the near and distant future.

Health so conceived is an ideal, presupposing notions of design or purpose and of optimal—or at least adequate—function with respect to purpose. Different sorts of beings have different purposes, hence different standards of health. Growth, for example, is essential to the health of children but not of adults; photosynthesis is essential to the health of green plants but not of fungi. Yet, for any living organism or system, health is the ability to accomplish the functions established by its particular evolutionary design. Humans, for example, evolved specifically as walkers, talkers, thinkers, and tool-users, but we also share many functions with other organisms: respiration, reproduction, resistance to infection, elimination of waste, procurement of nutrients, and so on. The ability to perform all such functions well is the ideal of *human* health. For oak trees and opossums, too, there are standards of health, different in detail but obviously related.

The concept of health applies not just to organisms but also to various other levels of biological organization. It is just as meaningful, or nearly so, to speak of the health of a cell, an organ, a population, an ecosystem, or a bioregion as it is to speak of the health of an organism.[2] What is denoted in each case is the wholeness—that is, the functional integrity—of the relevant entity.

Because the various levels of biological organization are interdependent, the health of any one component cannot be entirely separated from the health of the systems in which it is contained and those which it contains. A healthy body cannot be made of unhealthy cells, nor (except for certain "weedy" species) is a population likely to thrive if it inhabits a degraded ecosystem. Because of this interdependence, opportunities for health at all levels are maximized by health writ large—the harmonious integration of all life.

Total harmony, however, is a romantic fantasy. Most organisms live only by the deaths of others. Health writ large does not, therefore, imply an end to predation, parasitism, decay, disease, or even extinction. But it does imply limits to them all. The main contention of this book is that in the Southern Appalachian bioregion, as in much of the rest of the world, those limits are being exceeded.

The concept of health, like any other concept, is a means for organizing thought and focusing attention. Useful concepts are productive of insight; they help us think in ways that enhance appreciation and understanding. "Health" in the sense just defined is useful in this way. One who employs the concept to categorize, for example, woodlands can distinguish, perhaps at a glance, stands of forest that are rich and wholesome from those that are poor, injured, diseased, or degraded—and appreciate the wide significance of the distinction. The broad notion of health also suggests helpful analogies. It is illuminating to observe, for example, that wetlands are to a landscape as kidneys are to a body: they are organs for removing contaminants and wastes from the fluid (water) that flows through the landscape's "circulatory system" (its creeks and streams). Similarly, to see that the spread of invasive organisms in a forest is analogous to the spread of infectious organisms in the human body frames the problem of invasives in a way that both reveals its significance and suggests measures for prevention and cure. And to reveal significance is already an accomplishment; the bland fact that TVA is pumping oxygen into Douglas Reservoir, for example, becomes striking when we realize that this portion of what used to be a healthy, self-aerating river is now so sick that it must be kept on artificial respiration.

The fact that much of our land is unhealthy is not always easy to appreciate, partly because health is often confused with the condition of being average or normal. What we are used to seldom strikes us as dysfunctional—and we are used to a disintegrating environment. Health, however, is an ideal, not a kind of average. It would be no small blunder, for example, to seek the elements of a healthful diet by examining what the *average* American eats. (A healthful diet is far from what is now average; see chapter 6.) It would be a blunder of similar magnitude to take what today constitutes an "average" environment as a baseline of environmental health; for the Southern Appalachian environment today falls short of functional integrity, not only for the region's flora and fauna, but for its people. As with diet, what constitutes the health of this land and its people are far from what is now normal.

Consequently, the first step in characterizing a bioregion's health is to understand how its living systems evolved to function, not how they happen to be functioning (or malfunctioning) today. Thus we begin, in chapter 1, with a survey of the region's biological and cultural history.

We do not imagine, of course, that health can only be realized in the aboriginal life of some idyllic past. The ideal of health is dynamic, evolving as the systems to which it applies evolve—and, besides, the past has been neither idyllic nor always healthy. (This is not the first time, for example, that our region has seen catastrophic climate changes or rapid extinctions.) The dynamism of the ideal of health is readily illustrated by historical examples. Some of the Paleo-Indian groups of Southern Appalachia, for instance, maintained a healthy relationship with the land by fishing, hunting, or farming out an area and then moving elsewhere, allowing the depleted ecosystems to regenerate themselves. With larger populations and more powerful technologies, however, such behavior is utterly dysfunctional (witness the role of the westward march of settlement in creating the dust bowl). Thus a pattern of life that is healthy at low-population densities may become unhealthy as technology develops and densities increase.

It is therefore reasonable to assume that if we ever return to health, it will be very different from the forms of health that evolved here in the past. Nevertheless, Southern Appalachia's history provides a long, deep perspective from which to understand what overall functional integrity (however dynamic) has meant for the region's biotic systems. From this perspective, as subsequent chapters will show, the current state of life and the land appears pathological. That recognition is useful, for detection of pathology is the essential first step toward cure.

Valuing Health Writ Large

But why *value* health in so wide a sense? Nearly all of us, of course, care about our own health and the health of those we love. Some worry about the health of broader *human* communities. But few are much concerned with the health of the land or of the community of all living things. Why should we be?

One reason—obvious and alluded to already, but not wholly adequate—is that our personal health is inextricable from the health of the systems that sustain us. We cannot expect to be healthy, for example, unless the air we breathe is pure, our diet is healthful, and we exercise regularly. Yet our air is polluted (see chapter 2), our diet is often poor (see chapter 6), and our propensity to design cities and suburbs around the needs of automobiles rather than people discourages biking and walking (see chapters 5 and 9). Consequently, many people are afflicted with obesity, heart disease, lung disease, diabetes, asthma, or cancer. This book's subsequent chapters will detail various ways in which the health of the macrocosm (including air, water, soil, biological systems, and the built environment) affects the health of the microcosm (us). Given that we care about ourselves, the more deeply we understand these connections, the more widely we will value the health of the systems that supply our needs.

Yet because not all of life involves us directly, interest in our own health is ultimately too narrow a basis for valuing health writ large. It is highly unlikely, for example, that anyone except a few naturalists would be much affected by the extinction of the spruce-fir moss spider, a small tarantula now endangered by the demise (from imported insects and air pollution) of the high-elevation spruce-fir forests of Southern Appalachia. Why should we be concerned with this species? Or, to take a more human case, why should those of us who are past middle age care about global climate change? Though people a generation or two hence are likely to be much discomforted by it, we'll be dead by then. What is it to us? We might, of course, worry about our children or grandchildren and so concern ourselves with climate change for their sakes, but that would still leave much of the future unconsidered. The high-level nuclear waste that TVA produces in generating our electricity remains dangerous for tens of thousands of years. Somewhere over that vast expanse of time, it is likely to do great harm. But even our children and grandchildren will most likely be dead by then. So why should *we* care?

For many people, the ultimate answer to all these questions is religious. Love for the Creator seems to entail respect for the creation—all of it—and, consequently, a desire for its wholeness, its wholesomeness, its hallowedness . . . its *health*.[3]

But because religious beliefs are varied, contentious, and grounded in divergent faiths, there is need, if many are to cooperate in the restoration of a health larger than our own, to articulate reasons for caring that can appeal to people of any faith—or none.

Such reasons emerge when we reflect carefully and deeply on moral principles that most people already accept. Simple consistency, combined with a stolid resistance to prejudice, can take us a long way toward valuing health writ large. This point is probably best appreciated in historical perspective. Moral understanding has, roughly speaking, progressed from bigotry and provincialism to widened objectivity and care. One measure of this progress is expansion of the moral community—the group of those whose welfare we deem worthy of moral consideration.[4]

The earliest moralities were tribal; their moral community was, in its human dimension, essentially just the tribe. (For many tribal peoples the moral community had nonhuman dimensions as well, encompassing a panoply of gods, spirits, animals, and even plants. Over time, however, these nonhuman dimensions tended to fade, contributing less and less to the development of subsequent moralities.) So far as human beings were concerned, if you were a part of the tribe, you counted morally; if not, you did not—and from the tribe's point of view there was no moral objection to raping, robbing, or murdering you. Consider, for example, how in the book of Joshua, God orders the Israelites to invade the land of Canaan, seize its cities, and put its inhabitants—men, women, and children (in some cases, even livestock)—to the sword. No one in the narrative questions the propriety of this slaughter of people who were minding their own business, or at worst trying to defend their homes. The moral assumptions are tribal, and the victims are not members of the victorious tribes. Morally, they count for nothing.

As tribes drew together into federations and civilizations expanded from city-states to nation-states and empires, law and moral understanding followed suit, and the moral community expanded accordingly. Once a member of a group saw that people who formerly had belonged to an alien group were really not all that different from other members of the group, their inclusion in the moral community became a matter of course—but always there were other groups, more alien still, that remained outside. Offense against a full citizen of one's own social or political unit (city-state, nation, empire) was a crime and a sin; the same offense against a member of an alien group might be viewed with indifference, even mirth. Historical examples abound. Because many of the European newcomers to Southern Appalachia, for example, regarded the Cherokees as heathen savages, they had no moral objection to stealing their homes and herding them like cattle along the Trail of Tears. Likewise, many of these same settlers used Africans—whom they also excluded, or partly excluded, from their moral communities—as mere tools and commodities. From their perspective, Cherokees and Africans did not count—or at least they did not count for much.

During the twentieth century, however, as civilization globalized, the moral community gradually became global as well, expanding to encompass everyone on the planet. Today, at last, most reflective people recognize that tribe, race, gender, religion, nationality, sexual orientation, and other ways of partitioning humanity into those who count and those who count less or not at all are utterly arbitrary and irrelevant. *All* living people count—and with respect to fundamental rights we count equally.

But this may not be the final step in the progress of moral understanding, for we still generally fail to consider the vast majority of the human race—namely, those who are yet to live. The ways in which we are changing the climate, accumulating wastes, destroying forests, and extinguishing species demonstrate that failure. We have yet to extend full membership in the moral community to future generations.

We may wonder, moreover, whether the perimeter of the species *Homo sapiens* is a justifiable moral boundary. Is limiting moral consideration to this single species any less arbitrary than, for example, limiting it to the Caucasian race?

The effort to expand moral thinking in these two directions—toward the future and beyond the human species defines the discipline of environmental ethics. This book constitutes an essay in that discipline.

Extending moral consideration to future generations is simply a matter of consistency, for future people are in no morally relevant respect different from us. Time of birth has no more to do with how a person should be valued than do place of birth, tribe, nationality, religion, or gender. Since we agree that all currently living people are worthy of moral consideration, and since future people are in no morally relevant respect different from them, it follows that we ought to consider future people equally.

The moral irrelevance of time of birth is perhaps best understood via the realization that *we* are future people—to our predecessors. The distinction between past and future is, after all, not ultimate and absolute, but relative to temporal perspective. In that respect, it is like the designation "foreigner," which is relative to geographical perspective. Who counts as a foreigner depends on the country we inhabit. Likewise, who counts as a future person depends on the time we inhabit. *All* people are foreigners to people of countries other than their own. Likewise, *all* people belong to the future generations of their predecessors.

As it happens, a few of our predecessors were morally advanced enough to recognize this—and to include *us* in their moral community. The founders of the American nation, for example, designed its Constitution with future generations in mind; we benefit inestimably from their foresight. Similarly, the National Park Service Act of 1916 specified that the purpose of the parks is to "conserve the scenery and the natural and historic objects and the wild life therein and to provide for the enjoyment of the same in such manner as will leave them unimpaired for the enjoyment *of future generations*."[5] We are among those future generations. In Southern Appalachia, the Great Smoky Mountains National Park stands as a testament to the value of moral prescience. Decades, centuries, probably even millennia hence, there will likewise live people just as real, conscious, and valuable as we are. They may likewise remember us with gratitude—or with resentment and sorrow.

It is because future people are in no morally relevant respect different from us that they deserve full membership in our moral community. Later chapters, especially chapters 10 and 11, will explain more clearly *how* they ought to count in the decisions we make now.

But there are also reasons to enlarge our moral community in a second direction—beyond the human species. This process has been underway for some time. Early in the nineteenth century, small groups of reformers, primarily in England, began to question the treatment of certain animals— particularly horses, which were at that time the main engines of transportation. Thus the humane movement was born. In the twentieth century, as horses became less common, the movement focused mainly on dogs, cats, and other companion animals—with considerable success. Today most people agree, for example, that beating a horse or starving a dog are wrong, not merely because these acts offend our sensibilities or promote violence against humans, but because they cause the animal itself to suffer. We enforce this agreement with anticruelty laws. Since what we regard as wrong is the suffering of the animal, it follows that we have already extended moral consideration beyond the human species. These animals are members of our moral community, though obviously not on the same terms as humans.

It may seem objectionable to include nonhuman animals in the moral community, since, unlike us, they are not capable of acting morally. But not all of *us* are either. Very young children and the mentally impaired are likewise incapable of acting morally, yet they are members of our moral community, for we agree to grant them moral consideration. Philosopher Tom Regan clarifies this point by introducing a distinction between moral agents and moral patients.[6] A moral agent is one who can act morally (and so, of course, deserves moral consideration); a moral patient is an individual who cannot act morally, yet deserves moral consideration nonetheless. Both moral agents and moral patients are members of the moral community. The animals we include in our moral community are moral patients—like young children or mentally impaired adults.

Such a view need not make us the custodians of all animals everywhere. Nature is rife with amoral suffering, for which we clearly are not responsible. But suffering becomes a moral issue when moral agents are deliberately responsible for it—as we are, for example, when we destroy habitat or raise animals in factory farms (see chapters 4 and 6).[7] Hence inclusion of animals in the moral community does increase our moral responsibility.

What legitimately qualifies a being (whether moral agent or moral patient) for membership in the moral community is a matter of intense contemporary debate. Regan suggests a complex criterion for full possession of rights, which he calls being a "subject-of-a-life."[8] Peter Singer, following the utilitarian tradition of Jeremy Bentham, offers a simpler and, in the end, more plausible criterion of membership: ability to suffer. Suffering, reasons Singer, has in itself (that is, aside from its consequences) a negative value—regardless of the sufferer's tribe, nationality, gender, race, *or species*. To suppose that only the suffering of one species matters is an elementary prejudice, like thinking that only the suffering of one tribe or race matters. Singer dubs this prejudice (by analogy with racism and sexism) *speciesism*. On Singer's view, all suffering matters, and anything that is capable of suffering should belong to our moral community, if only as a moral patient.[9]

Such an expansion of the moral community would, as Singer shows, require substantial changes in what we eat, how we conduct research, and how we seek recreation and entertainment—prospects which make many people defensive. A common reaction is to take refuge in the venerable, widespread, and self-serving prejudice that nonhuman animals do not suffer at all—or at least not much. But there has never been any good reason to believe this, and the available evidence is overwhelmingly to the contrary.[10]

This evidence of nonhuman animal suffering is of three kinds: behavioral, evolutionary, and neurophysiological. The behavioral evidence is familiar to us all. When a dog has an injured paw, it whimpers and limps; when it misses a human or animal friend, it whines; when it is severely hurt, it howls in pain; when it is ill, it becomes unresponsive and withdrawn. We understand these signs of suffering sympathetically, because we know by personal experience what they express.

The second line of evidence is evolutionary. It is clear that the ability to experience pain and suffering serves two important adaptive functions: to limit possibly injurious movement when an animal is sick or wounded and to prevent repetition or continuation of behavior that is likely to lead or has led to injury. Because other animals evolved in the same way we did, it would be miraculous if only we were capable of experiencing pain and suffering, or that the etiology of their reactions to illness and injury was completely different from the etiology of ours.

Finally, there is the neurophysiological evidence. We know a good bit about why and how humans feel pain. Our nervous systems are equipped with specialized pain receptors and specialized neural fibers that transmit pain signals to our brains. Some anesthetics work by blocking these signals, enabling us to turn them on and off at will. We find these same receptors and neural fibers in a wide variety of animals, and the same anesthetics that quell pain sensations in us also quell pain behavior in animals. Suffering is more complex than pain, since it involves emotion. But the parts of our brains that are involved in emotional experience are also present in many other animals, and we have no reason to believe that their function is radically different in animals than in us. It is highly probable, therefore, that many animals experience both pain and suffering.

Admittedly, we cannot strictly *prove* that nonhuman animals suffer. The only way to be *absolutely* sure would be to feel their suffering directly, which is impossible. But likewise I cannot absolutely *prove* to myself that you are capable of suffering. For all I know, it is just barely possible that your reactions to injury, loss, or disease are programmed responses, unaccompanied by suffering of any kind. (This is an instance of the philosophical "problem of other minds"—roughly, the problem of determining whether any being apart from oneself experiences anything.) But it would foolish for me seriously to doubt that you can suffer, for all the evidence indicates that you can. Likewise, it is foolish seriously to doubt that nonhuman animals can suffer, for the evidence for their suffering is essentially the same as the evidence for human suffering—behavioral, evolutionary, and neurophysiolgical. It is true that for humans the behavioral evidence is more complex, since some of our behavior is verbal; we can describe what we feel, whereas other animals cannot. But our descriptions—especially of pain—are usually not very articulate; and, though nonhuman animals cannot describe their pain or suffering, many can vocalize it vividly. Our evidence for the suffering of some nonhuman animals, then, is nearly as strong as our evidence for the suffering of other humans, which we rightfully take for granted.

Pain and suffering are obviously of moral concern in human beings. Most people also agree that they are of moral concern in companion animals, such as dogs and cats. Why, then, do we not extend this same moral concern to *all* animals that are sentient (i.e., capable of suffering)? Unless there is some genuine moral difference (and not a mere speciesist prejudice on our part), are we not committing the same elementary mistake we have historically made over and over again? If the ability to suffer makes a creature morally significant (and why should it not?), then all sentient creatures ought to belong to our moral community. That, in a nutshell, is the most influential argument for expanding the moral community beyond the human species.[11]

Such an expansion has its limits, however, for it is very unlikely that all animals suffer. Protozoa, for example, like plants, have no nervous systems and so are almost certainly not sentient. And many "primitive" animals have such rudimentary nervous systems that it is unlikely that they can either sense pain or suffer.

Since, however, at least some animals suffer, their health matters not just to morally perceptive humans but *to the animals themselves*—and so has a value independent of any human valuing. Moreover, because the health of all animals depends intimately upon the health of the natural systems that sustain them, expanding the moral community to include all sentient animals entails concern for the health of these natural systems. Ecosystems, of course, are not sentient, so an ecosystem's health cannot matter to it. But the health of each ecosystem supports the health of many sentient creatures to whom their own health matters. Thus, the health of each ecosystem matters to these sentient creatures—and would still matter even if *we* failed to value it. If we, however, expand the moral community to include all sentient animals, then we cannot neglect the health of ecosystems—even those that do not have much to do with us.

Some philosophers (known as ecocentrists) hold that even ecosystems have intrinsic value—value, that is, apart from the value they have for us and for the sentient creatures that inhabit them. Others (called biocentrists) maintain that all living things (including even plants and bacteria) have intrinsic value. These thinkers aim to expand the moral community to include all living organisms or systems of life. They generally do not, of course, think that plants or ecosystems matter to themselves, but they recognize that even plants or ecosystems can be healthy or unhealthy and so can be benefited or harmed. Whether the moral community ought to be expanded so far beyond the human species need not be discussed here. Nor can we hope, in such short scope, to survey the many nuances of difference and controversies among the various schools of environmental ethics.[12] Yet enough has been said, I think, to show that there is reason to value health writ large—the health of the natural systems in which all life is embedded.

Spiritual Health

The reasoning of the previous section appeals only to the intellect. It bypasses a deeper motive for valuing health, one closer to the full and original meaning of the term and not so readily amenable to reason. That motive might appropriately be characterized as "spiritual"—by which I mean a motive "pertaining to the relation of human values and purposes to values and purposes that are greater and more enduring than anything human." Rather than appealing to the intellect, this motive appeals to the universal human need for hope.

Because life is finite and tragic, humans have always sought a context of meanings that transcends us. Aboriginal peoples express this urge in ritual and myth; more developed cultures in the teachings of religions. Both typically regard the realm of transcendent meanings partly as supernatural, but partly also as natural. Job 38:39–39:4 (New English Bible) provides a case in point:

> Do you hunt her prey for the lioness and satisfy the hunger of
> young lions, as they crouch in the lair or lie in wait in the covert?
> Who provides the raven with its quarry when its fledglings croak
> for lack of food? Do you know when the mountain-goats are born
> or attend to the wild doe when she is in labor? Do you count the
> months that they carry their young or know the time of their
> delivery, when they crouch down to open their wombs and bring
> their offspring to the birth, when the fawns grow and thrive in
> the open forest, and go forth and do not return?

The supernatural is prominent in this passage, since the thesis is that God provides for all nature. But implicit in the text, too, is the idea that nature has purposes apart from ours, for which we should arrogate to ourselves no responsibility or credit. However religions conceive the relation of spirituality to the *supernatural*, they all to some degree regard our attitude toward the transcendent meanings and purposes of *nature* as spiritually significant.

To speak of nature's purposes and meanings is, of course, neither to affirm nor deny that life (human or otherwise) has one ultimate meaning or purpose. The meanings and purposes that we can observe are invariably local and temporary: the salamander's stalking of the cricket, the flash of the firefly seeking a mate, the burrowing of the copperhead, the emergence of the gray bat from its cave at dusk. Even the grand natural phenomena that coordinate life's multiple purposes into intricate patterns—the alteration of night and day, the subtle changes wrought by the phases of the moon, the variations of weather, the seasonal transformations of the landscape—though meaningful in unimaginably many ways to unimaginably many creatures—impose no single purpose upon life as a whole. Nevertheless, life's local and temporary meanings and purposes constitute, collectively and in relation to the rhythms of these natural phenomena, a vast and ancient context of significance that transcends all merely human creations. In this sense, nature is imbued with transcendent meaning.

This conclusion stands, even if "meaning" is defined strictly in terms of interpretation or information processing, for human brains are not the only brains that interpret and process information. The brains of indigo buntings, for example, somehow manage to interpret the north circumpolar stars as navigational beacons that guide migrational flights of thousands of miles.[13] Examples of this sort abound, as all naturalists know.

Because we, however, live in increasingly artificial environments whose purposes are ever more rigidly dictated by humans, we have become progressively more dissociated from nature's transcendent meanings. Contemporary global culture (which is now the dominant culture of Southern Appalachia) urges us to organize our lives exclusively around human-centered purposes: comfort, entertainment, money, sex, power. But dissociation from all but human-centered purposes is a prescription for alienation, shallowness, and, in the end, despair: for alienation, because it sets us against the natural world, reducing everything beyond our narrow interests to mere obstacle or raw material; for shallowness, because merely human meanings lack grandeur of scope; and for despair, because if the only meanings we acknowledge are those we or other humans create, then when we (individually or collectively) come to die, all the meanings we know die with us.[14]

Such self-centered dissociation from transcendent purposes and meanings is a paradigm of spiritual dysfunction. Spiritual health, therefore, is by opposition: the wholesome integration of an individual's purposes and meanings into the grand context of transcendent purposes and meanings—supernatural (if there are such), but also natural. The spiritually healthy person has regard for health writ large—for the wholeness, wholesomeness, and holiness of life. Such a person strives to comprehend in relation to all life who and what we are—and to act accordingly. We are, pretty clearly, neither gods nor the crowns of creation, but instead tiny, temporary newcomers in a lush, mountainous bioregion of a watery blue planet that spins in lethal emptiness. Life emerged on this earth several billions of years before we did. It may yet, despite our short-sighted meddling, flourish for several billions more, continually generating wonders. In that grand prospect, spiritual dysfunction lacks interest, but spiritual health finds hope.

Chapter 1

HISTOry

Stan Guffey

When William Bartram traveled in and marveled at the splendor of the Southern Appalachians on the eve of the American Revolution, he found a pristine land that thrilled his naturalist's eye and elevated his Quaker spirit. Bartram's rhapsodic descriptions provide a baseline against which to consider the changes two and a quarter centuries have brought:

> I traveled some miles over a varied situation of ground, exhibiting views of grand forests, dark detached groves, vales and meadows . . . flowering plants, and fruitful strawberry beds . . . I entered a spacious forest . . . a pretty grassy vale . . . through which my wandering path led me, close by the banks of a delightful creek, which sometimes falling over steps of rocks, glides gently with serpentine meanders through the meadows . . . the land rises again with sublime magnificence, and I am led over hills and vales, groves and high forests, vocal with the melody of the feathered songsters . . . rested on the highest peak; from whence I beheld with rapture and astonishment a sublimely awful scene of power and magnificence, a world of mountains piled upon mountains.[1]

At the time of Bartram's journey, humans had inhabited the region for at least 13,000 years; still, he describes a magnificent wilderness. The Native American peoples left their mark on the landscape through their hunting, gathering, and horticultural modes of subsistence, but the ecological changes they made were different in kind and magnitude from those that have characterized the last three centuries. The ecology of Native American groups is a story of cultural adaptation to Holocene environments, changing subsistence modes, increasing population, and (after the arrival of Europeans) decimation and displacement. By the time of Bartram's visit in 1775, European hunters, traders, soldiers, and missionaries had been through and in the region for 235 years, and active settlement had been underway for at least 25 years. The arrival of the Europeans began a period of rapid, and to the Native Americans, catastrophic, change. They had lived *in* the land. We merely live *on* it.[2]

Before Human Settlement

By about 20,000 years ago, the topographic texture of the upper Tennessee valley that humans were soon to enter was much like it is today. Over tens and hundreds of millions of years, folding and faulting, uplift and erosion had cut the valleys and river channels and shaped the mountains and ridges. Life had changed even more profoundly. Millions of species had come and gone. Fishes had given rise to amphibians, then reptiles, dinosaurs, birds, and mammals. Life had survived mass extinctions, the last of which had ended the dinosaur's reign, 65 million years earlier. Climates had fluctuated. Flowering plants had replaced the once-dominant conifers, which had, in turn, replaced the ferns, lycophytes, and horsetails that formed the Pennsylvanian coal beds. Late in the tale, humans had evolved in Africa, and biologically modern humans spread from Africa throughout Asia and Australia, but not yet to the Americas.

The earth was in the midst of the last of at least seventeen glacial episodes occurring over the past two million years or so, and in North America glaciation was at its maximum. Most of northern North America, an area of about 18.5 million square kilometers (including the Greenland ice sheet), was covered in ice up to three kilometers thick, but glacial ice did not reach the Tennessee Valley, or even the Blue Ridge (the Southern Appalachian mountains). The glaciers began to retreat about 20,000 years ago, and waxed and waned until about 13,000 years ago.[3] After that, the retreat was steady and rapid. Although warmer than during the full glacial period, the climate was still strongly seasonal, with winters much colder than today's and a marked seasonality in precipitation. The climate was also wetter than at full glacial, but still considerably dryer than today.

The biotic landscape would have been unfamiliar to present-day inhabitants.[4] Mountaintops and some ridge crests were above the tree line, and the very highest peaks may have been devoid of vegetation entirely. At elevations below about 800 meters, tundra graded into savannas with scattered spruce, and to the southeast jack pine. Tree density in the valleys may have been higher, but not high enough to form a closed canopy. Wetter locations supported scattered oaks and elms, which, here and there, may have formed a closed canopy. Most of the woody and herbaceous species of the modern Southern Appalachians existed not here but well to the south. Eastern hophornbeam, American hornbeam, and other understory species of the contemporary deciduous forest were more abundant than today.

The fauna, too, was unfamiliar. The drier, cooler climate was generally hostile to amphibians; the rich salamander fauna of Southern Appalachia would come later. Reptiles were likewise limited by the cold. And there may have been fewer resident bird species than today—among them such current year-round residents as juncos, cardinals, blue jays, doves, cedar waxwings, ravens, an assortment of woodpeckers, and great horned, screech, and barred owls. Other birds that today live farther north and visit only seasonally were probably then residents, including crossbills, grosbeaks, pine siskins, and long-eared, saw-whet, and boreal owls. Ground birds, such as pheasants and grouse, and perhaps ptarmigans and prairie chickens were probably at home in the open forests. Glacial-period summers in the Southern Appalachians would have attracted migratory songbirds from the lower latitudes, but probably fewer species than today. Migratory raptors of the same species as today may have been seasonally present, along with now extinct eagles, hawks, and the teratorn, a very large, flesh-eating species with no modern representatives. Several species of old world vultures and the giant condor (known today as the California condor) filled the scavenging niche.[5] Turkey buzzards and black buzzards probably arrived later, from South and Central America.

The mammal fauna was much richer than it is today, especially the large herbivores. Several large and small mammals that are (or should be) found in the region today were likely present—including white-tailed deer, elk, black bears, gray wolves, cougars, and beaver. Probably, as with birds, a number of species that now live farther north were also present, including moose, musk oxen, and caribou. Also present were peccaries and jaguars (which occur today from the Southwest southward), tapirs (that today are forest species in Central and South America), and capybaras (which today are limited to a small area of the northern Andes). The savannas and open forests of the region may have supported many now extinct species of large mammals, including woolly and Columbian mammoths, mastadons, horses, bison, llamas, camels, stag-moose, musk oxen, four-horned pronghorns, mountain deer, sloths, giant beaver, short-faced bears, lions, dire wolves, and sabertooth and scimitar cats—to name a few.[6]

Discovery of America

Columbus, of course, did not discover the Americas, nor did Norse explorers or any of a large number of less historically justified or mythical candidates; at least they were not the first. The Americas were discovered and settled by peoples coming from northeastern Asia, sometime between 14,000 and 40,000 years ago. These Asian immigrants populated both American continents with rich cultures. One of the legacies of Columbus and the flood of Europeans that followed him was these first people's near total destruction.

When and how these Asian migrants entered the Americas is one of the most enduring controversies of American prehistory. Humans were definitely living in what is now the western United States by at least 14,000 years ago, the earliest firm date for archaeological sites with clear and indisputable evidence of human activity. Earlier dates from sites in North and South America are not generally accepted by archaeologists, nor are even earlier sites with putative assemblages of "Paleolithic" stone tools.[7] Conservatively setting the date of the discovery of America at around 14,000 years ago is consistent with our understanding of glacial environments at the time, and with the rapid expansion of populations throughout the Americas after 13,500 years ago. Humans may have been in the Americas before 14,000 years ago—perhaps as much as 20,000 or even 40,000 years ago,[8]—but things really got rolling with the retreat of the glaciers.

Settlement of the Upper Tennessee Valley

When and why people first entered the Southern Appalachian region, and exactly where they came from, is not known. Only a few sites with reliably dated cultural remains from the time of presumed first settlement exist in eastern North America; none are in the Southern Appalachians. Paleo-Indian sites from the Shenandoah, Susquehanna, and Delaware valleys and in southern Florida are dated to around 13,000 years ago, setting the latest possible date for settlement of the Southern Appalachian region.[9] Although excavated and dated sites are few, there have been a large number of surface finds in Appalachia of distinctive fluted stone spear points akin to those of the Clovis hunting culture, whose sites in the western United States have been reliably dated back to about 14,000 years. Indeed, these surface collections are so numerous in the mid-south (especially east of the Blue Ridge), that some prehistorians have suggested the area as a center of Palo-Indian flourishing.[10] Clovis-style points from the Southeast show only minor differences from Plains Clovis points, suggesting a rapid spread of Clovis people from the northern plains into the Southeast.

The signature Clovis artifact, the Clovis point, shows the cultural (and in all likelihood the demographic) continuity between Plains and eastern Clovis peoples. The Clovis point is a large, stout tool that could have been hafted for use as a knife or as a spear point. Archaeological evidence from the Plains and from the East clearly show that one of its important functions was as a spear point for hunting large herbivores. Because there is no evidence of the spear thrower in Clovis assemblages, the spear was probably used for thrusting into prey at close quarters, much as some African peoples hunted elephants down to modern times.[11]

The disappearance or possible extinction of Pleistocene megafauna may attest to the success of this tool and to the hunting strategies of the people who employed it. Transformation of the Clovis culture on the North American Plains coincides with the disappearance of the large herbivores that seem to have been the culture's primary resource base. The role of Clovis hunters in the extinction of the Pleistocene megafauna remains a topic of academic debate that has spawned an enormous literature. Populations of large grassland herbivores may have been declining in response to changing, post-glacial climates, but there is general consensus that Clovis culture hunters at least hastened the megafauna's demise.[12] Before the transformation of the Clovis culture on the Great Plains, perhaps at least in part because of a declining resource base, people had spread from this area into other parts of North and South America, including the Southern Appalachian region.

Shenandoah, Susquehanna, and Delaware valley sites have yielded sufficient evidence to attempt a reconstruction of eastern Clovis lifeways.[13] The basic social and economic unit was probably a band of about forty people who exploited a large area in a seasonal round. Primary habitation sites, occupied in the summer, were river floodplains and terraces, which yielded annual and early successional plants, such as goosefoot, blackberries, and grapes. The rivers provided fish and perhaps mussels, and surrounding ridges were sources of other plant foods and game. Several bands may have camped together, socializing, trading, and sharing in spiritual life. In the fall groups would split up, moving up the valleys and ridges in search of game. As oaks, hickories, and American chestnuts became established beginning in the early Holocene, about 10,000 years ago, nuts from these trees would become an important autumn food source for the people, and for the deer and other game they hunted. Winter encampments were probably in sheltered valleys. Fall and especially winter groups may have been smaller than summer bands, perhaps no more than an extended family, because of the difficulty of procuring food in those seasons. In the spring extended families and bands would return to favored summer sites.

Settling In: Adapting to Southern
Appalachian Holocene Environments

The disappearance of the large herbivores and changes in the composition of the biotic community that accompanied warming and increased rainfall rendered the Clovis point obsolete; it was unsuitable for hunting the mammals that inhabited the new environments. Clovis points disappeared earliest from the Plains. This fact may reflect the earlier disappearance there of the Pleistocene megafauna or the migration of peoples from the Plains in pursuit of the game that persisted longer to the south and east. In the East smaller points replaced Clovis tools about 12,000 years ago.

The first manifestation of this regional adaptation to changed and changing environments was the Dalton tradition.[14] The signature artifact of Dalton is a point that is both smaller and of a different design than the Clovis point. Again there is no evidence of a spear thrower, suggesting that the weapon was still thrust, or because of its smaller size, perhaps thrown. The point also functioned well as a knife and saw.[15] The hunting and gathering rounds developed earlier continued, with a shift from hunting large, now extinct herbivores, to the deer and elk of forest edges and interiors.[16] The Dalton tradition persisted until about 9,000 years ago, and it was replaced by a larger variety of localized tool-making traditions and subsistence modes adapted to developing Holocene environments.

Several archaeological sites in the Little Tennessee River Valley, including Ice House Bottom, Bacon Farm, and Rose and Calloway Islands, have provided a treasury of information on native peoples and environments from the Paleo-Indian settlements of 9,500 years ago to the Overhill Cherokee towns of the nineteenth century.[17] It is sad irony that this information was obtained as a salvage operation, an attempt to learn as much as possible as quickly as possible, before the gates of the Tennessee Valley Authority's Tellico Dam were closed. The sites went under the waters of the Tellico Reservoir in 1979.

The lowest occupation levels observed at Ice House Bottom date from about 9,500 years ago, a cultural time period that archaeologists refer to as the Early Archaic. The stratigraphy reveals that after about 10,000 years ago, erosion and runoff from formerly frozen slopes of the Blue Ridge deposited silt and sand in the river valley, creating bars, islands, and bottomlands. These newly formed alluvial landscapes became favorite seasonal campsites, providing access to a wide diversity of habitats and resources. Food remains indicate the sites were used from summer through early winter as a part of a seasonal round of habitation and resource utilization begun by earlier populations. Spring rains carried additional sediment into the valley, covering the remains of previous settlements and preserving the remarkable record of occupancy that archaeologists have been able to reconstruct. In the late spring the people returned to the newly rejuvenated alluvial landscape, probably occupying the same sites used during the previous summer.

Their diet was diverse. From the river they obtained suckers, drum, catfish, and probably other fish species. In the valley and uplands, they hunted white-tailed deer, elk, black bear, rabbits, raccoons, squirrels, foxes, opossums, turkey, and passenger pigeons. Mussels and aquatic and terrestrial gastropods were also consumed. During the spring and summer, they ate berries, shoots, leaves, and seeds of annual and early successional plants. They obtained hickory nuts and later acorns from the forests in the fall, as these species became established in the warmer and somewhat wetter Holocene climate.

Valley sites also provided access to chert outcrops for making stone tools on the lower slopes. From the chert they manufactured scrapers, knives, saws, spear points, and a variety of tools for working wood. Early Archaic projectile points in the Little Tennessee Valley were smaller than the earlier Dalton points, reflecting a change in hunting weapons from the thrust spear to the atlatl, or spear thrower, which propels lighter spears longer distances with greater accuracy.

The archaeological record suggests that settlement and subsistence patterns in the Little Tennessee Valley did not change greatly between about 9,000 and 7,000 years ago. An abundance of projectile points from this period, as well as faunal remains from habitation sites, indicates that people still relied on white-tailed deer and small game. However, it also appears that after about 8,000 years ago, plant foods and fishes were being more widely exploited than in the preceding millennium. Increased abundance of nutting and grinding stones indicates greater reliance on nuts and seeds. Grooved stones, interpreted on the basis of ethnographic analogy as fishing-net weights, first appear during this period. Grinding stones and fishing nets both indicate not only increased reliance on plants and fishes for food, but also an increased technological sophistication in their harvest and preparation.

Elsewhere in Southern Appalachia, and in the eastern woodlands generally, more dramatic subsistence changes occurred. During the Middle Archaic, as archaeologists designate the period from about 8,000 to about 6,500 years ago, use of riverine resources increased substantially. Shell middens throughout the Southeast and Mississippi valley attest to the enormous numbers of freshwater mussels consumed.[18] An abundance of fish bones and fishing implements from these sites indicates an increase in fishing. Use of plant foods, seeds, and nuts also increased. Small game and white-tailed deer remained important in the diet, but the relative subsistence contribution of hunting appears to have declined, especially after about 7,000 years ago.[19]

Although the Canadian glaciers had not yet entirely disappeared, the climate of the Southeast from 8,000 to 6,000 years ago was warmer and dryer than the last 2,000 years have been. Deciduous forest trees, especially oaks, became more abundant than in previous millennia, but dry conditions, and natural and human fires, may have fostered an open canopy in many areas.[20] Because of the low rainfall, the Tennessee River and its tributaries were lower than today. This may have created large shoals and riffles, habitat for most of the region's mussel species.[21]

About 6,000 years ago temperatures began to decrease in the valley, and rainfall increased. Forests took on a more modern composition and character.[22] Concurrent with these changes was a shift of culture to the Late Archaic period. Archaeologists observe an increase in the number of habitation sites in the Late Archaic, indicating a substantial population increase over the preceding millennium.[23] The Late Archaic was a period of increasing regional specialization in resource use, with at least six distinctive cultural traditions identifiable in eastern North America. The period also saw increased trade in high-value commodities, including copper and hematite from the Great Lakes region, shells from the Gulf and Atlantic coasts, soapstone from the eastern Blue Ridge, and a wide variety of fine-grained cherts, flints, and other rocks from various locales.[24]

Late Archaic cultural traditions in the Southern Appalachians, and throughout much of the Central Mississippi River drainage, are given a distinctive name, Central Riverine Archaic, because of their continued and intensified exploitation of river and valley resources.[25] Between 6,500 and 3,000 years ago, life became more sedentary for these peoples, fish and plants were more widely used, and mussel consumption declined (perhaps reflecting a decrease in mussel habitat brought about by increased rainfall). Fish were taken by nets, spears, hook-and-line arrangements, fish traps, and fish toxins obtained from hickory nut husks and other plants.[26] More use was made of berries and seeds from early successional plants of floodplains and disturbed habitats. Fire was probably used to maintain early successional plant communities for this purpose, as well as to keep upland forests open.[27] Some annual plants were becoming domesticated, heralding the dawn of horticulture in eastern North America.[28]

Pots and Cultivars

Pottery and domesticated plants are hallmarks of the Woodland cultural period in eastern North America. The earliest North American pottery is from sites along the South Carolina coast dating to about 4,500 years ago.[29] By 3,000 years ago pottery making had spread throughout the lower Mississippi valley and the Southeast as far north as northern Georgia.[30] It was probably about this time that pottery entered the cultural inventory of people in the Southern Appalachian region. Pottery permits the storage of food in watertight containers and facilitates food processing. More secure food storage increased reliance on annual plants, stimulating the development of horticulture and agriculture.

Maize, beans, and squash were domesticated somewhere in central or extreme southern North America about 7,000 years ago.[31] These domesticated plants would later become central to subsistence in the river valleys of the eastern United States, but before they arrived, a number of native plant species had already been domesticated or semi-domesticated. By 3,500 years ago people in the upper Tennessee valley relied on wild plants for a substantial part of their subsistence. In additions to nuts from the forests, a small number of annual, disturbed-habitat species had become important, including marsh elder, chenopodium (goosefoot), and knotweed.[32] The process of harvesting seeds from these plants probably promoted these plants' persistence in alluvial habitats and their spread to lands burned or otherwise disturbed by humans.

Domestication is a process rather than an event. Harvesting, purposeful creation of suitable habitat, and land disturbance impose new selective pressures on plants. Eventually humans began to select desirable traits purposefully. One of the most important traits, and one that leaves a recognizable signature in the archaeological record, is increased seed size. By at least 2,500 years ago, Early Woodland people in the upper Tennessee valley were cultivating, in addition to many indigenous plants, gourds and sunflowers that had probably arrived from the western and southwestern margins of the deciduous forest region.[33]

Other trends that began in the Archaic also mark the Early Woodland cultural period, including burials in mounds, ceremonial earthenworks, fabrication of ornaments and other socially valued artifacts, and continued trade in commodities and finished artifacts throughout eastern North America.[34] These cultural patterns suggest a widening scale of social organization and increased social stratification, arising and attaining its greatest manifestation in the Ohio valley. The Adena complex, as archaeologists call this group, did not directly penetrate the Southern Appalachians, but elements of it did.

In the Illinois and Ohio valleys, maize and beans entered the eastern North American woodlands about 2,200 years ago, presumably via the cultures of the Southwest. Squash may have arrived somewhat earlier. However, these Mesoamerican crops did not contribute much to the subsistence of the people until several centuries later, and societies based on the full-scale cultivation of maize did not develop until about 1,200 years ago.[35] Subsistence still involved a mixed economy of hunting, fishing, and harvesting of nuts and grains—including domesticated or semi-domesticated indigenous species. Nuts, especially acorns, may have actually increased in importance prior to the adoption of full-scale agriculture. In the Illinois and Ohio valleys, Adena gave way to (or developed into) the society known as Hopewell, which flourished from about 2,200 to about 1,600 years ago. Many of the elements that would characterize later cultural and demographic developments in the Southeast were present in Hopewell (and earlier in Adena), including denser populations, social stratification, and year-round occupancy of relatively large villages in the river valleys.[36]

The Adena and Hopewell cultures constituted the beginnings of complex, socially stratified societies, organized on a much larger spatial scale than earlier societies. These chiefdoms, as anthropologists call them, consisted of groups of villages in a river valley, or perhaps in several valleys, under the control of or tribute to a single ruler. Larger groups shared cultural patterns than in earlier periods, reflected in a decrease in the number of distinct local styles of pottery and other elements of material culture.

The Middle Woodland peoples in the upper Tennessee valley and the Southeast differed significantly from the Hopewell, although societies throughout the East were engaged in regular trade. Population densities may not have been any lower here than in the Hopewell heartland, but social groups were not as organized or centralized. Villages were larger than in previous centuries and were most likely occupied year-round, but chiefdoms involving the organization of people over a large area into a single society appear not to have developed. Mortuary practices in the Middle Woodland period of the upper Tennessee valley were less complex than in the Hopewell core, and earthworks were smaller and less numerous.[37]

Maize and Chiefdoms

Between about 1,100 and 500 years ago, the upper Tennessee valley became one of the centers of a cultural flourishing that occurred throughout the Midwest and Southeast.[38] This so-called Mississippian period is characterized by the development of large-scale maize and bean agriculture, a population increase, and larger-scale social stratification and organization. Elements of the Mississippian cultural tradition appeared first in the Midwest, where they represent a continuation of developments seen in Adena and Hopewell cultures, and in the Southeast in the lower Mississippi valley and Gulf Coast. The sudden and almost simultaneous appearance about 1,100 years ago of Mississippian societies throughout the Southeast is remarkable. It is not surprising that earlier generations of archaeologists interpreted the Mississippian as a radically new cultural tradition imposed on pre-agricultural societies by people arriving from elsewhere.

Yet, the development of distinctive Mississippian cultures was a continuation of trends that had been occurring for 6,000 years. Societies whose subsistence focus was located in the fertile river valleys and who had already developed an indigenous horticulture using native annual species adopted maize and bean agriculture. The result was revolutionary. Maize and beans increased the potential yield of calories and protein from land that had been the source of subsistence for millennia. The increased carrying capacity afforded by agriculture resulted in population increases and the concentration of populations into larger villages. Large-scale agriculture with wood and stone implements is not a solitary activity, and anthropologists hypothesize that the organization of agricultural production and the storage and distribution of harvests fostered the development of stratified societies and chiefdoms.

Maize, as we noted earlier, spread into the Southeast and Midwest about 2,000 years ago from the Southwest, where it had been grown for at least 3,000 years. The first varieties cultivated in the East were small-cob, low-yield 12- and 14-row hard flints. People in the Ohio, Illinois, and lower Mississippi River valleys, and valleys of Gulf-draining rivers in the Southeast grew this corn, but it was at best a minor contributor to subsistence. About 1,000 years ago a hardier and higher-yielding eight-row flint, *Maiz de Ocho,* or northern flint corn, entered the eastern woodlands, where it rapidly became a major source of nutrition. Beans also appeared about this time and were as readily adopted as maize.[39] Hunting, fishing, and gathering of wild foods continued during the Mississippian, but the valley agriculture was central. In addition to maize, beans, and squash, Mississippian peoples were probably tending "orchards" of fast-growing trees, such as mulberry and walnut, and may have engaged in the pen raising of turkeys.[40]

Between about 1,100 and 800 years ago, Mississippian societies in the Southern Appalachians developed a series of statelike chiefdoms centered in large valley villages.[41] Smaller villages were politically connected and paid tribute to the rulers of one or more larger villages. Chiefdoms were hierarchical, with some receiving tribute from a group of nearby smaller villages, and at the same time paying tribute to one of a few larger and more powerful villages in the region. Tributary alliances were dynamic, shifting as the power of different leaders and villages waxed and waned, and were established and maintained by warfare. Villages, often palisaded, were centered on a large platform mound that held the residence of the principal ruler of the village. Nearby were granaries and the houses of the common people. Villages were in close proximity to the fields of maize that provided the bulk of Mississippian subsistence. In addition to the field crops, small garden plots were located adjacent to houses, and semiculti- vated orchards of mulberries, walnuts, and butternuts may have occupied the higher terraces. In addition to maize, the river and creek bottoms in many areas supported dense stands of river cane, an important raw mate- rial used for making houses and baskets.[42] Cane stands were probably maintained by the use of fire.

By about seven hundred years ago, on the eve of the period of Euro- pean contact, Southern Appalachian Mississippian civilization was at its height. Primarily on the basis of pottery styles, archaeologists distinguish three cultural groups. In Tennessee and southwest Virginia, the Dallas people lived in the upper Ridge and Valley; the Mouse Creek culture inhabited the lower French Broad River and Mouse Creek in the southern Ridge and Valley; and the Pisgah group lived in the Blue Ridge of North Carolina. Pisgah, the primary group ancestral to the post-contact Cher- okee, may have been a different linguistic group from Dallas and Mouse Creek. Some archaeologists suggest that the size and power of Southern Appalachian Mississippian chiefdoms were in decline at the time of European contact.[43] This may be the case, but the only European eyewit- nesses of Mississippian culture in the upper Tennessee valley, the mem- bers of de Soto's expedition, report a dynamic and sophisticated agricul- tural society. What is certain is that on the heels of this initial contact European diseases decimated Native American populations in the region and radically altered the Mississippian way of life.

Contact

Our brief survey of 13,000 years of Native American history in the South- ern Appalachians does not do justice to the richness of the cultures that developed and flourished in the region, nor does it give us any feeling for the undoubtedly rich social, intellectual, and spiritual life of the people. We might be left with the terribly mistaken impression that prehistoric Native Americans were simply intelligent actors responding to the need for subsistence, to environmental change during the transition from glacial to interglacial environments, and to cultural innovations emanating from outside the region. In part this is due to the brevity of our account. Mostly it reflects the limits of the data available in the archaeological record.

With the arrival of de Soto's conquistadors we have the first written account of Native American societies in the Southern Appalachians and our first glimpses into the deeper elements of their culture. But subsequent accounts by Europeans describe cultures vastly different from those de Soto observed, demonstrating that the consequences of even this initial contact were catastrophic for the area's inhabitants.

Spain, through the persistence of the Italian mariner Christopher Columbus, claimed credit for the discovery of the Americas. Enterprising men from Spain, Portugal, France, England, and Holland had been vying to tap the riches of Asia, and their royal sponsors hoped to prolong their favorite pastime, one-upmanship. The Americas got in the way. On his initial voyage Columbus thought he had reached Asia, new islands of the Indies that Europeans had recently been exploiting by voyages around Africa. Columbus's geographical error survives in the name he "gave" to the inhabitants of the lands he encountered: Indians. Of greater significance for subsequent history is the fact that some of the "Indians" Columbus encountered were wearing gold ornaments.[44] In the fifteenth century the Portuguese had set the model for colonization with their establishment on the Canaries, Azores, and Madeiras of slave-based plantations that produced agricultural commodities for European markets. Novel commodities generated by slave labor made money, but gold was another matter: gold was direct wealth, and the initial Spanish expedition efforts aimed to find and seize it.

Spanish adventurers wasted little time. Initially gold was obtained from the native inhabitants of Cuba and Hispaniola, the Tiano, and then from mines that were opened on the islands. Shortly the Spanish needed additional slaves to work the mines and to manage the cattle and sugar operations on the islands, because too many of the enslaved Taino died. Early in the sixteenth century, Spanish raiders began capturing slaves from Central America. Because many of these new slaves died as well, slave raiding became a regular activity. From their Central American captives the

Spanish learned of the Aztec empire, and were excited by stories of its fabulous wealth and gold. In 1521 Hernan Cortés, with a small band of soldiers and native allies who had been tributaries to the Aztec empire, and with the aid of smallpox, captured the Aztec capital and quickly gained control throughout the empire. During the 1530s Fransisco Pizarro defeated and gained control of the Incan empire in the Andes. Spanish efforts north of Mexico had yielded no gold, and one expedition into Florida, that of Cabeza de Vaca, ended with the deaths of all but four of the conquistadors. When de Vaca and his three associates reappeared in Mexico in 1536, eight years after the start of the Florida expedition, he did not have gold, but he had tales of gold to be found somewhere in the North American interior. De Vaca's stories and the hope born of lust for gold were sufficient for Hernando de Soto, who had been an officer in the conquests of both Cortés and Pizarro. He now longed for his own empire to conquer. From 1539 to 1541, de Soto led his six hundred men on what Taylor describes as "a violent rampage through the carefully cultivated and densely populated heartland of the Mississippian culture."[45] De Soto's chroniclers provide us with the first and last eyewitness accounts of the flourishing Mississippian cultures of the Southern Appalachians.

De Soto and his men began their "rampage" from Tampa Bay, eventually traversing parts of present-day Florida, Georgia, South Carolina, North Carolina, Tennessee, Alabama, Mississippi, Arkansas, and Texas (and perhaps also parts of Kentucky, Indiana, Illinois, and Missouri).[46] In the Mississippian villages they found signs of wealth: large fields of maize, densely populated villages with central temple mounds and powerful rulers, but little gold. The expedition was well equipped with accoutrements of war and enslavement, including three hundred sets of iron collars, but carried little food. Instead, they used force, capture, torture, and execution to extract food from what the chroniclers describe as abundant stores in the Mississippian villages. Looting of villages and graves and torture of captives still did not turn up gold. In most villages de Soto and his men were told that what they were looking for might be found farther north or west: somewhere else.

De Soto's wanderings in search of gold took him through the Blue Ridge and into the upper Tennessee valley around the end of May 1540, but his exact route remains uncertain. Davis, following Hudson and Duncan, describes a route through the mountains along the French Broad River, with a stop at Conasoga, a large village thought to be near present-day Hot Springs, North Carolina. Following their transit through the mountains, the Spaniards next came to the village of Chiaha, which Davis and others place in the vicinity of present-day Dandridge, Tennessee. Sheppard, on the other hand, places the route through the mountains along the Little Tennessee River, with Conasoga to the south of present-day Asheville, North Carolina, and Chiaha near the site of today's Fontana Dam. In any case, de Soto's marauders rested and availed themselves of the hospitality of the inhabitants of Conasoga and Chiaha, and took what was not offered.

Eventually, by either account of the route, de Soto's band reached the villages and hamlets along the lower reaches of the Little Tennessee River. In the large village of Tali the conquistadors mistreated the inhabitants as usual, before again moving southwest through the valley. In mid-July they entered the town of Coosa in northern Georgia, near present-day Eton. Coosa, the largest village the Spanish had encountered since crossing the Blue Ridge, was a major agricultural and ceremonial center, and the power center of a large tributary area. Chiaha, itself a major agricultural and tributary center, may have been tributary to Coosa. De Soto and his men stayed in Coosa for three weeks, enslaving many of the inhabitants. Around August 20, the force left Coosa, with a number of its inhabitants as captives. Though they had left the Tennessee valley, de Soto's conquistadors would continue their depredations through the Southeast and Midwest for another three years. De Soto died on the banks of the Mississippi River in 1542, his body consigned to the river with a ballast of sand to conceal his death from the natives. Friar Bartolomé de Las Casas, a sixteenth-century critic of Spanish depredations in the America's, voiced a humanitarian judgment of de Soto and his savage treatment of the peoples, declaring "We do not doubt but that he was buried in hell . . . for such wickedness."[47]

De Soto's four-month rampage through the upper Tennessee valley was disastrous for its native people and cultures. We can only begin to imagine the immediate cultural shock, but we have a better grasp of the effects of European diseases brought by the conquistadors. Even prior to de Soto, epidemic smallpox, measles, and influenza had stricken the native populations of the Southeast. In the high country of South Carolina, de Soto's chroniclers had noted that maize was more scarce and the people weaker than in the recent past.[48] Diseases contracted from Spanish and French trading posts on the Atlantic and Gulf coasts had spread from native group to native group. People in the western Blue Ridge and Ridge and Valley had thus far escaped the epidemics, but de Soto's band brought that isolation to an end. Twenty years after de Soto, when soldiers from a Spanish colony on the Gulf entered the Coosa sphere again, the changes were so marked that the Spanish thought they were in a different area from that described by the de Soto expedition. The chiefdom had been so depopulated and its power so reduced that former tributary villages were refusing to pay tribute. By 1600, the Native American population of the Southeast, including the upper Tennessee valley, had been reduced by as much as 90 to 95 percent.[49]

The sixteenth century was a period of cultural loss and reconfiguration among the Native Americans of eastern North America. European diseases decimated populations and altered power relations among surviving groups. Trade and conflict with Europeans changed economies and introduced new needs and wants into the cultures. Most important, the presence of Europeans and the necessity of dealing with them became an overriding element in the day-to-day affairs and futures of native peoples. Some surviving groups migrated out of the area; others formed new coalitions or cultures.

Mississippian Mouse Creek and Dallas survivors of the epidemics probably migrated to the Southwest, joining with similarly fragmented groups in northwestern Georgia and Alabama and developing into the Creek Confederacy of historical times. Others in the upper reaches of the valley likely amalgamated with the late prehistoric Fort Ancient people to the northwest, emerging as the historic Shawnee. The surviving Pisgah peoples of the western Blue Ridge, along with survivors of Mississippian and other groups from the eastern Blue Ridge and the South Carolina Piedmont, coalesced as the Qualla culture, probably centered along the upper reaches of the Little Tennessee River and its tributaries. Early in the seventeenth century, some of the Qualla settled in the region of the lower Little Tennessee River valley, establishing the westernmost group of the four historic Cherokee population centers.[50] Elements of pre-contact Mississippian lifeways continued among the reconfigured surviving groups, but many cultural elements were certainly lost. In the narrative of his 1775 journey into the region, William Bartram observed that the Cherokee appeared to have as little knowledge as the Europeans of the origin and purpose of the platform mounds that once formed the center of Mississippian villages.[51]

European trade transformed the native cultures. Through most of the sixteenth century, the Spanish traded with Native Americans in the upper Tennessee valley from settlements on the Gulf and Atlantic coasts. Usually the trade was conducted through intermediaries, or by journeys of native groups to the Spanish settlements. However, on several occasions, but only for brief periods, the Spanish established forts and trading installations in northern Georgia and western North Carolina to foster and protect trade. Native Americans traded deer hides, and sometimes the skins of other animals, for iron tools, firearms, gunpowder, and bullets, iron and brass pots, fabrics, and other goods. Native groups in the Southeast also adopted crop plants and domesticated animals brought to the Americas by the Spanish. Cultivated plants appropriated earliest by the Cherokee were peaches, sweet potatoes, cowpeas, and watermelons. Horses were adopted as soon as they could be obtained; cattle, sheep, goats, hogs, and chickens came later, in the eighteenth century.[52]

Depopulation and trade with Europeans produced ecological changes as well. Entire villages were wiped out or abandoned, and thousands of acres of bottomland maize fields underwent succession to forest. Fire-maintained river cane fields decreased so substantially that large stands were extremely rare by the late eighteenth century.[53] There was less demand for firewood and for timber for building and palisade construction in the large Mississippian villages; hence, timber density increased around the villages. Bison re-entered the Southern Appalachians, or at least became more numerous, after about 1600, as did elk. Bison may have been present before de Soto's incursion, but they could not have been numerous, for de Soto's chroniclers, who were careful to note many other animal species, did not mention seeing them, although they did acquire one buffalo blanket from a Mississippian village. The arrival of bison upon the decline of Native American populations suggests that hunting may have kept them from becoming established in earlier times. Herds were resident on the Cumberland Plateau and in the Nashville Basin and Blue Grass region before 1600, but hunting pressures may have prevented their expansion into the Ridge and Valley.[54]

Because of the commercial value of deer hides to the Europeans, and because of the Native Americans' desire to obtain goods in trade, deer hunting may have increased with the gradual increases of Native American populations after 1600. However, the destruction and demographic collapse of the Mississippian cultures probably reduced white-tailed deer habitat, which consisted of open forests, edges, and early successional communities. Use of fire for hunting, a cultural practice of great antiquity, may have increased as hunting for trade replaced the subsistence hunting of precontact times; this would have tended to increase the abundance and quality of deer habitat, particularly on dryer ridges and slopes.

Trade with the Europeans and adoption of European crops, material goods, and practices were also important to the growth of Cherokee population and power, and to their two centuries of resistance to Euro-American expansion. As the heirs to 10,000 years of living in the eastern North American forests, the Cherokees knew the biota, the geology, and the seasons and could build a society from the land. Yet, they had to react, too, to the presence of the Europeans. The Cherokee readily adapted European tools and crops to suit their own culture. But by the nineteenth century, the Cherokees' genius for responding to the inevitability of white presence—adopting American cultural elements and practices and adapting them to their own still vigorous culture—became their downfall. The Cherokee, by their own standards as well as by the standards of their envious white neighbors, were better at living and prospering in the land than were the whites. In the end, their presence could no longer be tolerated.

Cherokee Land, Euro-American Settlement

The Cherokee inhabited four distinct areas on both sides of Blue Ridge divide. The Lower Towns, scattered along the river valleys of northeastern Georgia and northwestern South Carolina, and the Valley Towns, located in valleys to the southwest, were east of the divide. West of the divide, the Middle Towns dotted the upper tributaries of the Little Tennessee and the Overhill Towns populated the lower reaches of the Little Tennessee, Tellico, and Hiwassee.[55] All of the towns were involved in extensive trading contact and conflict with the South Carolina colony, and to a lesser extent with the colony of Virginia, and for a time they blocked settlement from the south. Initial Euro-American settlement was by way of the Great Valley into the Ridge and Valley section of southwest Virginia and upper East Tennessee. Traders and frontier adventurers preceded Euro-American settlers, and along with the settlers came soldiers.[56]

Limited trade between the Virginia colonies and the Cherokee began as early as the 1670s, but with the English defeat of the Spanish on the Atlantic Coast south of Florida around 1700, the English colonies, especially South Carolina, dominated trade thereafter. By the first decade of the eighteenth century, a few colonial traders were residing in the Lower Towns and periodically visited the Middle Towns and Overhill Towns. Whitetailed deer hides remained the primary commodity exchanged by the Cherokee for the same goods that had been obtained from the Spanish. Munitions—firearms, gunpowder, and lead—increased in importance as items of trade as the Cherokee came to rely on this technology for trade hunting, subsistence, and defense. By the 1730s trade with the Cherokee was so lucrative and the traders so numerous that colonial governments had begun to regulate and partition it. By the 1750s many Cherokee towns had become almost entirely dependent on European trade goods, forcing them to obtain hides for exchange. The resulting deer kill was staggering. Donald Davis, who analyzed surviving trade accounts and export records from the time period 1739–61, conservatively estimates a deer kill of at least 25,000 animals per year, and suggests that the total figure for this twenty-two-year period may be close to a million animals.[57]

Although white-tailed deer were the most abundant game hunted for trade, other large mammals were also taken. Bison and elk hides made up a small percentage of the total, reflecting the low densities of these animals in the Southern Appalachian region. With the decline of beaver populations in the Northeast, beaver pelts became an important trade item in the Southeast. Introduction of steel traps among the Cherokee dramatically increased the take and also hastened the decline of southeastern beaver populations. By the 1740s beaver had become so rare in the region that they were no longer important in trade, and by the late eighteenth century beaver had been largely extirpated from the Southern Appalachians. Deer kill peaked in the 1750s, as deer populations fell, and by the 1760s elk and bison had largely disappeared from the Southern Appalachians. In addition, deer, bears, wolves, and mountain lions were all decreasing, as settlers from the western fringes of the Virginia and Carolina colonies joined in the melee. Long hunters from the frontier settlements—so-named for their long hunting forays into Cherokee and Shawnee hunting grounds—began to compete with the Cherokee in pursuit of the same valuable game. Complaints registered with colonial officials by Cherokee leaders suggest that the long hunters were depleting game and making Cherokee subsistence more difficult.[58]

Deer herds were also reduced by competition from livestock, particularly hogs. Many of the traders living among the Cherokee had brought livestock with them. Hogs and cattle were rarely penned or fenced in this period; rather, they were turned loose in the forest to forage on the same mast (fruit and nuts) and browse (herbaceous plants) consumed by native grazers. By the mid-1740s Euro-American settlement of the northern upper Tennessee valley had begun, and the settlers, too, brought their free-ranging livestock with them.[59]

In 1745 the royal governor of Virginia began the sale of land in the upper Tennessee valley of Virginia. Settlers began arriving by 1748. One of the first was Stephen Holston, who settled along the middle fork of the river that still bears his name. Holston did not persist in the region for very long, but he was followed by others, including settlers brought by the Loyal Land Company, which received a grant of at least 800,000 acres from the colony of Virginia. The grant did not officially involve lands south of the Virginia border, but by 1750 lands were surveyed for settlement in the vicinity of Long Island in the Holston in present-day Sullivan County, Tennessee.[60] Although British colonial policy prohibited settlement beyond the Blue Ridge, by the beginning of the French and Indian War (or Seven Years War) in 1754, hundreds of settlers, their livestock, and their European perspectives and lifeways were established in the Holston and Watauga River valleys.[61]

The war stalled and reversed settlement, at least for a brief time. Many of the settlements in the upper valley were abandoned during the war, and others were destroyed. Initially the Cherokees were allies of Great Britain, and they sent warriors to participate in British campaigns along the Virginia frontier and acquiesced to the construction of Fort Prince George (1753), east of the Lower Towns, and Fort Loudoun (1756), at the confluence of the Little Tennessee and Tennessee rivers in the midst of the Overhill territory. But ill-treatment of Cherokee warriors by British officers whom they had joined in Virginia, and murders by settlers in the Holston and Watauga valleys, precipitated a schism in the Cherokees' attitudes toward the British. Cherokee warriors attacked and forced the abandonment of most of the settlements in the upper valley, and warriors from the Middle and Lower towns pushed back encroachment on the Carolina Piedmont. In the summer of 1760, a British force attacked and destroyed the Lower Towns and their crops, driving the survivors to refuge in the Middle Towns. The pursuit was repulsed at the Middle Towns, and the British were forced to withdraw. Two months later a Cherokee siege of Fort Loudoun forced its abandonment. But defense of the Middle Towns and the capitulation of Fort Loudoun were the Cherokees' last successes in stemming the tide of white encroachment.[62]

In the Spring of 1761, a British force attacked the Middle Towns, eventually destroying all fifteen villages and fifteen hundred acres of crops. Most of the inhabitants escaped over the mountains to the Overhill Towns. But epidemics spread through the villages, diminishing the people's ability to defend themselves or to tend their crops. Moreover, the three-year interruption in trade with the British forced the Cherokee to return to the use of bows and arrows for hunting and defense. Creek, Chickasaw, Catawba, and Iroquois war parties, well armed with European weapons, began to attack the Overhill Towns, and a large British force descended the Holston River from Virginia. Faced with the possibility of destruction by well-armed attackers, and the certainty of famine in the coming winter, the Cherokee sued for peace in August.

The terms of the peace agreement were remarkably mild, given that the Cherokee were severely weakened. The Cherokee agreed to relinquish about half of the Lower Towns' hunting territories, but retained the remainder of their lands—for a time. Most important, they kept their identity as a nation. Whites continued to move into Cherokee lands after the peace settlement of 1761, especially from the north and northeast. Anxious to maintain peace with the Cherokee and other native groups, the British issued a proclamation prohibiting settlement west of the Appalachians. This official action and a series of treaties that followed slowed settlement but did not stop it.

During the revolutionary period, the Cherokee were nominal allies of the British, and their villages frequently suffered attacks by frontier militia. Throughout the war and after, white settlers continued to flow into Cherokee lands. In 1784 the new republic opened negotiations with the Cherokee, culminating in the 1785 Treaty of Hopewell. But many white settlements remained in Cherokee territory, and the settlers, according to the new government, were now too numerous to remove. As settlers continued to arrive, negotiations with the Cherokee became almost continuous.[63] By 1788 there were at least 25,000 whites in the upper Tennessee valley. There were, by contrast, perhaps 2,000 Cherokee warriors at that time, and probably a total Cherokee population of less than 15,000.[64]

Concessions continued. In 1809 some Cherokee moved west of the Mississippi River, a trek that most of the remaining people would be forced to make three decades later. In 1816 the Cherokee gave up claim to the

Lower Towns and their hunting territories in South Carolina, and in 1817 yet another treaty ceded additional Cherokee lands. Kituah, the land of the Cherokee, was fading. The Cherokees responded to the inevitable by assimilating into Euro-American culture, adopting a written language, developing a civil government modeled on those of the whites, and successfully participating in the growing economy of the new republic's southwestern frontier.[65]

But pressures from white settlers were unrelenting. The governments of Tennessee and Georgia, proceeding independently of the federal government, acted to seize the remaining Cherokee lands. The Cherokee appealed unsuccessfully to Congress and the U.S. Supreme Court. In 1838 they were rounded up at gunpoint and forced to relocate to inhospitable lands in present-day Oklahoma. Refusing to ride in government wagons, many walked, and thousands died along what became known as the Trail of Tears. Except for a small group that hid in the wilds of the Great Smoky Mountains and was later permitted to purchase a small reservation in the mountains, Kituah was gone. As H. H. Jackson aptly observed in his 1885 account of relations between the government and Native Americans, "Never did mountaineers cling more desperately to their homes than did the Cherokee."[66] The Cherokee homeland had become a region of American settlers, more numerous than the Cherokee had ever been, and with entirely different designs on the land.

From Frontier America to American Landscape

The trickle and then flood of arrivals from the British colonies and the American states brought people from a variety of social and ethnic backgrounds, and with a diversity of aspirations and expectations, into the land of the Cherokee. Most were of English ancestry, but there were also Scots-Irish, Irish, Scots, Germans, and Scandinavians. Settlers of means (wealthy landowners from the settled areas, lawyers, military officers, and land speculators), of whom there were a surprising number, also brought slaves of African ancestry. In addition to the yeoman farmers, who made up the majority of free arrivals, immigrants also included indentured servants and people who were fleeing misfortune or the law in the older settled areas. Nearly all were seeking to improve their economic fortunes.[67]

Their methods were not derived entirely from European customs. Two hundred years of experience in North America had modified Euro-American approaches to living on the American landscape. Farmers adopted Native American crops and techniques that were better suited to the New World than many of the European crops they had brought with them. The abundant game in the forests had lured the Euro-Americans to hunting, an activity practiced exclusively (and only recreationally) by the aristocracy in the Old World. Hunting was an important part of the subsistence and trade economies during the settlement period, and in isolated areas of the Appalachians it would remain so until modern times. Euro-Americans readily adopted the Native Americans' use of fire as a tool for clearing land, controlling pests, improving browse, and hunting.[68] The expansive forests, new to most Europeans, were an obstacle to farming, but the vast supplies of timber tempted them to consume and waste wood at a prodigious rate. Mast-producing trees—oaks, chestnuts, hickories, and beech—permitted methods of animal husbandry that were very different from those practiced in Europe, and on a larger scale. Because swine were particularly well suited to free-range husbandry, hogs came to dominate as commodities for trade and for local consumption.

But Euro-Americans also imposed Old World techniques on their new environment. The Scots and Scots-Irish brought livestock transhumance (seasonally changing pasturage). They and the English also brought in-field and out-field systems of farming, while the Germans brought systems of continual use of agricultural fields. Most important, the Euro-American arrivals imported their vision of what a settled landscape should look like and how its economy should function. Land should be cleared. Forests should fringe settled areas. Within cleared and settled areas, forests should be confined to woodlots sufficient to meet domestic fuel needs. Agriculture was to supply local subsistence needs and provide commodities for participation in a wider economy. Livestock and agricultural surpluses were for trade, either as barter or for money, to obtain manufactured goods. Local manufacturing, using available resources, was to provide goods both for local needs and for trade.[69]

Trade required transportation, and soon roads began to scar the landscape. The earliest roads developed along the well-worn routes of Native American trails. Trails from the Carolina Piedmont were expanded into rough wagon roads into the Cherokee Lower Towns in the seventeenth century. Long hunters from North Carolina and Virginia used the Native American trail network, and later often used what they learned of it to direct the construction of wagon roads into the region. For example, Daniel Boone's Wilderness Road from southwest Virginia through the Cumberland Gap was an expansion of a segment of the Great Warrior Path that connected the Cherokee lands in Tennessee and Kentucky with the Shenandoah Valley. The Great Warrior Path, expanded as a road for the passage of wagons and livestock, was also the main conduit for settlers into the upper Tennessee valley in the mid-eighteenth century. Early settlements in the Watauga and Nolichucky valleys were connected with other settled areas by existing trails expanded into rough wagon roads. The road from the East into the Nolichucky settlements, perhaps the first transmountain road in the upper Tennessee valley, was an expansion of an ancient trail used by the Native Americans. Reflecting a pattern that extends to the present, one of the first actions of the Watauga Association in 1772 was the construction of a new road connecting the Watauga settlement with the settlements and road networks of southwestern Virginia.[70]

Wendell Berry, in an essay musing on the 1797 construction of a road in eastern Kentucky, precisely captures the meaning roads had for Euro-Americans:

> The Indians . . . and peasants were people who belonged deeply and intricately to their places. Their ways of life had evolved slowly in accordance with their knowledge of their land, of its needs, of their own relation of dependence and responsibility to it. The road builders were a *placeless* people. . . . The difference between a path and a road is not only the obvious one. A path is little more than a habit that comes with knowledge of a place. It is a sort of ritual of familiarity. As a form, it is a form of contact with a known landscape. It is not destructive. It is the perfect adaptation, through experience and familiarity, of movement to place; it obeys the natural contours; such obstacles as it meets it goes around. A road, on the other hand, even the most primitive road, embodies a resistance against the landscape. Its reason is not simply the necessity for movement, but haste. Its wish is to *avoid* contact with the landscape; it seeks so far as possible to go over the country rather than through it. . . .[71]

The earliest Euro-Americans settled first in the river valleys, where alluvial soils were richest and easiest to cultivate. These valleys were in close proximity to the developing road networks and contained native vegetation that was easiest to clear. Forests had replaced the cleared fields of the Mississippian peoples, but floods maintained early successional plant communities on some floodplains and terraces. During the early settlement both rich and poor settlers obtained lands in the valleys, although the well off and powerful usually obtained the best and largest tracts. The wealthy and powerful purchased or obtained grants from civil government. Poorer settlers, the majority, obtained smaller tracts from wealthy purchasers or grant holders, buying outright or on the promise of future payment after successful clearing and development. Some poor settlers obtained tracts under the rule of "corn rights," a commitment to clear the land, produce a crop, and establish a permanent habitation. Other settlers, especially those who moved beyond the lines established by treaties with the Cherokee, were simply squatters on land that had not come under the control of colonial or federal authority. As settlement continued, later arrivals were obliged to settle the less-desirable, less-productive lands in the narrower and higher valleys and on ridges.

The first tasks of settlers included the clearing of land and the construction of a habitation. Cleared land was of course an absolute necessity for field agriculture, but perhaps equally important was the cultural psychology of the white arrivals: cleared landscapes represented civilization, or at least its beginnings; forests in close proximity were a threat. Like the Native Americans, settlers used fire to clear early successional plant communities and to maintain landscapes in early successional states. Clearing forests of large trees was a longer and more time-consuming process. Trees were killed by girdling with an axe, severing the cambium along the full circumference of the tree. Girdled trees died within a season. In the early settlement period, in the absence of sufficient labor or even of saws, the dead trees were left standing. Slash and understory vegetation were burned, and corn could be planted in the newly exposed, ash-enriched soil. Wind would eventually topple the trees in a few years, a process that could be hastened by burning slash heaped at the base of the trees. Toppled trees were used for fuel or for the construction of dwellings, barns, and fences, or burned in place. Later, saws and labor made it possible to clear land by removing trees outright, but girdling continued in isolated parts of the Southern Appalachians into the modern period.[72]

Maize, or Indian corn, was an early mainstay of agriculture during the settlement period and afterward. Varieties obtained from the Native Americans were better suited to the climate than were most European grains, and they could be more easily cultivated in newly cleared land. Wheat and rye cultivation required the plowing of a field free of stumps, roots, and rocks, and their harvest required enough open space to swing a scythe. Ground for planting corn was prepared with a hoe, and the grain and stalks were harvested by hand. Early settlers cultivated wheat, rye, and buckwheat on the best bottomlands and terraces in the upper valley, expanding their fields as the area of plowable lands increased. However, corn was the predominant crop on upper terraces and ridges, and the sole grain crop of later arrivals, who could obtain only marginal lands. Grains were used for human sustenance and livestock feed, and a considerable percentage of the corn crop (at different times and places as much as 25 percent) was made into whisky for local consumption and trade.[73] As transportation networks expanded, corn became an important commodity for trade outside the region.[74]

In addition to grains, the early Southern Appalachian farmers grew flax, and later hemp and cotton, for local use in clothing and cordage. Many farms produced tobacco, primarily for local consumption. Tobacco would not emerge as an important cash crop until after the Civil War. Every farm also had at least one large garden that produced vegetables and medicinal herbs. Beans, turnips, potatoes, sweet potatoes, winter squashes, and cabbages were major vegetable crops, because they could be easily stored for the winter, but a wide variety of other vegetables and greens were grown for immediate consumption during the growing season. Most farms also had at least a few apple, peach, and plum trees, and many grew native persimmons and pawpaws.

Wild animals and free-ranging livestock were a continual threat to crops and gardens. Stacked rail fences, made possible by the abundance of timber, were used to exclude cattle and draught animals, but were ineffective against hogs and a variety of wild species. Squirrels, in particular, increased in abundance as the amount of edge habitat increased and as populations of their predators declined. White-tailed deer also benefited from the increase in edge habitat and were drawn to the browse available in cultivated landscapes. Wattle and picket fences and fences built from field stones were constructed to protect gardens in some areas, but for most farm families protection of their crops required the continual vigilance of people and dogs.

Squirrel, deer, raccoon, bear, and possums were hunted as threats to produce and as sources of food and pelts. Turkeys, grouse, and other birds were also hunted for meat. The earliest settlers could include elk and bison among potential game, but by about 1780 both had all but disappeared from the area, and deer populations had declined substantially. By themselves, settlers would probably have eventually eliminated bison and elk, but these species' demise was hastened by commercial hunters serving the market for hides and for tallow rendered from bison. Large predators were also hunted, as actual or perceived threats to livestock. The earliest game laws enacted by colonial and state governments in the region were bounties on wolves and cougars. Laws regulating the taking of deer came much later. Large predators were quickly eliminated from the Ridge and Valley section, but wolves persisted in the Great Smoky Mountains into the early twentieth century, and cougars may have lingered there and in the Cumberlands even longer.[75]

Livestock grazing may have also contributed to the decline of large herbivores. Early settlers released hogs and cattle into the forest to browse on the mast and herbaceous vegetation. Cattle did not exactly thrive as forest dwellers, but they did survive and increased at little cost to the farmer. Hogs, on the other hand, flourished on the abundant mast, growing fat and reproducing at an amazing rate. Environmental consequences aside, hogs were the perfect Southern Appalachian livestock.[76] Competition for mast between the settlers' hogs and native white-tailed deer and black bears may or may not have contributed to the decline of the native species, but the effects of feral hogs on Southern Appalachian environments were undoubtedly substantial. Hogs root in the forest floor, feeding on mast, as well as the roots and tubers of perennial plants. Hog "plowing" and feeding encouraged the growth of pioneer annuals, including European weeds, at the expense of native perennials. Feral hogs, hybrids between domestic varieties and European wild boars, are troublesome even today in parts of the Southern Appalachians, especially the Great Smoky Mountains National Park and surrounding national forests, affording us an unwanted opportunity to study their effects on native communities.[77]

As the region became more densely settled and more land was cleared, especially in the Ridge and Valley, cattle were raised in pastures sown with non-native grasses and perennials. The higher quality of pasture browse for cattle was sufficient to keep cattle in or near these grasslands. Salt licks, used to supply necessary micronutrients to livestock, also tended to keep cattle from straying. In areas adjacent to the high elevations of the Blue Ridge, cattle were pastured at low elevations during the fall and winter, and then driven to prepared high-elevation pastures in the spring. Mountaintops and ridges above four thousand feet from northwest Georgia through southwest Virginia are dotted with the remnants of grassy balds that were cleared and used for summer pasturage of cattle and, to a lesser extent, sheep. Some of these summer pastures were used by farmers in the more remote and less densely settled regions into the early twentieth century.[78]

Methods for raising hogs did not change as rapidly, primarily because hogs thrived in the Southern Appalachian mast forests, requiring little time or resource investment by farmers. Hogs could be difficult to round up from the forests for driving to market, but the loss was made up for by the natural increase of populations. Increasing human populations and declining forests necessitated changes in hog management, but several variants of forest pasturage continued into the modern period. In some areas, forest pasturage became a seasonal activity: hogs were held in fenced lots from late winter through early summer and fed on corn; in late summer the hogs were turned out to subsist on forest browse and mast. Other farmers confined hogs to fenced lots on forested lands year-round, supplementing the mast and browse with corn feed as necessary. Yet another variant involved the planting of fruit trees in hog lots, with apples and persimmons taking the place of forest mast. In many places today, remnant orchards of apples and persimmons stand as signatures of former hog lots.

Hogs, cattle, and, in the upper valley, sheep were raised for domestic use and for sale in distant markets. Pork quickly became the predominant flesh consumed in the Southern Appalachians, even by descendants of the pork-adverse Scots. Pork remains a significant part of the diet among rural Southern Appalachians today and an element of a distinct regional cuisine. Beef was less frequently consumed locally, because of the value of cattle for trade: Farm families usually kept one or more milk cows, but most of the stock was raised for market. Mutton was not an important food item in the Southern Appalachians; rather, sheep were raised for wool, which was used locally and sold to distant markets.

By the late eighteenth century and continuing up to the Civil War, the upper Tennessee valley was a major livestock-producing region. Road and river transportation developed as much to allow for the transport of live-stock and other agricultural products out of the region as to facilitate the movement of people and goods into it. Expansion of the slave-based plantation economy of the Deep South, and the growth of populations east of the Southern Appalachians, provided ready markets. By 1800 Knoxville had developed as a river-transportation hub, connecting upper Tennessee valley farms with markets in the lower Mississippi valley.

During the nineteenth century Chattanooga also developed and grew as a Tennessee River transportation hub, and later Asheville developed as the center of a network of roads connecting the Southern Appalachians with the Piedmont and Costal Plain. During the first half of the nineteenth century, grain was carted and cattle and hogs were driven to Knoxville and Chattanooga in seasons of river transport, to be shipped as far south as New Orleans. Ginseng, harvested from the Southern Appalachian forests, was also shipped to New Orleans, the first leg of a long journey to markets in China.[79]

With the completion of the Buncombe Turnpike in the 1820s, Asheville became the principal transit point for the movement of livestock to markets in the more populous Eastern Seaboard. Asheville would remain the main point for moving cattle and hogs to external markets into the 1840s, when meat-processing plants and steam-powered transportation reduced the need for transporting livestock on the hoof.

By 1800 the number of free white settlers and slaves in the region totaled over one hundred thousand.[80] Settlers continued to arrive during the early nineteenth century, purchasing smaller or less-desirable tracts, or in the remote areas of the Blue Ridge arriving as squatters. As successive treaties coerced more lands from the Cherokee and made them available for official settlement, more lands to the south of Knoxville and in western North Carolina were opened for settlement. By 1810 the Ridge and Valley section had a population density of almost six persons per square mile, and by the 1820s the valley population may have been as high as thirty persons per square mile.[81]

Although some regions, particularly the Blue Ridge and the higher-elevation Cumberlands remained thinly settled, by 1830 the region had ceased to be a settlement frontier, and less than half of the land in the Ridge and Valley section remained forested. Bristol, Knoxville, Chattanooga, and Asheville had become large towns, as well as transportation hubs and centers of commerce, manufacturing, and civil administration. However, the population of Southern Appalachia remained largely rural and agricultural, as it would up to modern times. Most farm families were self-supporting in food production and other needs, and many produced surpluses for consumption by nonfarmers in the growing towns and for export.

Apart from agriculture, economic activity included market hunting for hides, furs, and tallow, and the collection of ginseng for export through middlemen to China. Within a decade or two of the arrival of the first settlers, early processing and manufacturing enterprises also sprang up to serve the needs of people in the region. Water-powered gristmills were established in most valleys within a few years of settlement, and, after the Revolutionary War and the decline of Cherokee resistance, they spread throughout the region. Sawmills, often in conjunction with gristmills, were also established, particularly in proximity to the developing transportation and commercial centers. Only later would sawmills become common in the more thinly settled, forested regions. Early in the settlement period, sawmill location was dictated by proximity to demand rather than proximity to timber supply, because timber was relatively abundant almost everywhere.[82]

Perhaps the earliest industry in the Southern Appalachians was the chemical industry. Salt, niter, and possibly other compounds, such as alum, were extracted and processed in the late eighteenth century from deposits in the upper valley of the North Fork of the Holston River in Smyth County, Virginia. The town of Saltville owes its name to this early industry, and it became the site of a major chemical-manufacturing complex in later years. Salt, the primary means of food preservation for subsistence farmers, was a valuable commodity. Gunpowder was almost as important as salt, and niter from the Saltville works was an essential ingredient. By 1806 as many as four powder mills may have been operating along the Holston River in Kingsport, Tennessee (originally named King's Mill), using water power to mill and blend the ingredients of gunpowder. Other early industrial activities operating at Kingsport and elsewhere at about this time were the extraction of linseed oil from flax seed, leather tanning, and the manufacture of iron goods.

Hundreds of small iron smelters and forges were established in the region prior to the Civil War, and a few, those in proximity to the nodular iron ore and coal deposits of the Cumberland Plateau, became major industrial operations, persisting into the twentieth century. Most of the operations were small, producing iron goods for local consumption, although a few firms on the navigable rivers sold goods to distant markets. Iron manufacturing involves two basic processes, smelting ore to obtain pig iron and forging the refined iron into functional items, such as nails, wheel rims, kettles, tools, and the like. Both processes required water power to operate bellows and hammers, and both required large amounts of carbon. Charcoal obtained from the low-oxygen combustion of hardwoods was the primary source of carbon for iron manufacture in the Southern Appalachians prior to development of rail-transportation networks. Forges required less carbon than furnaces, and when it could be obtained, even relatively low-quality bituminous coal could be used. Not so for smelting; this required clean carbon, free of metal, sulfur, and other contaminants.

Prior to the Civil War, charcoal was almost the exclusive source of carbon for smelting, and large amounts of charcoal needed demanded enormous amounts of wood. Davis estimates that the two tons of iron produced per day from the average nineteenth-century furnace required the woody biomass of about 300 acres of mature hardwoods annually to produce the necessary charcoal.[83] He calculates that twelve small works operating in upper East Tennessee, collectively producing considerably less than two tons per day, together consumed 12,000 cords, or more than 500 acres of hardwood in 1820. Extending the annual operation of these smelters over thirty years yields a figure of 15,000 acres. And these were small smelters operating in only two counties in one year; there were many more smelters operating in the Southern Appalachians before and after 1820, some with a larger capacity. In 1840, around the peak of iron production in upper East Tennessee, furnaces turned out 3,124 tons, consuming about 1,500 acres of forest.

Probably the most environmentally destructive industrial process before the Civil War was copper smelting. Beginning in 1850 with the opening of the first mine and smelter near Ducktown, Tennessee, dozens of operations opened to obtain copper from the "Copper Basin" of southeast Tennessee and northwest Georgia. Copper mining, like all mining, degrades the land. Copper smelting, however, is much more destructive still. Ores in the basin, which contain a relatively low concentration of copper, require processing, which in turn requires enormous amounts of timber.[84]

Fig. 1. *The still devastated landscape of Copper Basin near Ducktown, Tennessee. In the foreground is the historic Burra Burra Mine Pit. Copyright by Mignon Naegeli.*

Processing involved two steps. First, the ore had to be roasted by heating it for several weeks on top of a large pile of continuously burning wood. The roasted ore was then placed in iron furnaces heated with charcoal. Heating liquefied the copper, freeing it from the ore. By the end of the 1860s, more than forty square miles of timber had been removed from the vicinity of Ducktown to roast and smelt copper ores. Those forests could not re-establish themselves, because roasting and, to a lesser extent, smelting of the ores releases sulfur and metallic contaminants as gases and particulates. The gases from the Ducktown operations killed all vegetation for several miles around the processing sites. The bare soil eroded down to and into deep horizon clays. The sulfates increased the acidity of the soil above what seeds and seedlings can tolerate, and the acid mobilized metals to toxic levels. Copper smelting created a wasteland, a chemical desert in a region of abundant rainfall and an extraordinarily rich biota. Only in the middle of the twentieth century, through active restoration efforts, did vegetation begin to return to this landscape.[85]

The Upper Tennessee Valley in the
Mid-Nineteenth Century

By the middle of the nineteenth century, three hundred years of Euro-American contact with the landscape and its native inhabitants had brought changes that neither the Native Americans nor the European arrivals could have conceived. Except for a small, persistent, and determined group that managed to survive in the Great Smoky Mountains, the Native Americans were gone, their 12,000 years of cultural continuity living *in* the Tennessee valley brought to an end. In one hundred years of settlement, the population density of Euro-Americans and their unwilling slaves exceeded by an order of magnitude the highest Native American population densities the valley had ever seen.

But population density is not the whole story, or even the biggest part. The new arrivals, the Euro-Americans, brought a different attitude to the continent and to the bioregion. The landscape, its living inhabitants, soils, minerals, waters, even its air were to be used and manipulated to suit the needs, wishes, and seemingly insatiable desires of the new owner-inhabitants. Most of the native vegetation of the Ridge and Valley section had been altered by the time of the Civil War; the old forests were removed and replaced with simpler, biotically poorer communities. Still, most of the forests of the Blue Ridge remained unaltered, and the rivers, though polluted with human and animal wastes and the effluents of growing industry, ran free. One hundred years later the Blue Ridge forests were recovering from the first round of cutting, Ducktown was still a chemical wasteland, and the rivers were contained behind dams. Another fifty years later, in the present, these Euro-American values have altered the land almost unimaginably. But perhaps we are also initiating a new tradition of respect, humility, and sustainability. Perhaps.

Chapter 2

Air

Ulla-Britt Reeves and John Nolt

Always before the European invasion—and, indeed, for a long time after—the air of Southern Appalachia was healthful and pure. But as industry and automotive transportation expanded in the years following World War II, air quality plummeted. Though the passage of the first Clean Air Act in 1970 brought significant improvements, continued reliance on coal-fired power plants and polluting industries and the rapidly increasing use of internal-combustion engines still render the sultry summer air unfit to breathe. Regional air pollutants (see list below) contribute to asthma, chronic bronchitis, lung cancer, and other lung diseases, heart disease, diabetes, and neurological disorders—sometimes with lethal results—and their effects on the environment are multiple and profound.

Major Types of Outdoor Air Pollution

Ozone—produced from nitrogen oxides and volatile organic compounds in the presence of sunlight. Ozone is a toxic, colorless gas that oxidizes (in effect, burns) living tissue, causing respiratory diseases and damaging plants.

Fine particles (PM2.5)—(or particulate matter) consist largely of sulfate ions (from sulfur dioxide), nitrate ions, ammonia, organic chemicals and soil dust. They contribute to visibility-impairing haze and are harmful to human lungs when inhaled.

Toxic pollutants—these include mercury and about two hundred other toxic substances. They cause a variety of health problems, including cancer, neurotoxicity, and learning disabilities.

Greenhouse gases—chiefly carbon dioxide (CO_2), but also methane and other gases. Global warming, or climate change, increases temperatures, raises sea levels, increases the spread of tropical diseases, and damages ecosystems.

Acid precipitation—produced as sulfur dioxide and nitrogen oxides combine with water to create sulfuric or nitric acids. These precipitate out as acid rain, snow, fog, or dry particles and damage streams, soils, and plants.

Tropospheric (Ground Level) Ozone

Ozone is involved in two momentous but entirely different air-pollution problems: stratospheric ozone depletion and tropospheric ozone pollution. We have, ironically, too much ozone near the ground (in the "troposphere," the lowest six miles of the atmosphere) and too little higher up (in the stratosphere, from six to thirty miles above the ground). This section deals with tropospheric ozone pollution. Stratospheric ozone depletion will be discussed later.

Ozone (O_3) is a highly reactive form of oxygen the molecules of which consist of three oxygen atoms instead of the usual two. Because of its reactivity, it can oxidize (that is, sear or burn) sensitive respiratory tissues—in plants, animals, and humans. Small amounts of ozone occur naturally in the troposphere, but human activities, especially in the summer months, may increase ozone concentrations to harmful levels. Though some electrical equipment creates ozone directly, nearly all humanly generated ozone pollution is formed indirectly in the air by reactions between two other pollutants, nitrogen oxides and volatile organic compounds (VOCs), in the presence of sunlight. Ozone is therefore a secondary pollutant, derived from these two primary pollutants, which we do produce directly.

The nitrogen oxides, chiefly NO and NO_2, are collectively designated, using the variable "x," by the chemical formula NO_x. In conversation, this formula is pronounced like the name "Knox"—prompting the wry suggestion that, in view of its bad air, Knoxville should be renamed "NO$_x$ville." But NO$_x$ pollution extends all across Southern Appalachia.

NO$_x$ is a byproduct of burning, created when the heat of combustion causes oxygen and nitrogen molecules that occur naturally in the air to combine. Its main sources are power plants, industrial processes, and internal-combustion engines. TVA's coal-fired power plants are responsible for about one-third of the total NO$_x$ emissions in the region TVA serves. Cars, trucks, boilers, furnaces, and incinerators account for much of the rest, though a considerable quantity of NO$_x$ is also released by the application of nitrogen fertilizers.[1]

VOCs, which react with NO$_x$ to produce ozone, are an extremely broad class of chemicals that consist of millions of different hydrocarbon gases and vapors. Many VOCs, such as those responsible for the scent of pine trees, occur naturally. But others are released by the burning of fossil fuels or the evaporation of solvents in paints, adhesives, cleaning fluids, and many other chemical products. Some VOCs are in themselves relatively benign; others (such as gasoline or asphalt fumes) are toxic or carcinogenic.

During summer months, when sunlight is most intense, the ozone-forming reaction of VOCs with NO$_x$ is at its peak. Ozone levels in the Tennessee valley are highest on late summer afternoons, as sunlight and traffic combine to maximize ozone production, and lowest at night, when ozone formed during the day breaks down again into ordinary oxygen. But ozone levels in the Smokies are not so directly tied to traffic patterns. In the summer they remain constantly high for long periods, so that hikers and mountain vegetation receive unhealthy exposures both day and night.[2]

Ozone can irritate eyes and sear lung tissue even at low levels in sensitive populations. It scars lungs, promoting asthma, chronic bronchitis, and other respiratory problems—some of which can be fatal, especially to the weak and elderly. Its effects are greatest among children, asthma sufferers, athletes, outdoor enthusiasts, and the elderly. Some recent studies even link ozone exposure to cardiovascular disease, strokes, and lung cancer.[3]

Natural ozone concentrations in the air are between 20 and 40 parts per billion (ppb).[4] According to the EPA, concentrations exceeding 80 ppb for an eight-hour average are harmful to human health. The American Lung Association has advocated a lower figure of 70 ppb for an eight-hour average.

Ozone levels in much of the Southern Appalachian bioregion often exceed these standards in the summer months, but the full extent of the

Table 1

HEALTH EFFECTS OF TROPOSPHERIC OZONE IN THE BIOREGION
BETWEEN 1999 AND 2001

	Total at Risk Population	High Ozone Adult Asthma Cases	No. Unhealthy Ozone days	Air Quality Grade
Tennessee[a]				
Anderson	71,330	3,846	30	F
Blount	105,823	5,635	61	F
Hamilton	307,896	16,242	30	F
Jefferson	44,294	2,331	39	F
Knox	382,032	20,117	51	F
Meigs	11,086	567	—	—
Roane	51,910	2,830	—	—
Sevier	71,170	3,756	84	F
Sullivan	153,048	8,361	23	F
Total	1,198,589	63,685	—	7 F
North Carolina[b]				
Avery	17,167	872	1	C
Buncombe	206,330	10,171	10	F
Haywood	54,033	2,708	40	F
Jackson	33,121	1,655	10	—
Swain	12,968	619	0	A
Yancey	17,774	884	19	F
Total	341,393	16,909	—	3 F
Virginia[c]				
Georgia[d]				
Fannin	19,798	1,125	—	C
Overall Total	**1,559,780**	**81,719**	—	**10 F**

NOTES: [a]24 counties in Tennessee in the bioregion are missing monitors; [b]9 counties in North Carolina in the bioregion are missing monitors; [c]all 7 counties in Virginia in the bioregion are missing monitors; [d]5 counties in Georgia in the bioregion are missing monitors.
SOURCE: American Lung Association, State of the Air 2003.

problem is unknown. Of the 61 counties discussed in this book, only 16 have ozone monitors. Of these, 10 received a failing grade for dirty air from the American Lung Association's "State of the Air Report 2003" (see table 1). (Cities and counties are assigned grades ranging from "A" through "F" based on how often their air quality exceeds the EPA's 80 ppb standard.) It is likely that many counties with unhealthy air were not listed as failing simply because they lack monitors. The American Lung Association estimates that over 1.5 million people in our region are at risk from unhealthy air and over 81 thousand asthma attacks are directly attributable to high ozone levels. Knoxville consistently ranks among the 25 most polluted cities in the nation with respect to ozone. It was eighth in 2002, ninth in 2003.[5]

Ozone levels at high elevations in the Great Smoky Mountains National Park also frequently exceed the EPA health limit for ozone. From 1998 to 2002, the Smokies violated the 80 ppb standard 185 times during the summer and fall months.[6] Often Smoky Mountain air is more dangerous to breathe than the air of the cities from which visitors come. The National Park Service has recently begun posting health risk warnings for visitors and employees on days when ozone exceeds the EPA health standard.[7]

Health risk warnings may help people, but they cannot help plants. Ninety plant species in the park show signs of ozone-like damage, ranging from leaf injury and loss to reduced growth.[8] The visible symptoms include flecking (small colored areas, metallic or brown, fading to tan, gray, or white); stippling (tiny white, black, red, or red-purple spots); pigmentation (bronzing), in which the entire leaf turns brown or reddish-brown; chlorosis, a total death of tissue which turns leaves yellow or white; and early loss of leaves or fruit.[9] Over 50 percent of the black cherry, yellow poplar, and sassafras trees sampled near park ozone monitors in recent years have exhibited visible ozone damage—70 percent at the highest monitoring station. Among the black cherry seedlings surveyed along trails in 1992, 47 percent showed damage, and 88 percent of the tall milkweed plants surveyed in 1994 exhibited ozone-induced leaf damage. In general, the higher the elevation, the greater the damage. Continued exposure could extirpate some sensitive species, such as white pine or black cherry, from the mountains.[10]

Ozone also damages crops. The EPA has estimated that crop losses due to current ozone levels range on average from 10 to 15 percent. For some crops grown in the Southern Appalachian region—particularly spinach and tobacco—losses may exceed 15 percent.[11] In Tennessee alone, ozone damage to crops is estimated to be costing farmers between $38 and $65 million annually.[12]

It was once believed that the best way to reduce ozone formation was to lessen emissions of VOCs. More recently, however, it has become clear that because of the inevitable presence of natural VOCs, limiting human-produced VOC-emissions would improve the ozone levels only slightly. Therefore, the only practical way to address the ozone problem is to significantly reduce emissions of NO_x.

With respect to the power plants, at least, improvements are underway. TVA's annual NO_x emissions reached an all-time high of over 500,000 tons in 1994. Upgrades to the power plants reduced them to 261,812 tons in 2002. Even bigger improvements are planned. TVA has pledged to put selective catalytic reduction control technology on eighteen of its plants, reducing NO_x emissions by 75 percent during the summer by 2005.[13] While this commitment should help to reduce the summer ozone peaks, the technology is expensive to operate and will not be used at other times of the year. Year-round NO_x reductions are needed, however, to solve another NO_x-related problem, acid deposition (discussed in chapter 3). In the meantime, NO_x emissions from traffic continue to rise.

$PM_{2.5}$

Sulfur dioxide (SO_2) is a contributor to three important problems: regional haze, acid deposition, and fine particulate matter. Haze reduces visibility, and acid deposition harms plants and aquatic animals, but fine particulate matter directly impairs human health. This section deals with particulate matter. Haze and will be discussed later in this chapter and acid deposition in chapter 3.

Like NO_x sulfur dioxide is primarily a product of combustion, but unlike NO_x it is created only if the fuel itself contains sulfur. Gasoline, propane, natural gas, and other hydrocarbon fuels typically contain little or no sulfur and so do not contribute much to SO_2 pollution. But there is often a good bit of sulfur in coal. Thus coal-fired power plants are by far the largest source of sulfur dioxide pollution, accounting for about 75 percent of it in the Southern Appalachian region.

The primary culprit in SO_2-related health problems is fine particulate matter (often called $PM_{2.5}$ or PM fine), which is formed when sulfur dioxide mixes with sulfate and nitrate particles, ammonia, organics, and soil dust. The body can defend itself against larger particles, but when these fine particles are inhaled, they may pass through the cillary apparati and nose hairs and become permanently embedded deep in the lungs, where they promote a variety of maladies.

Particulate matter has been linked to asthma attacks, chronic bronchitis, emphysema, lung cancer, heart disease (ischema), strokes, birth defects, impaired fertility, and premature death.[14] Particulate matter exposures for a healthy, non-smoking adult have effects similar to those of living with a smoker and regularly breathing in secondhand smoke.[15] Healthy adults living in areas with high pollution levels exhibit faster and more significant decreases in lung function over the course of their lifetimes and in general are more susceptible to the host of illnesses associated with breathing polluted air. Long-term exposure to particulate matter is believed to account for up to 4 percent of all U.S. deaths—the equivalent of a one- to three-year drop in life expectancy.[16]

Children and the elderly are at a heightened risk of health problems from airborne particles.[17] Children are more vulnerable because they breathe in 50 percent more air per pound of body weight than adults. Fine particle pollution has been linked to increases in Sudden Infant Death Syndrome (SIDS) in cities with high pollution levels. Children are also more likely to suffer respiratory illnesses because, on average, they are active outdoors more than adults are and tend to be exposed to more air pollution.[18] Across the United States, childhood asthma has increased 55 percent between 1982 and 1996 and is now the leading cause of hospitalizations and lost school days, and the number-one health care cost for children.[19] Of all asthma cases 40 percent occur in children, yet children make up only 25 percent of the overall population.[20]

According to a recent study on the health impacts of fine particulate matter, in the 61 counties represented in the Southern Appalachian bioregion, over 950 deaths, 16,825 asthma attacks, and 143,339 lost work days annually are directly attributable to particulate matter from power plants. The average annual death rate from power plant pollution per 100,000 people in this region is 47.6—one of the highest in the nation (the highest estimated rate is 59 per 100,000 in Gadsden, Alabama).[21]

Fig. 2. TVA's Bull Run Steam Plant. A thermal inversion condenses water vapor in the plume. The water vapor acts as a tracer for invisible SO_2, NO_x, VOCs, particulate matter, and other pollutants. Copyright by Mignon Naegeli.

The many studies linking human health impacts to fine particulate matter pollution provide compelling reason to reduce sulfur dioxide pollution significantly and immediately. Over 75 percent of the SO_2 in the Smoky Mountains comes from coal-fired power plants. Not all the sulfate that affects the region comes from TVA, however. One study concludes that about 37 percent of the sulfate originates from power plants in the Ohio River Valley.[22] Scrubbers, which pass waste gases through a spray of finely pulverized limestone mixed with water, are the most efficient means of reducing sulfur dioxide emissions. While scrubbers efficiently remove sulfur dioxide from waste gases, they do not destroy the pollution but merely capture it in a toxic and acidic sludge—creating a waste-disposal problem.

TVA's sulfur dioxide emissions peaked in 1976 at 2,376,000 tons per year and by 2002 were reduced to 547,167 tons through the use of new technologies and fuel switching mandated by the Clean Air Act. However, SO_2 emissions from the two TVA coal plants closest to the Great Smoky Mountains National Park (Bull Run and Kingston) have continued to increase since 1995—up to a combined total of 133,340 tons.[23]

The 1990 Clean Air Act, though it mandated overall reductions in air pollution, gave polluters (including TVA) a certain amount of flexibility to pollute at reduced levels. This flexibility takes the form of pollution credits, which polluters may buy, sell, use, or bank. TVA promised not only to comply with the 1990 act but to do so without buying pollution credits outside its own system. In the spring of 1994, TVA achieved a surplus of credits by adding scrubbers to its largest sulfur dioxide source, the Cumberland power plant, about fifty miles west of Nashville. This retrofit reduced SO_2 emissions by about 278,000 tons per year.[24] But for many years thereafter, this surplus enabled the agency to avoid retrofitting controls at other plants. Gradually, however, TVA's pollution credits have dwindled, and, in order to remain in compliance with the law, the agency has had to purchase pollution credits from other utility companies. In addition, in the fall of 2002, TVA signed a contract for five additional scrubbers on four power plants (Bull Run, Kingston, Paradise, and Colbert). This equipment should reduce SO_2 emissions by 200,000 tons each year starting in 2006, eventually bringing the total, systemwide emissions down to around 300,000 tons. Still, if TVA were to reduce SO_2 emission to meet new plant standards, the total output would only be 159,404 tons.[25]

Regional Haze

To the Cherokees these mountains were the "land of the blue mist." European settlers named them, more prosaically, "Smoky Mountains"—which later generations glorified to "*Great* Smoky Mountains," presumably in order to entice tourists. These names refer to the ethereal blue mist that once surrounded the mountains like a nearly transparent veil. Today, most often in the summer months, the "smoke" of the Smokies is supplanted by a white, brown, gray, or yellowish haze. The original blue mist consisted mainly of water vapor and volatile organic compounds released by vegetation. But the haze so evident today consists of sulfate particles from the burning of coal at power plants in the Ohio and Tennessee valleys, smoke, dust, nitrate aerosols, ammonia, and the volatile organic compounds that billow from the exhaust pipes of the millions of cars and trucks moving ceaselessly through the lands below.

Until power plants and traffic began to fill the valleys—until the late 1960s, in fact—both the skies and the mountain mists were a healthy blue, and summer was the season of clearest air. From the high vantage points of Thunderhead Mountain or Clingmans Dome, the haze appears as a murky gray or featureless yellowish white sea that fills the entire Tennessee Valley, hiding it from view. From the valleys below, it is frequently visible at sunset as dirty brown or gray streaks near the horizon. At night it reflects back the light of cities, paling the sky and obscuring the stars.

Estimates suggest that fine sulfate particles account for 60 to 85 percent of the visibility loss; volatile organic compounds account for most of the rest. Chemical reactions in the atmosphere, accelerated by heat, humidity, and sunlight, transform the sulfur dioxide into sulfate particles, which scatter light, creating the visible haze. The average summer sulfate concentration in the park is ten to forty-two times higher than natural levels.[26] Nitrate aerosols, derived by chemical reactions from NO_x, account for only about 2 percent of the summer visibility impairment in the Smokies. Since 1948, pollution haze has decreased visibility ranges in the Great Smoky Mountains by 80 percent in the summer and 40 percent in the winter. The natural annual average visibility range is 77 miles, and the natural summer visibility range is 113 miles. Annual average visibility is now 25 miles and the summer visibility range is 14 miles. Often in the summer months, visibility drops as low as 5 miles, leaving the beautiful vistas of the mountain range completely obscured.[27]

Volatile organic compounds (VOCs) are responsible for about 11 percent of the visibility impairment. The final component of the haze, accounting for about 12 percent of the visibility loss, is elemental carbon, in the form of soot or smoke.[28] Open fires, diesel exhausts, cigarettes, wood stoves, fireplaces, small engines, and poorly maintained automobiles are common sources of smoke. Smoke particles are themselves harmful to the nose, throat, and lungs. But smoke often contains other dangerous substances—such as the tar and nicotine in cigarette smoke.

In addition to the health impacts and visibility impacts, nitrogen oxides and sulfur dioxide contribute to the formation of acid rain, acid snow, acid fog, and dry acidic particles—which, collectively, are known as "acid deposition."

Mercury

Mercury has the dubious distinction of being an air pollutant that does much of its damage in the water. Mercury is a toxic heavy metal that is emitted into the air by power plants and from the combustion of municipal waste, medical waste, and hazardous waste. Becoming water soluble through oxidation, it is then precipitated into lakes and streams where it may harm both human health and the environment. Mercury bioaccumulates through the food chain, becoming most concentrated in the top predators, which in lakes and streams are the sport fish. Moreover, because mercury is an element, it never biodegrades; mercury pollution from fifty years ago is still a problem today.

Airborne mercury comes predominantly from human-made sources, although it is also produced naturally by volcanoes and forest fires. The EPA estimated in 1997 that the combustion of fossil fuels or wastes produced 87 percent of the total mercury emissions (about 158 tons per year) in the United States.[29] In the bioregion, TVA power plants are by far the largest contributor to mercury emissions.

Mercury, a neurotoxin, was the cause of "Mad Hatter's disease" in the nineteenth century when it was widely used in the felt industry. Exposure to even small amounts can damage the nervous system, and it is particularly dangerous to pregnant women and children. Mercury poisoning causes irreversible effects that include developmental delays, learning disabilities, birth defects, and damage to the brain, kidneys, and other organs.[30] These days, consumption of mercury-contaminated fish is the most common form of exposure.

Mercury contamination not only affects health; it also threatens the fishing industry. In Tennessee the economic benefit of freshwater sport fishing in 2001 was over $480 million. Neighboring states also benefited, including Virginia ($517 million), Georgia ($544 million, 1,086,000 anglers), and North Carolina ($1 billion, 1,287,000 anglers), with a total economic benefit to this area of over $2.6 billion dollars. As consumers and anglers become more aware of the threats posed by mercury contamination, the fishing and tourism industries may suffer.[31]

Across the Southeast there is considerable variation in the number of state fish advisories that are issued for lakes and rivers. In 2003, for example, the state of Georgia issued advisories against fish consumption on 42 lakes and 77 rivers, whereas Tennessee issued only two advisories (North Fork of the Holston River and the East Fork of Poplar Creek).[32] The discrepancy is not due to the fact that Tennessee has less mercury pollution than Georgia. Rather, Georgia issues advisories for lakes and rivers with mercury levels above .23 parts per million (ppm), while Tennessee does not issue advisories until mercury levels exceed 1 ppm, over four times the level Georgia allows.[33] If Tennessee were to post advisories at the same level as Georgia, it would have *at least* 18 more "contaminated" waterways.

Other Toxic Industrial Air Pollutants

Mercury is only one of many toxic air pollutants, though an especially important one. Though toxic industrial air pollution has, as noted above, diminished overall since the passage of the Clean Air Act in 1990, it has not been eliminated. Our bioregion, unfortunately, has done worse than many other areas of the country in reducing industrial toxins. In 2001, North Carolina reported toxic air emissions totaling 115,130,332 pounds, making the state the second worst in the nation for toxic air pollution. Tennessee ranked seventh at 79,573,558 pounds.[34] But these figures are lower than actual emissions, since they represent only releases that industries report to the EPA. Emissions from some government sources and individual households are not reported, and nobody is actually checking most

Fig. 3. *Industrial pollution still exists in Southern Appalachia. An industrial plant on Blair Bend, off Route 11 in Loudon, Tennessee. Copyright by Mignon Naegeli.*

of the industry measurements. More ominously, Tennessee ranks first in the nation for developmental toxins, second for reproductive toxins, and third for suspected neurological toxin releases.[35] Yet there has been considerable progress. There were no reporting requirements in the 1950s and 1960s—when clothes hung out to dry in Knoxville turned black with soot, the nuclear weapons plants at Oak Ridge routinely released radioactive gases, and Chattanooga had some of the most poisonous air in the entire nation. Toxic releases in those days were certainly worse. Most of the progress can be credited to regulations mandated by the Clean Air Act.

The single most profound improvement has probably been the elimination of leaded gasoline. Lead accumulates in the body and is especially toxic to children, causing hyperactivity, learning disabilities, and decreased intelligence. In adults it may contribute to hypertension, heart attacks, and strokes. It also increases the likelihood of birth defects and stillbirths. Airborne lead has decreased dramatically over the past two decades.

Still, Southern Appalachia has more than its share of big industrial air polluters. The biggest (in sheer mass of toxic emissions to the air) is Lenzing Fibers in Lowland, Tennessee. Lenzing had the dubious distinction of ranking fifth in the nation for total toxic air emissions in 2000.[36] But a more recent report from the U.S. Public Interest Research Group in 2003 shows Lenzig Fibers as number one in the nation for both reproductive and developmental toxicant releases for air and water emissions. Lenzig also ranks third in the nation for suspected neurological toxicant releases.[37] The plant's main air pollutant is carbon disulfide, a widely used solvent. Inhalation of carbon disulfide has produced abnormalities of the genitourinary and skeletal systems, as well as disturbances of ossification and blood formation and dystrophic changes in the liver and kidney, in test animals.[38] Lenzing reported emitting 2,640,422 pounds of carbon disulfide, plus a smattering of other air pollutants, in 2001.

The bioregion's second largest air polluter is the TVA Kingston Fossil Plant in Harriman, Tennessee. TVA's Kingston plant reported 5,926,225 pounds of toxic air emissions in 2001.[39] Hydrochloric acid made up approximately four-fifths of the toxic air releases from the plant.[40] Hydrochloric acid can irritate eyes, nose, throat, airways, mouth, and skin of animals and people. It is also a biocide and is capable of killing plants, bacteria, algae, and fungi.[41]

Tennessee Eastman in Kingsport is the bioregion's third largest air polluter. Eastman reported releasing 4,735,421 pounds of toxic chemicals into the air in 2001.[42] This may appear to be a great improvement from Eastman's earlier emissions. In 1993, for example, it reported 29,027,863 pounds of toxic air releases. But the numbers are deceptive. In 1993 the EPA required polluters to report emissions of acetone, which is suspected of enhancing the carcinogenic effects of some other chemicals, such as benzene. But objections from the chemical industry ended the reporting of acetone emissions in 1994. As a result, while Eastman reported emitting 22,200,000 pounds of acetone into the air in 1993, acetone amounts are absent from subsequent reports to EPA, reducing the total emissions figure dramatically.

The chemicals listed in table 2 affect human health and the environment in widely varying ways; and, though all are toxic, some are much more toxic than others. It is not possible to list all their effects, some of which are in any case unknown. Some—glycol ethers, styrene, and toluene, for example—are reproductive toxins. Others—such as chloroform, dichloromethane, formaldehyde, and (again) styrene—are carcinogens.[43]

Many industries not listed in table 2 because their reported emissions are under a hundred thousand pounds still do considerable damage. Horse-

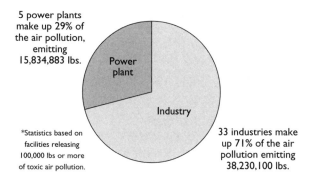

5 power plants make up 29% of the air pollution, emitting 15,834,883 lbs.

Power plant

Industry

*Statistics based on facilities releasing 100,000 lbs or more of toxic air pollution.

33 industries make up 71% of the air pollution emitting 38,230,100 lbs.

Fig. 4. *Toxic Air Emissions of Power Plants Compared to Other Industries in the Bioregion, 2000.*

head Resource Development, a hazardous waste-processing plant in the small community of Clymersville, in Roane County, Tennessee, for example, reported only 4,953 pounds of air emissions in 2001.[44] Yet neighbors have been plagued by dust containing toxic metals, some of which have been found in elevated levels in their blood. The dust is also killing vegetation.[45]

Table 2 reveals that the bioregion's five coal-fired power plants are all among the top ten toxic air polluters.[46] These plants are all significant sources of airborne mercury. They also emit relatively small amounts of other toxic metals, including arsenic, cadmium, chromium, beryllium, nickel, lead, manganese, selenium, vanadium, and radioactive uranium, radium, and thorium, which may become water pollutants as they settle out of the air.[47] The power plants' largest toxic emissions, of course, consist of sulfur dioxide and nitrogen oxides (NO_x), discussed above. And, of course, their largest emissions of all consist of carbon dioxide, which, not being considered toxic, is not included in the toxic-release inventory figures, though it is the most important contributor to global climate change.

A historically important source of toxic and radioactive air pollution was the U.S. Department of Energy plants in Oak Ridge. For a long time, little was known of the effects of their emissions, but some quantitative research is at last beginning to emerge. One study concerns the old graphite reactor at Oak Ridge. From 1944 through 1956, this reactor released radioactive iodine 131 into the air. The iodine, falling out on the

Table 2

INDUSTRIES EMITTING MORE THAN 100,000 POUNDS OF
TOXIC CHEMICALS INTO THE AIR IN 2001

Facility Location	Main Air Pollutants	Total Air Emissions (pounds)
A. E. Staley Mfg. Co. Loudon, TN	Acetaldehyde, Hydrochloric acid, Hydrogen fluoride, Sulfuric acid	530,784
Acupowder TN L.L.C. Greenback, TN	Copper, Manganese	155,542
Ahlstrom Engine Filtration LLC. Chattanooga, TN	Methanol	1,108,748
Alcoa Alcoa, TN	Carbonyl sulfide, Hydrochloric acid, Hydrogen fluoride	1,520,109
American Electric Power Clinch River Plant Cleveland, VA	Hydrochloric acid, Hydrogen fluoride, Sulfuric acid	2,152,283
Blue Ridge Paper prods. Inc. Canton, NC	Acetaldehyde, Ammonia, Cresol, Hydrochloric acid, Hydrogen fluoride, Methanol, Sulfuric acid	2,782,668
Bowater Newsprint Calhoun Ops. Calhoun, TN	Acetaldehyde, Ammonia, Chloroform, Hydrochloric acid, Methanol, Phenol, Sulfuric acid, Zinc compounds	1,035,498
Chattem Chemicals Inc. Chattanooga, TN	Ammonia, Methanol	391,967
Carolina Power and Light, Asheville Plant Arden, NC	Hydrochloric acid, Hydrogen fluoride, Sulfuric acid	2,412,342
Day Intl. Arden, NC	Methyl ethyl ketone, Toluene	311,187
DTR Tennessee Inc. Midway, TN	Methyl isobutyl ketone, Toluene	315,500

Facility Location	Main Air Pollutants	Total Air Emissions (pounds)
Eastman Chemical Co. Tennessee Ops. Kingsport, TN	Acetaldehyde, Ammonia, Barium compounds, Biphenyl, Bromine, Bromo-methane, Butyraldehyde, Chlorine, Ethylbenzene, Ethylene, Ethylene glycol, Hydrochloric acid, Hydrogen fluoride, Methanol, Methyl isobutyl, Ketone, Propional-dehyde, Propylene, Sulfuric acid, Toluene, Xylene, o-Xylene, p-Xylene	4,735,421
Foamex L.P. Morristown, TN	Dichloromethane	447,782
Harris-Tarkett Inc. Johnson City, TN	Methyl Ethyl Ketone, Methyl Isobutyl Ketone	103,072
Holliston Mills Inc. Church Hill, TN	Toluene	442,287
Johns Manville Intl. Inc. Etowah, TN	Formaldehyde, Methanol	243,270
Lenzing Fibers Corp. Lowland, TN	Carbon disulfide, Dioxin and Dioxin-Like compounds, Hydrochloric acid, Hydrogen fluoride, Lead compounds, Mercury, Sulfuric acid	2,640,422
Malibu Boats West Inc. Loudon, TN	Acetone, Styrene, Toluene	112,736
Marley Mouldings L.L.C. Marion, VA	Certain glycol ethers, Methanol, Methyl ethyl ketone	131,880
Mastercraft Boat Co. Vonore, TN	Dimethyl phthalate, Methylenebis (phenyliso-cyanate), Styrene	149,448
Merillat Corp. Atkins, VA	Xylene, n-Butyl alcohol	102,630

Facility Location	Main Air Pollutants	Total Air Emissions (pounds)
Nu-Foam Prods. Inc. Chattanooga, TN	Dichloromethane	374,415
RFS Ecusta Inc. Pisgah Forest, NC	Ammonia, Chlorine, Chlorine dioxide, Dibutyl phthalate, Hydrochloric acid, Hydrogen fluoride, Manganese, Mercury, Methanol, PCBs, Sodium hydroxide, Sulfuric acid, Toluene, Xylene	254,124
Sea Ray Boats Inc. (Riverview Facility) Knoxville, TN	Methyl methacrylate, Styrene	254,790
Sea Ray Boats Inc. (Knoxville Facility) Knoxville, TN	Styrene	308,401
Sea Ray Boats Inc. Vonore, TN	Methyl methacrylate, Styrene	341,920
Stanley Furniture Co. Robbinsville, NC	Acetone, Isopropyl alcohol, Lead, Methanol, Toluene, Xylene	101,127
Steelcase Inc. Fletcher, NC	Methanol, Toluene, Xylene, n-Butyl alcohol	112,556
Tennessee Watercraft Inc. Vonore, TN	Styrene	208,189
U.S. Army Holston Army Ammunition Plant Kingsport, TN	Acetone, Ammonia, Ammonium nitrate, Nitrate compounds, Nitric acid, PCB, Sodium hydroxide, Sulfuric acid	107,940

Facility Location	Main Air Pollutants	Total Air Emissions (pounds)
U.S. DOE Oak Ridge Y-12 Natl. Nuclear security Complex Oak Ridge, TN	Chlorine, Copper, Dioxin and Dioxin-like compounds, Freon 113, Hydrochloric Lead, Lead compounds, acid, Mercury, Mercury compounds, Methanol, Nitrate compounds, Nitric acid, Ozone, PCBs, Sulfuric acid, Tetrachloro ethylene, Zinc	187,107
U.S. TVA Bull Run Fossil Plant Clinton, TN	Hydrochloric acid, Hydrogen fluoride, Sulfuric acid	4,305,815
U.S. TVA John Sevier Fossil Plant Rogersville, TN	Hydrochloric acid, Hydrogen fluoride, Sulfuric acid	989,168
U.S. TVA Kingston Fossil Plant Harriman, TN	Hydrochloric acid, Hydrogen fluoride, Sulfuric acid	5,926,225
Viskase Corp. Loudon, TN	Carbon disulfide, Hydrochloric acid	2,268,148
Weyerhaeuser Co. Kingsport Paper mill Kingsport, TN	Acetaldehyde, Acetone, Ammonia, Barium compounds, Chlorine, Chlorine dioxide, Chloroform, Dioxin and Dioxin-like compounds, Formic acid, Hydrochloric acid, Hydrogen fluoride, Lead compounds, Manganese compounds, Mercury compounds, Methanol, Phosphoric acid, Polycyclic aromatic compounds, Sodium hydroxide, Sulfuric acid, Zinc compounds	313,188

land, was ingested by cows downwind in Bethel Valley and became concentrated in their milk. Once ingested by humans, iodine 131 is taken up by the thyroid gland, where it is especially carcinogenic to young girls. In one of the first quantitative assessments of the health effects of air pollution from the bomb plants, the Tennessee Department of Health concluded that the lifetime risk of thyroid cancer to girls who regularly drank milk from cows grazing two miles from the reactor was increased by about 0.6 percent. The increase in cancer risk would have been less for those who drank milk from cows farther away, for males, and for adults.[48] This is the estimated risk for only one pollutant. There are as yet no quantitative data on cumulative risks from the panoply of contaminants released at Oak Ridge.

The effect of outdoor air pollution depends to a large extent on the wind. The predominant airflow in Southern Appalachia is from the west, but local topographic features influence wind flow. The Tennessee Valley, with its northeast to southwest orientation, channels air currents so that surface winds tend to blow either up the valley from the southwest or down the valley from the northeast. (In Asheville, on the North Carolina side of the mountains, by contrast, surface winds blow most often directly from the north or from the south.) The Smokies are a formidable barrier to airflow, and Tennessee Valley weather almost never comes over the mountains from the southeast.

Unfortunately, our bioregion has nearly the highest incidence of stagnant air in the eastern United States,[49] and because industries tend to locate in sheltered valleys, toxic emissions often linger near the site of release or slightly downwind, prolonging exposures.

Indoor Air Pollution

Ironically, the air pollutant responsible for the most deaths in the bioregion is the one easiest to avoid: tobacco smoke. Studies show that tobacco smoke kills some 3,000 people annually from lung cancer and another 62,000 people annually from heart disease across the United States. In addition, it contributes to asthma attacks and lower respiratory-tract infections, as well as increasing the risk of Sudden Infant Death Syndrome.[50]

Second to tobacco smoke is radon, a naturally occurring radioactive gas given off by rocks and soil that can seep into buildings from the ground.

In open air, radon disperses rapidly and harmlessly, but in tightly closed buildings it may accumulate to dangerous concentrations. Inhalation of radon can cause lung cancer. Radon is prevalent in some of the soils of Southern Appalachia—especially in upper East Tennessee and northwestern North Carolina.[51]

Because radioactive particles adhere to smoke particles in the lungs, simultaneous exposure to radon and tobacco smoke results in a lung cancer risk much greater than the sum of the risks for each. This is a classic example of synergy: the multiplication of effect when two or more pollutants are present together. For many combinations of pollutants, the potential for synergistic multiplication of health effects is unknown.

Other common sources of indoor air pollution include mold, bacteria, wood and natural gas stoves and fireplaces, home cleaning products, scented products, foam insulation, plastic products, synthetic fibers, adhesives, artificial wood products, polishes, paints, pesticides, photocopiers, dust, and asbestos. Some indoor air pollutants, such as carbon monoxide, can kill quickly. Some, like radon and asbestos, are carcinogenic. Many, either singly or in combination, also cause headaches, eye irritation, nausea, allergies, memory loss, and suppression of the immune system—a suite of symptoms known collectively, in severe cases, as "sick building syndrome."[52] Because indoor pollutants are trapped and repeatedly recirculated, indoor air pollution may be more harmful to human health than air pollution outside. In fact it is, according to the EPA, the single most important environmental cause of cancer.[53]

The easiest and least expensive way to reduce indoor air pollution is to remove the source—for example, by prohibiting smoking indoors or by stopping leaks that contribute to mold growth. The next most effective way is to increase ventilation—for example, by opening the windows or by running exhaust fans in the bathroom more frequently. But because Southern Appalachia is cold in the winter and hot in the summer, many buildings have heating and cooling systems that require windows to be closed through much, if not all, of the year. Heating, of course, is necessary in our climate. But air-conditioning, widely available only in the past few decades, is (for most healthy people, at least) a luxury. There are alternatives: open windows, appropriate architecture, fans (preferably solar-powered), and proper placement of trees can do much to make buildings without air-conditioning breezy and tolerable even when the temperature is sweltering. It is ironic that in our effort to escape the summer heat through air-conditioning, we consume electricity generated by combustion, which releases greenhouse gases that have the undesirable side-effect of intensifying the summer heat.

Climate Change

When Hurricane Opal roared across the Smokies on October 5, 1995, flattening whole stands of trees and blocking trails, she provided a portent of things to come. The trees of the high forests, already thinned by acid deposition, ozone damage, and infestations of imported insects, could not adequately shield one another from the blast, and whole stands fell together. Red spruce in the transition zone from spruce to fir suffered heavy losses.

The Smokies are no strangers to wind and storm; the stunted and twisted trees clinging to rocks on the high peaks attest to that. But wind damage is likely to increase, not only because of the progressive thinning of the trees, but also because of the increasing ferocity of storms. The earth's atmosphere is heating up; and, just as water heated in a pan roils faster and faster as the heat increases, so the atmosphere will churn as it warms.

The atmosphere is warming for the same reason that a parked automobile gets hot on a sunny day: the heat is trapped by a greenhouse medium—glass in the case of the automobile, greenhouse gases in the case of the atmosphere. Other things being equal, the denser the greenhouse gases in the atmosphere, the greater the heating. And each year humans inject trillions of pounds of greenhouse gases (especially carbon dioxide) into the atmosphere, mostly by the burning of fossil fuels. In 1850, before widespread industrialization, the atmosphere contained 250 parts per million of carbon dioxide. Today it contains about 370 parts per million—an increase of 48 percent. The increase is accelerating.

The quantities of carbon dioxide are enormous. TVA's power plant system releases over 109 million tons of carbon dioxide annually (nearly 0.4 percent of the world's total), mostly from the burning of coal. At room temperature and sea-level pressures, this amount would fill a volume of over 1.8 trillion cubic feet. A box measuring a mile square and ten miles high (as high as wispy cirrus clouds) could not contain it all. TVA's Cumberland facility alone puts out over 19 million tons every year, placing it eleventh in a list of top one hundred power plants emitting quantities of CO_2 in 1999. And while it has reduced other pollutants, TVA has not reduced CO_2 emissions at all; in fact, emissions have risen by 6 million tons since 1999.[54]

Carbon dioxide emissions of cars, trucks, furnaces, stoves, and other combustion sources are also significant contributors to climate change. Other greenhouse gases, including methane, NO_x, and chlorofluorocarbons, are even more effective than carbon dioxide at trapping heat, though they are emitted in smaller quantities.

Such gargantuan emissions, together with actual global temperature measurements, have convinced nearly all climatologists that the average global temperature is rising and will continue to rise throughout this century. Because many other factors—including cloud cover, ozone depletion, sulfur dioxide emissions, and the extent of forests—also influence global temperature, the exact amount of the increase is unknown. Using a wide array of the best available climate models, the Intergovernmental Panel on Climate Change, probably the world's foremost authority on the subject, projects a global average increase of 1.4–5.8 degrees Centigrade (2.5–10.4 degrees Fahrenheit) between 1990 and 2100, with nearly all land areas warming more than the global average. The latest studies take into account both warming and cooling effects of pollutants. For example, emerging research indicates that sulfate particles might actually have a cooling effect, though this will not be sufficient to offset global warming.[55]

The effects of climate change are diverse. The increasing fluctuation of air currents produces extreme and violent weather—including not only prolonged heat waves, but abnormal cold snaps. The warmest year on record worldwide was 1998. 2002 and 2001 were second and third, respectively. Yet in Southern Appalachia, the winter of 2002–3 was unusually cold. Increases in ground-level ozone from hotter temperatures will threaten air quality and subsequent respiratory problems. Increases in rainfall, temperatures, and air pollution could worsen the threats to sensitive animal and plant species, like the salamanders and spruce fir forests, which are unique to our region and are very sensitive to climatic changes. Temperature changes will affect forests by driving out the native eastern hardwoods and replacing them with pine and scrub oaks and invasive plant species that thrive in hotter climates.[56] Rainfall is also likely to increase, producing more frequent flooding. Unfortunately, more precipitation also means increased siltation, which makes reservoirs shallower and thus floods harder to control.

Rivers, lakes, and streams are also affected by increasing temperatures. Higher temperatures can produce thermal stress and promote disease in such cool-water fish as sauger and walleye. Fish populations may change significantly. Higher temperatures alter the growing season of wetland vegetation, increasing the decomposition and nutrient cycling rate. Summer heat can dry up soils and decrease runoff, lowering dissolved oxygen levels. Some aquatic plants may be eliminated and possibly replaced by species that can withstand higher temperatures, higher (or lower) pH, and increased siltation.

On land, an increase in the minimum winter air temperature would extend the range of subtropical species, bringing invasions of exotics that would compete with indigenous species, rendering the existence of certain threatened or endangered species still more precarious. Growing seasons are likely to change, so that crop varieties now suited to this bioregion may no longer be successful. The instability of climate and weather patterns may make both long- and short-term agricultural investments increasingly risky. Flooding, hail, and high winds may damage crops and orchards. New diseases or exotic insects may afflict crops and livestock.

Global warming may reduce or eliminate some tree species, altering composition of forests. If there are long droughts, forest fires will increase in frequency and intensity. Increasing temperatures may strike the final blow to the cold-adapted spruce-fir forests of the high Southern Appalachians—forests already falling to exotic insects, ozone, acid deposition, and wind.

On a larger scale, polar melting and thermal expansion of the seas will probably raise ocean levels, flooding vast areas of coastland, especially in Florida, within a century or two. While this will not affect Southern Appalachia directly, it may increase population pressure as refugees flee inland.[57]

Human health impacts are also a concern of the changing climate. As the temperatures rise, so do ozone levels and thus asthma attacks, allergies, and other respiratory problems. Heat-related diseases like heat stroke may become more prevalent, as may waterborne diseases that infect humans through fish and shellfish. Warmer temperatures might expand the range of certain infectious diseases, such as malaria, hanta virus, and dengue fever, into Southern Appalachia.[58]

Even if we were to stop releasing all greenhouse gases immediately, the atmosphere would not return to normal for centuries. Moreover, there is a long lag time between elevated greenhouse gas levels and increased temperatures, so that the effects of the gases we introduce into the air today may not be felt for decades. At present, we are doing virtually nothing to slow carbon dioxide emissions, and two historic events—global deforestation and the industrialization of less-developed nations (China in particular)—will accelerate the increase still further. Expect rough weather.

Stratospheric Ozone

Though excess ozone in the troposphere is harmful, ozone in the stratosphere is vital to most living things. Stratospheric ozone acts as a shield that absorbs ultraviolet radiation from the sun. Without the ozone layer, more of this radiation would reach the earth's surface, causing rapid sunburn, cataracts, skin cancer, and immune-system deficiencies in humans and animals—and cell death, mutation, and growth inhibition in plants. Ultraviolet radiation can also accelerate the breakdown of plastics, such as the plastic siding now used on many houses.

Stratospheric ozone is not appreciably increased by ozone production near the earth's surface. Ozone molecules are too unstable to survive long enough to reach the stratosphere in significant quantities. (Nor does there seem to be any technically feasible artifice for moving large quantities of ozone from the troposphere into the stratosphere—which would be an elegant solution to both ozone problems were it feasible.)

Ozone in the stratosphere is generated by the interaction of ultraviolet radiation with ordinary oxygen molecules. Under natural conditions, this interaction occurs at a steady rate, which has kept the amount of stratospheric ozone stable for millions of years. But this equilibrium can be broken by ground-level release of less-reactive substances—particularly chlorofluorocarbons. Chlorofluorocarbons, which are extremely stable, do last long enough to migrate to the stratosphere. There each chlorofluorocarbon molecule can catalyze the destruction of many thousands of ozone molecules. During the last few decades of the twentieth century, tens of millions of tons of chlorofluorocarbons were manufactured, incorporated into consumer products, and ultimately released into the air. These chemicals are now eroding the ozone layer. International agreements have already phased out the manufacture of most chlorofluorocarbons, and there is hope of achieving a total worldwide ban by 2006; but because of their persistence in the atmosphere, the ozone layer is not expected to return to normal until about the middle of this century. Over the mid-latitudes of the Northern hemisphere, the ozone layer has thinned about 4 percent from pre-1980 levels in the winter and spring and about 2 percent in the summer and fall. The result has been a 6- to 14-percent increase in ultraviolet-B radiation at ground level.[59]

Ultraviolet-B radiation harms plants, animals, and humans. Although most terrestrial plants have some defenses against ultraviolet radiation, these are energy consuming and usually decrease the amount of light that enters leaves for photosynthesis. Excessive ultraviolet radiation changes the pigmentation, thickness, and anatomy of leaves and may repress growth and flower formation.[60] Exposure to intense ultraviolet rays may also make plants more vulnerable to diseases or insect infestations.

It is possible that ultraviolet radiation has already damaged Southern Appalachian forests. Throughout July of 1995, the ozone layer thinned to record or near-record lows over southern West Virginia. In the last ten days of July, the leaves of yellow poplars, sycamores, red maples, and redbuds turned brown and began falling along a fifty-mile stretch of the Coal River Valley, west of Charleston. In August the foliage of shagbark hickories and three species of oak—white, chestnut, and chinquapin—curled and turned brown. White pines, too, began to exhibit needle damage of a kind associated with excessive ultraviolet exposure. By August 20 damage was evident at all elevations. Though there was little rain in July and August, and drought undoubtedly contributed to the defoliation (similar drought-related damage was observed in the Tennessee Valley and on the Cumberland Plateau), the combined rainfall for May and June at the Coal River was over three inches above normal.[61] Researchers reported observing blotchy and wrinkled burns on the sunward sides of deciduous trees, and deformed growth of needles in white pines, possibly due to the intense ultraviolet radiation.[62]

Noise

Noise, though it travels through the air, is not a substance and hence not a pollutant in the ordinary sense. But it is a common, and sometimes harmful, concomitant of many human activities. Extremely loud sounds, such as the noise of machinery, high-decibel music, or low-flying jet aircraft can cause traumatic or progressive hearing loss. At lower levels, noise increases stress, producing such symptoms as headaches, nausea, and high blood pressure.[63]

Until the European invasion, the most intrusive noises in the Southern Appalachians (apart from an occasional thunderstorm) were probably the buzzing of insects and the squawking of crows and jays. Bird song was prominent, along with the sighing of wind in the trees. Near rivers, the rushing of water was audible. On summer evenings, choirs of frogs, crickets, and katydids could be heard, punctuated now and again by the lonesome call of an owl. At night, rarely, the mountain silence might be broken by the howling of wolves—or the heart-stopping wail of a mountain lion. Sometimes in winter—or in the dark, calm hours before the dawn—one might experience utter silence—silence so intense that within it could be heard the bombardment of air molecules against one's eardrums, playing a subtle music.

Things are different today, as the reader—if he or she is now anywhere within the bioregion—can verify simply by taking a moment to listen. Even on the peaks of the Smokies, one is seldom entirely out of hearing of traffic noises, gunshots, distant chainsaws, and aircraft. In most other locations, possibly excepting some isolated coves, traffic noise (the roar of engines, the blasting of horns, the squeal of brakes) is prominent—as is the noise of lawn mowers, leaf blowers, sirens, slamming doors, weed eaters, sound systems, and heavy machinery. Near interstates, the bellow and growl of diesel trucks is audible day and night.

These pervasive sounds do not harm people in any obvious way. They may be ugly, of course; but most people, most of the time, simply ignore them. So if we were concerned only with physical health, we might dismiss most of the noise as trivial. But though common noise pollution may do little apparent physical harm, it may degrade the health of the spirit. Seekers of silence in the Smokies, for example, are often disturbed by the invasive tree-top roar of sightseeing helicopters flying up from Pigeon Forge. This would be less troublesome if silence could be found elsewhere. It cannot.

What Will They Breathe?

Given the degree to which we are assaulting our lungs with ozone, particulate matter, NO_x, and a panoply of other toxic chemicals, it is doubtful that we ever experience a full, pure, healthy breath of air. But to these facts must be added one more—something that does not affect us significantly now, but raises a stark question for the future: for each atom of carbon added to the atmosphere by combustion, two atoms of oxygen are removed. As a result, the oxygen content of the atmosphere is decreasing by about thirteen parts per million per year.[65] This is a very small loss, since the current oxygen concentration is about 210,000 parts per million. But we should not ignore the fact that while injecting all these harmful gases into the air, we are gradually and ever more rapidly removing the one gas that is absolutely essential for all animal and human life.

Chapter 3

waтer

John Nolt and Keith Bustos

The Old River

From the Cumberland Plateau to the Smokies, the topography of the upper Tennessee valley is dominated by a remarkable series of parallel ridges that run from northeast to southwest. Four major rivers—the Powell, the Clinch, the Holston, and the Tennessee—and many smaller streams traverse the rich bottomlands between these ridges. Eastward into the Smokies, though the same southwest-to-northeast orientation prevails, the land is more jumbled and the rivers run westward to the Tennessee through deep gaps and gorges. The waters of the Nolichucky, French Broad, Pigeon, Little Tennessee, and Hiwassee all penetrate the mountains to reach the Tennessee.

Or, at least, what is called the Tennessee. The name rests on a mistake. The Cherokees called the river the *Hogohegee,* which in their language means simply "the big river."[1] The name "Tennessee," or, more properly, "Tanasi," was first applied by white settlers to the Little Tennessee River, which joins the Hogohegee just below Knoxville. But "Tanasi" was not the Cherokee name for the Little Tennessee either; the Native Americans called that river the Settico. Tanasi was, rather, an important Cherokee town on the banks of the Settico. Ironically, the site of this town, which gave the state and the river their names, now lies with other lands sacred to the Cherokee in a watery grave beneath TVA's Tellico Reservoir.

Moral reform begins, says Confucius, with the rectification of names. But, for the Hogohegee at least, it is too late; the Old River is lost. The sparkling blue current that once rushed through deep forested gorges and over wide rocky shoals now lies buried by hundreds of thousands of acres of deep, permanent flood. Beneath the murky, sluggish water, its banks and bottomlands, with their formerly fertile and lively farms, are entombed in toxic silt. Where once the riverbanks resonated with the chirping of birds and the croaking of frogs, there reigns a watery silence, disturbed only by the occasional heavy rumble of towboat engines or the frantic whine of a speedboat far above. Those old banks were long lined with dark forests and damp vegetation—sycamore, tupelo, and willows. The wider banks of the new Tennessee are often denuded and ever more frequently armored with bare, sun-bleached limestone, called riprap. It is dumped there to keep the fertile mud of the old forested hillsides or bottomland, now constantly pounded by speedboat wakes, from sloughing off and adding to the silt load.

Nowhere along the new Tennessee can the old Hogohegee be seen. From the confluence of the Holston and the French Broad just above Knoxville, where the Tennessee nominally begins, to Kentucky Dam just above its mouth on the Ohio, the water impounded and pooled by each successive downstream dam laps against the base of the next dam upstream, keeping the Hogohegee totally submerged. Where the old river made a continuous, often rocky descent, these colossal dams—nine of them on the main stem—divide the modern Tennessee River into a series of discontinuous steps. Behind each, the river swells out to inundate the land. Viewed from above, the new, human-made river is wider over much of its course than even the Mississippi, of which it is a mere tributary of a tributary.

Yet remnants of the Hogohegee survive, for the Cherokees did not define its bounds as we do. The upper end of our Tennessee lies just east of Knoxville, where the river branches into the Holston and the French Broad. But for the Cherokees the French Broad, though carrying a larger flow, was a tributary and the Holston a continuation of the Hogohegee, which thus has its origin far up in the mountains of southwestern Virginia. It is here along parts of the Holston that the old river may still be seen.

However we define the river, it is ultimately the destination to which all the waters of the bioregion flow. But the waters themselves arrive by air.

Water that Falls in the Mountains

The air that brings water to Southern Appalachia flows, when warm, from the tropical waters of the Gulf of Mexico or the Atlantic Ocean (sometimes in violent storms) and, when cold, from the frigid depths of the Canadian arctic. The predominant air flow, however, is from west to east: over the Cumberland Plateau, down into the Tennessee Valley, and up again over the Southern Appalachians. The lift of the air as it streams up over the mountains cools the air and often condenses its moisture. Then rain falls in the highlands—or, if it is very cold, snow. Hikers on the heath balds at the crest of the Smokies often see sinuous fogs rising up and over the peaks from the west and hear the rush of the air against the silence, eloquent testimony to the forces that shape Southern Appalachian weather.

Because they force the air to rise, condensing the moisture it contains, the Smokies constitute the wettest part of the bioregion. Some parts of the Smokies are temperate rainforest, with annual rainfalls exceeding a hundred inches per year. But about fifty inches, still an ample amount, is more typical for most of the region.

The rain, however, as was shown in chapter 2, is not pure. Dotting the Ohio and Cumberland valleys to the northwest and along the Tennessee Valley are the tall stacks of coal-burning power plants—Widows Creek, Kingston, Bull Run, Gallatin, John Sevier, and others. The effluent from these stacks is nearly invisible. It is not the thick black smoke of early industrialism, but a hot diaphanous billow of oxides of carbon, sulfur, and nitrogen. Moved by the prevailing winds, this pollution trails eastward in the lee of the stacks like the wake of a flotilla of boats, spreading out in widening plumes. As it approaches the mountains, this polluted air gathers other effluents rising, mostly unseen, from the stacks of thousands of industries, from a multitude of commercial incinerators, and from the exhaust pipes of millions of cars and trucks: oxides, once again, of carbon, sulfur, and nitrogen, hydrocarbons, and, in smaller amounts, a host of such industrial toxics as modern chemistry has devised. The stories of the toxic chemicals, the carbon compounds, and other contaminants have been told in chapter 2. But the fate of the sulfur and nitrogen oxides lies with the water.

Acid Deposition

The chemical reactions that seal that fate occur in the clouds. Sulfur dioxide and nitrogen oxides, combining with moisture in the air, form sulfates, nitrates, and dilute sulfuric and nitric acids. Then the moisture cools and condenses to form acid rain, acid ice, acid snow, or acid fog. Dry acidic particles may also settle directly out of the air. All these forms of acidity are known collectively as "acid deposition." Acidity is measured on the pH scale. Lower numbers correspond to greater acidity. Natural precipitation has a pH of 5.0 to 5.6. But the scale is logarithmic, so that a reduction of pH by one unit indicates a tenfold increase in acidity. Acid precipitation is harmful to terrestrial plants and many forms of aquatic life. The *average* pH of precipitation in the Great Smoky Mountains National Park (GSMNP) is 4.5—about ten times greater than normal.[2] Exposure to acid cloud water and ice damages plants directly by leaching nutrients from their leaves.[3] Clouds in the Smoky Mountains have an average annual pH of 3.5, and their pH sometimes dips as low as 2.0.[4] Rime ice on Mount Mitchell in western North Carolina—at 6,684 feet the highest eminence east of the Mississippi—sometimes has a pH as low as 2.1.[5]

The Great Smoky Mountains National Park receives some of the highest sulfur and nitrogen deposition rates in the nation. Between 1981 and 2000, annual wet nitrate deposition increased 16 percent.[6] Soil at the high elevations of the Smoky Mountains is so saturated with nitrogen deposited by the polluted air that it cannot absorb any more.[7] This means that it no longer has the capacity to neutralize nitric acid before it runs off into streams or percolates down to the water table. As a result rain flushes pulses of acidity into the high streams, which have little ability to maintain their natural pH level.[8] Though sulfate deposition has recently been reduced, thanks largely to the acid-rain provisions of the 1990 Clean Air Act (which targeted coal-fired power plants), increasing nitrate deposition has only made the problem worse. Moreover, two separate ecosystem models have concluded that sulfate reductions of 70 percent are necessary just to prevent more damage to sensitive sites.[9]

The average streamwater pH in the most sensitive streams of the Great Smoky Mountains has dropped by almost half a unit over the last twenty years, indicating an approximately threefold increase in acidity. Some of the higher-elevation streams have the highest nitrate concentration of any streams draining undisturbed watersheds in the United States.[10] During heavy rains, stream acidity increases by a factor of ten.[11] Water samples from streams in the Smokies have a median pH level of 5.6.[12] Changes in stream chemistry directly affect aquatic organisms, including the native brook trout. Though brook trout evolved in naturally acidic high-elevation streams, they are, paradoxically, very sensitive to acid fluctuations. If pH levels fall below 5.5, brook trout may die; and even a pH of 5.6–5.9 can jeopardize young fish and cause reproductive problems in adults.[13]

The acid also takes its toll on the forests. Long-time visitors to Mount LeConte (6,593 feet), which looms picturesquely above Gatlinburg, or to Clingmans Dome (6,643 ft., the highest peak in the Smokies) recall the red spruce and Fraser fir forest that occupied their higher flanks and summit as recently as the 1980s. Now, especially near their summits, only the bleached skeletons of these trees stand above fields of acidophilic blackberry brambles. Over 95 percent of the mature trees at the top of Mount LeConte have died. These dead forests leave even the most casual viewer with an overwhelming impression of malady. And these are not just isolated instances. Summarizing three decades of research on Appalachian forests, one recent survey concludes that "nearly all the forests in the region above 4,500 feet are in deep trouble."[14]

Acid is not the only problem here. Ground-level ozone and intense ultraviolet radiation from the thinning stratospheric ozone layer (see chapter 2) have also stressed the highland forests. And it was infestation by the balsam woolly adelgid (an aphid-like insect resembling a small bit of white fuzz), which was unintentionally imported from Europe, that finished off the Fraser firs. But just as an AIDS victim usually dies not of AIDS itself, but of secondary infections, so the adelgid was probably not the sole cause of the dying of the fir trees.

The degree to which acid precipitation is responsible for the loss of the Fraser firs has been a matter of bitter scientific and political debate for decades. But for the red spruce the issue is now beyond question. The red spruces are not infested by the adelgid, yet they are dying too. Research has shown that when nitrates and sulfates flow through the soil, they mobilize aluminum, which is naturally present in the soil. Normally, most of the aluminum is bound to other elements in the soil and is unavailable to plants. An overdose of free aluminum inhibits root growth and interferes with the spruce tree's ability to absorb calcium and zinc, which are essential nutrients.[15]

Fortunately, as acidified water moves down the mountain slopes, its acidity decreases and so does the damage. The limestone of the lower elevations and the valley floor is chemically basic, and it (together with the summer respiration of certain algae in the rivers and reservoirs) neutralizes the acid. By the time the water that falls as acid rain in the mountains reaches the Tennessee River, it is no longer abnormally acidic. Thus the damage done by acid deposition appears to be limited mostly to the highlands. It is, however, spreading downward and outward year by year from

the peaks, and the majestic spruce-fir forest of the high Southern Appalachians, the largest remaining forest of its type in the world, may already
be damaged beyond recovery.

Water that Falls on Fields

The rain that falls in the mountains falls elsewhere too—often on pasture
or croplands. The Southern Appalachian hills are much used for grazing,
and many pastures extend right down to a river or stream bank. When
they do, cattle consume the vegetation along banks, which then crumble
and erode, filling the water with muddy silt. Fecal material from cattled
washed or dropped into the stream contaminates it with viruses and bacteria, making the water unfit for drinking or swimming; swallowing the
water may cause gastrointestinal disease; getting it into an open cut can
cause infection.

The urine and feces of cattle are high in nitrogen and other nutrients,
and these are washed into the streams and rivers. There they are joined by
similar nutrients from other sources: leaking septic tanks, urban sanitary
sewers or sewage treatment plants, and runoff of fertilizer from agricultural
fields and suburban lawns. The combined influx of these nutrients into the
waterways is often excessive, sometimes causing extreme growth of species
of algae, which thrive on the nutrients. The algae cloud the water, making
it green, and preventing sunlight from penetrating deeply. When they die,
some algae release toxins powerful enough to irritate human skin and eyes.
More significant, the decay of dead algae that accumulate on the bottom
deprives the water of oxygen. Low oxygen levels, prevalent in many TVA
reservoirs, threaten fish and other forms of aquatic life.[16]

The effect of high nutrient runoff is dramatic where clear streams come
down from the mountains. The Little River, for example, where it leaves the
mountains near Townsend, Tennessee, is, even in the hot summer months,
cleat as crystal with a sandy and rocky bottom visible six feet below the surface almost as through unsullied air. Trout flit in and out of the shadows of
trees near the river bottom. But near Rockford, only a few miles downstream, its summer waters have the appearance of muddy pea soup—the
result of runoff of agricultural wastes and fertilizers and leakage from septic tanks—the latter a symptom of Blount County's rapid overdevelopment.

By the time the Little River reaches the Tennessee, it is a sad, sluggish, mucky stream—and (to add injury to insult) its thick sediments are laced with carcinogenic PCBs, dumped several decades ago by the Aluminum Corporation of America (ALCOA) and TVA's Singleton Marine Ways facility.[17]

Runoff of nutrients into streams is facilitated by the destruction of their riparian corridors—the borders of vegetation that under natural conditions line a stream or river. Healthy streams in Southern Appalachia do not have bare, mown, or closely cropped banks; they are lushly overhung by trees and bushes that provide shade and habitat for frogs, turtles, salamanders, herons, and other aquatic creatures. Riparian vegetation filters out contaminants and excessive nutrients before they reach the water, reduces streambank erosion, and traps much of the silt carried by surface runoff from fields during heavy rains. Such vegetation also keeps the water shady and cool, reducing the growth of algae. Healthy streams are normally clear, and they have rocky bottoms rather than the thick, oozy bottoms now so common, which are largely the result of disturbance by humans and livestock. While a winding stream through a cropped pasture may be picturesque, it is no more natural or healthy than a typical suburban lawn.

A particularly embarrassing instance of riparian corridor destruction may be seen on the south bank of the Tennessee River just west of the Alcoa Highway bridge in Knoxville. There on the agricultural campus of the University of Tennessee (where more enlightened practices ought to prevail), trees and bushes have been removed and fields are plowed almost to the river's edge, producing readily visible gullies and bank erosion.

Turbidity and siltation affect aquatic life. Silt can clog the gills and filter-feeding apparatus of many organisms. Turbidity may block vision and keep animals from finding their food. Many species of fish, insects, and mussels need rocky stream bottoms to reproduce, and many are, as a result, rare, threatened, endangered, or extinct. Most of these species, however, are small and secretive, so that their plight has scarcely been noticed. Their names are largely unfamiliar—Smoky madtom, blackside dace, yellowfin madtom, tippecanoe darter, southern brook lamprey, brook trout, flame chub, blue shiner, splendid darter, Tennessee dace, blotchside logperch, golden topminnow, redband darter, firebelly darter, spotfin chub, slackwater darter, Tuckasegee darter, Barrens topminnow—though one, the snail darter, once achieved a certain notoriety. Their stories will be told more fully in chapter 4.

These creatures are also assaulted by the chemical-laden run-off from fields, railroad rights-of-way, or roadsides that have been treated with pesticides and herbicides. Intensely contaminated runoff may kill aquatic life directly; more diluted, it may increase the incidence of cancer and other maladies.

Water that Falls on Asphalt

Because of burgeoning urbanization—particularly around Knoxville, Oak Ridge, Chattanooga, Pigeon Forge, the Tri-Cities, and Asheville—much of the rain that in former years would have watered the woods and fields now falls on asphalt. Of this, a large portion—especially when the asphalt is hot—evaporates back into the air. The remainder mixes with road grit and the toxic effluents of leaky automobiles—antifreeze, brake fluid, gasoline and diesel fuel, grease, windshield-washer fluid, transmission fluid, abraded rubber from tires, asbestos from brake linings, and oil—to form a slippery slurry that precipitates automobile accidents like an advancing wave at the forward edge of the rain system. As the rain intensifies, this poisonous slurry washes across the pavement of roads, parking lots, and gas stations. Traveling in iridescent slicks along gutters and curbs, it plummets into a dark underworld of storm sewers, where it is conveyed downward through dark concrete tunnels, into a nearby river or stream.

Stormwater is untreated. Unlike sanitary sewers, which transport the wastes from indoor drains and toilets (and from businesses and factories) to sewage treatment plants, storm sewers flow directly to the nearest waterway. The chemicals, filth, and trash that people throw, drop, or leak onto roads or parking lots are thus transformed directly into water pollution.

Also, toxic materials are illegally dumped into storm sewers by negligent or unscrupulous businesses. The Norco Metal Finishing Company in Knoxville, for example, has been caught discharging toxic chemicals into storm sewers that drain directly into the Tennessee River.[18]

Storm sewers are typically composed of many sections of large-diameter concrete pipe that are fitted, but not sealed, together. In many places these pipes are cracked and broken by tree roots, construction, ground movement, or the aging of the concrete itself. During dry periods, contaminants such as gasoline, paint, bleach, antifreeze, oil, and industrial effluents that are

Fig. 5. *Heavy development in west Knox County has increased flooding. Copyright by Mignon Naegeli.*

dumped into the storm sewers are not flushed through the system into the streams but lie in low spots in the pipes, sometimes leaking slowly into the soil and ground water.

Leaks in the pipes—and in some cases direct connections made by unscrupulous or ignorant plumbers—also permit the influx of sewage from sanitary sewers into storm sewers.[19] This is particularly common during heavy rains, when the soil is saturated and both sewer systems are full of rushing water, and it is one of the ways by which sewage and industrial wastes enter rivers and streams.

Not all the water that falls on asphalt evaporates or goes to the storm sewers. Much of it runs off directly into rivers and streams. A meadow, woodland, or a field absorbs rain, both in the soil and in the vegetation, but asphalt absorbs nothing. During a heavy rain, the expansive roofs and parking lots of industrial parks and commercial developments quickly shed powerful torrents of water into nearby rivers and streams, increasing the frequency of damaging floods. A flood stage, for example, that used to be reached once in twenty-five years on average before development may now be reached every ten years. Recent flooding along the West Prong of the Little Pigeon River has undoubtedly been exacerbated by the increase in runoff from extensive paving associated with rapid development in Pigeon Forge and Sevierville.

Siltation

Before the arrival of European settlers, the Tennessee Valley was forested. But since the middle of the eighteenth century, logging, plowing, mining, construction, and road building have repeatedly bared large swaths of earth. The resulting erosion has flushed much of Tennessee's rich soil into creeks and rivers, taking some of it as far as the Mississippi Delta and the Gulf of Mexico. However, with the advent of government conservation programs in the 1930s, as well as decreased farming in the Southeast, this source of sediment began to diminish.

Then in the 1930s and 1940s, TVA dammed the Tennessee River. Large dams are efficient sediment traps. A reservoir with a capacity that is one-tenth of its annual inflow retains 80 to 90 percent of the silt that enters it.[20] This silt settles out as sediment, and, as a consequence, the reservoir grows shallower, until in a matter of decades or centuries (depending on its size) it is reduced to a shallow lake or stream-crossed wetland behind a useless concrete cliff.

The siltation process is most advanced in TVA's Nolichucky Reservoir and at three TVA reservoirs on the Ocoee River. Nolichucky Reservoir, also called Davy Crockett Reservoir, about eight miles south of Greeneville, Tennessee, was formed by a dam built in 1913 and acquired by TVA in 1945. Clogged by tailings from kaolin, feldspar, and mica mines located near Spruce Pine, North Carolina, the dam has been unable to generate power since 1972.[21] What was once a reservoir has since become a waterfowl refuge.

Ocoee Number Three Reservoir in Polk County in the extreme southeastern corner of Tennessee silted up even more quickly. Here, over a century ago, clear cutting combined with acid deposition from intense copper

Fig. 6. *Mudflats on Ocoee Number One Reservoir, showing the effect of heavy siltation. Copyright by Mignon Naegeli.*

smelting operations at nearby Copper Basin reduced thirty thousand acres (about fifty square miles) of rich forest land to sterile desert and exterminated aquatic life in the Ocoee River. The barren land eroded into deep furrows and gullies as the acidified soil, laden with heavy metals, washed into the river. Decades-long reclamation efforts have recently succeeded in revegetating most of the land, and fish are returning to the Ocoee after being absent for over a century.[22] But the reservoir, created in 1943 by a nine-million-dollar dam 110 feet tall and 612 feet wide, is all but gone, having lost 98 percent of its 8,700-acre-foot storage capacity to siltation.[23] Downstream, Ocoee Number One Reservoir (also called Lake Ocoee or Parksville Lake) and Number Two Reservoir are also highly silted, as evidenced by extensive mud flats.

Lake Junaluska, near Waynesville, North Carolina, a small private reservoir owned by the Methodist church, is plagued with silt and can be kept open only by dredging. It took fifty years for the lake to fill with sediment for the first time. It was first dredged in 1964. It has had to be dredged again in 1973, then in 1982, and again in 1992 or 1993. The sediment seems to be coming from residential and urban development in the area. Influxes of chemicals and solid waste have occasionally killed fish in the reservoir.[24]

At present rates of siltation, the larger reservoirs will take a good bit longer to silt up—several hundred to several thousand years, by TVA estimates.[25] But the erosion caused by the current influx of paper pulp and logging industries into the Southeast and the intense road building and construction throughout the region will undoubtedly add to the sediment load. And though the larger reservoirs are in no immediate danger of reverting to wetlands or mud flats, siltation continually diminishes their capacity and enlarges the winter mud flats around their shores.

Mud-flat expansion is accelerated by the wakes of powerboats. Where shorelines are unprotected by rocks or vegetation, the wakes undercut them, leaving a rooted overhang. This overhang eventually collapses into the water, producing an outward-sloping cliff, which in turn is undercut again, perpetuating the cycle. In this way as much as two meters of shoreline may be lost in a single year. Reservoirs thus grow continually wider and shallower, the ecologically important summer riparian zone narrows, and turbidity increases as sediments cloud the water.[26] Powerboat wakes have other, more subtle effects as well. Great blue herons fishing along the riverbanks need clear water to see their prey. When a powerboat wake disturbs the shallows in which a heron is fishing, it usually flies to clearer water, since that spot, now filled with roiling silt, will remain unusable for fifteen or twenty minutes.[27]

Silt from any source creates other problems as well. Suspended densely in water, it decreases light penetration, reducing photosynthesis and sometimes even killing aquatic plants. The plants then no longer supply oxygen to the water, and their decay lowers dissolved oxygen levels still further. Dense turbidity can also harm fish and other aquatic animals by temporarily blinding them or clogging their gills.[28]

Individuals can help reverse the siltation problem by reporting commercial violators, especially construction projects, to regulatory agencies—e.g., the Tennessee Department of Environment and Conservation (TDEC)—which have the power to issue citations that may result in fines. Because such agencies often lack the staff to do adequate field inspections, regulatory violations—damaged or inadequate silt fences, for example, are common—citizen reporting helps these agencies to locate and cite offenders.

Such efforts, though they can alleviate the problem, will not eliminate it. Sooner or later—and even before the reservoirs silt up completely—the dams themselves will defeat one of their central purposes, for by trapping the accumulating silt they will diminish the reservoirs' capacity for flood control. Dredging is unlikely to solve the problem. The sheer volume of the silt contained in the big reservoirs is so large that to remove any considerable portion of it would be prohibitively expensive.

Toxic Sediments

In addition the expense of mere removal, any attempt to remove silt from many reservoirs would constitute an imposing hazardous waste-disposal problem. The silt has entrapped not only an enormous quantity of plain old garbage, but also deposits of toxic chemicals poured into the river through decades of industrial abuse. Most experts believe there is no practical method for removing or cleaning this silt and that the safest thing to do is to let still more layers accumulate over it until it is deeply buried.

Contaminants vary considerably from reservoir to reservoir and, within a given reservoir, from one creek embayment to the next, and the contaminants migrate from time to time as currents shift the silt. Yet there are relatively stable areas in which particular contaminants are known to occur in especially high concentrations.

One notable case, already mentioned, is the Fort Loudoun Reservoir, where silt contains low levels of chlordane (a pesticide previously used to control termites and crop pests), zinc, and significant concentrations of PCBs (polychlorinated biphenyls).[29] PCBs are oily liquids manufactured until 1976, when their production was banned, for use as insulators in electrical equipment, particularly transformers, and as hydraulic fluid. PCBs are now known to cause cancer, reproductive problems, and skin eruptions. However, EPA rules concerning their disposal did not take effect until 1978 or 1979. Prior to and during this lapse, PCBs were often simply dumped onto the ground or into the water.

Once released, PCBs tend to bind to particles of soil, which migrate into waterways, eventually sinking to the bottom as silt. There they are ingested or absorbed by microorganisms, worms, insects, or mollusks living in the sediments. Predators eat these small organisms, and they in turn are consumed by larger predators, and so on, up the food chain. Since PCBs tend to be stored in fatty tissues rather than excreted, and since each predator eats many times its own weight in prey species, the slight concentrations of PCBs in organisms low on the food chain are multiplied in their predators. This effect, called bioaccumulation, may produce dangerously high levels of PCBs in the largest aquatic predators, particularly catfish and bass.

PCB contamination is serious in many of the bioregion's waterways, particularly Fort Loudoun, Watts Bar, Tellico, Melton Hill, Nickajack, Boone Reservoirs, Chattanooga Creek, and East Fork Poplar Creek. Fish from all these locations contain PCBs at levels that may increase the cancer risk for people who eat them. The Tennessee Department of Environment and Conservation has issued advisories against the consumption of various species, primarily catfish and/or bass, from each of these places.[30]

Probably the bioregion's most dangerously contaminated silt occurs in the vicinity of the U.S. Department of Energy's Oak Ridge Reservation, a legacy of the production of nuclear weapons that began during World War II. White Oak Reservoir, which collects drainage from seven nuclear and chemical waste burial grounds on the reservation, is the most radioactive lake in the country.[31] Its sediments contain extraordinary levels of both chemical and nuclear contaminants, including mercury, cesium 137, cobalt 60, strontium 89, strontium 90, plutonium 239, and PCBs.

Fed by White Oak Creek and Melton Branch, the lake empties its overflow through a small dam into the Clinch River. Here, at the White Oak Embayment, in November 1990, Department of Energy (DOE) officials discovered a "hotspot" of radioactive cesium 137 in the sediments that posed a substantial hazard to people fishing or swimming there. (This discovery occurred after repeated public assurances that no such contamination existed.) DOE and the U.S. Army Corps of Engineers belatedly decided to construct a cofferdam across the mouth of the embayment to contain the sediments. The dam was completed in the spring of 1992 at a cost of seven million dollars.[32]

White Oak Creek and Poplar Creek, both of which drain the reservation and empty into the Clinch River, are the chief sources of outflow of radioactive and chemical contaminants found in sediments in many areas along the lower Clinch and in Watts Bar Reservoir. These pollutants consist primarily of PCBs, mercury, and cesium 137, all of which are associated with the manufacture of nuclear weapons. Fish sampled downstream from the reservation have shown measurable amounts of mercury and PCBs, as well as Cobalt 60, Cesium 137, Strontium 89, and Strontium 90.[33]

Mercury, in particular, was released into Watts Bar in large quantities. Like PCBs, mercury accumulates in the tissues of fish and in the human body—where in sufficiently high concentrations it may damage the brain, nervous system, and kidneys. The Department of Energy knew all this but kept it from public attention until May of 1990, when the Foundation for Global Sustainability obtained and released to the media government documents detailing the contamination of Watts Bar. A group of resort owners later filed a twenty-four-million-dollar lawsuit against Department of Energy contractors Martin Marietta and Union Carbide for business lost as tourists and anglers quite reasonably avoided the reservoir. Just before the trial was to begin in 1994, the plaintiffs accepted an out-of-court settlement for two million dollars.

The Foundation for Global Sustainability also learned that the Department of Energy had failed to notify the Army Corps of Engineers of the danger of disturbing the contamination by dredging, which the corps oversees. As silt accumulates in the reservoirs, older layers are buried beneath newer ones. Since mercury, PCBs, dioxin, chlordane, and radiological releases have been greatly reduced over the past few decades, cleaner layers of silt have gradually settled over the most heavily contaminated layers, insulating the water from these toxic substances. However, dredging operations or currents generated by the powerful engines of towboats may stir up the buried layers sufficiently to resuspend the contaminants. Moreover, as the silt slowly migrates downstream, some of it passes through the dams, where contaminants may re-enter the water as a result of the great turbulence within the turbines.[34] These disturbances may make portions of the water temporarily poisonous.

Silt along the Pigeon River and particularly in Waterville Reservoir in Haywood County, North Carolina, was for many years contaminated with dioxin, a product of the chlorine bleaching process used throughout much of the last century to whiten paper by the Champion International Corporation. The Pigeon runs along the northern border of the Great Smoky Mountains into the French Broad, which flows into the Tennessee at Knoxville. Dioxin, which is not a single chemical but a family of over seventy-five related compounds, is best known as the primary contaminant in Agent Orange, the toxic defoliant used during the Vietnam War. Its health effects are multiple and profound: it causes cancer, birth defects, lowered sperm counts, and damage to the immune, reproductive, and nervous systems.[35] In the Pigeon River the dioxin moved from the silt into the food chain in quantities sufficient to produce deformed jaws, skeletal defects, and death in fish.[36]

But in the late 1990s, Champion switched to a new incineration/ evaporation process, which removed all the dioxins from the plant's water emissions. Shortly afterward the plant was sold and became Blue Ridge Paper Products. Water quality in the Pigeon rebounded, and the Tennessee Department of Environment and Conservation subsequently removed its warning against consuming fish from the river, but North Carolina has retained its fish advisory for Waterville Reservoir.[37]

Toxic Pollution of Surface Waters

Industrial pollution of the bioregion's waterways peaked in the 1970s. Since then, as a result of the regulation initiated by the Clean Water Act in 1972 and associated legislation, it has on the whole significantly diminished. Much of the pollution that persists in the sediments of rivers and streams is the heritage of past industrial abuse. But repeated efforts, most recently those of the George W. Bush administration, to weaken the Clean Water Act and cut funding for its enforcement are an unnerving reminder that this progress is not secure. Moreover, though industries generally pollute less now, many still release hundreds of tons of toxic materials into the water— sometimes with the blessing of regulatory agencies, sometimes without.

According to one recent estimate, one in three facilities violated their pollution permits at least once between January 2000 and March 2001.[38] Moreover, Tennessee ranks among the top ten states with the greatest number of major facilities in significant noncompliance with their Clean Water Act permits. Tennessee is also ranked in the upper eight states in the nation for having the worst violators between October 1998 and December 1999.[39] More specifically, 84 of 155 facilities in Tennessee (including Oak Ridge National Laboratory, Eastman Chemical Company in Kingsport, and sewage treatment plants in Jefferson City and Harriman) discharged significantly more pollution than their permits allowed between January 2000 and March 2001.[40]

Tennessee has within its borders 60,187 miles of streams and 536,724 acres of lakes, and state agencies have been unable to assess them all.[41] As of December 2002, about 50 percent of Tennessee's streams have been assessed, and of this portion about 30 percent are jeopardized by pollution.[42] Roughly 155 river miles are posted due to bacterial contamination and almost 120 river miles are posted due to contaminated fish.[43] Almost all of Tennessee's lake acres have been assessed and about 78 percent of them are listed as "fully supporting" their uses, the remaining 22 percent being listed as "impaired."[44] Within the impaired lakes more than 94,000 lake acres are posted due to fish contamination.[45] The overall water quality in the mountainous areas of the bioregion is relatively good, but many Ridge and Valley streams are polluted by urban development and agricultural practices. The urban streams of Bristol, Chattanooga, Kingsport, Knoxville, and Johnson City suffer from urban stormwater pollution. Many streams in East Tennessee are regarded as "seriously impaired" due to high concentrations of bacteriological pollution that stems from urban runoff, municipal bypassing, and animal operations.[46] Several lakes and reservoirs have fishing advisories due to PCB contamination. (For a complete list, see appendix 1.) These are chronic conditions. Acute incidents also still occur.

On February 25, 2000, for example, toxic chemicals released by a fire at the North American Rayon Corporation caused a massive fish kill in the Watauga River, stretching from Elizabethton to Boone Lake (about ten miles). The water downstream was covered by a soapy white foam. As the toxic suds drifted to the Johnson City Water Treatment Plant, managers shut down the treatment plant to guard the city's drinking water from contamination. A few days after the release, mats of dead fish washed up onto the shores, and the beds of once dark green algae lining the river's bottom died, turning the river to a dismal brown. Such acute incidents, however, need not do permanent damage, and today the Watauga is regaining its health.[47]

The bioregion's largest industrial source of water pollution is the Eastman Chemical Company in Kingsport, Tennessee. In 2000, the most recent year for which EPA data are available, Eastman discharged 388,138 pounds of toxic chemicals into the Holston River. Table 3 lists toxic chemicals released by the highest-volume polluters. Among these chemicals are a number of carcinogens and reproductive toxins. The figures are those reported by the industries themselves; there is no independent verification of their accuracy.

Chapter 2 explained how mercury emissions into the air from TVA's coal-fired power plants contribute to mercury pollution in the water. These same plants contribute to water pollution more directly. At many plants fly ash is sluiced to a storage pond, the water from which, though treated, still contains some toxic metals when released.[48] In 2000, the three coal-fired power plants in our bioregion—Bull Run, John Sevier, and Kingston—released a combined total of 68,435 pounds of toxic metals—including arsenic, barium, chromium, cobalt, copper, manganese, mercury, vanadium, and zinc—into regional waterways.[49]

Table 3

INDUSTRIES EMITTING MORE THAN 100,000 POUNDS OF
TOXIC CHEMICALS INTO THE WATER IN 2001

Facility and Location	Main Water Pollutants	2001 Surface Water Discharge of Toxic Chemicals (lbs.)
Blue Ridge Paper Products Inc. Canton, NC	Acetaldehyde, Acetone, Ammonia, Arsenic Compounds, Barium compounds, Benzo (G, H, I), Perylene, Catechol, Certain glycol ethers, Chlorine, Chlorine dioxide, Chloroform, Chloromethane, Cresol, Dioxin and Dioxin-like compounds, Formaldehyde, Formic acid, Hydrochloric acid, Hydrogen fluoride, Lead compounds, Manganese compounds, Methanol, Methyl ethyl ketone, Phenol, Phosphoric acid, Polycyclic aromatic compounds, Sodium hydroxide (solution), Sulfuric acid, Vanadium compounds, Zinc compounds	114,252
Eastman Chemical Co. Tennessee Ops. Kingsport, TN	1,2-Dichlorobenzene, 1,4-Dioxane, Acetaldehyde, Acetonitrile, Ammonia, Aniline, Antimony compounds, Arsenic compounds, Barium compounds, Benzene, Bromomethane, Butyraldehyde, Certain glycol ethers, Chromium compounds, Cobalt compounds, Copper compounds, Crotonaldehyde, Cyanide compounds, Cyclohexane,	395,036

Facility and Location	Main Water Pollutants	2001 Surface Water Discharge of Toxic Chemicals (lbs.)
Eastman Chemical Co. Tennessee Ops. Kingsport, TN (cont.)	Di(2-ethylhexyl) phthalate, Dibutyl phthalate, Dimethyl phthalate, Dimethylamine, Ethylbenzene, Ethylene glycol, Formaldehyde, Formic acid, Hydroquinone, Lead compounds, Manganese compounds, Mercury, Mercury compounds, Methanol, Methyl iodide, Methyl isobutyl ketone, N, N-Dimethyl-formamide, Nickel, Nitrate compounds, Propionaldehyde, Quinone, Toluene, Triethylamine, Vanadium, Xylene, Zinc compounds, m-Xylene, n-Butyl alcohol, o-Xylene, p-Xylene	
Modine Mfg. Co. Knoxville, TN	1,1,1-Trichloroethane, Nitrate compounds, Nitric acid, Sodium hydroxide (solution), Sulfuric acid, Zinc compounds	223,078
Shakespeare Conductive Fibers L.L.C. Enka, NC	Certain glycol ethers, Chlorine, Ethylene glycol, Formaldehyde, Formic acid, Nitrate compounds, Phosphoric acid, Sodium hydroxide (solution), Sulfuric acid	189,000
U.S. Army Holston Army Ammunition Plant Kingsport, TN	Acetone, Ammonia, Ammonium nitrate (solution), BIS (2-Ethylhexyl) Adipate, Chlorine, Ethylene glycol, Hydrochloric acid, Methyl Ethyl Ketone, N-Butyl alcohol, Nitrate compounds, Nitric acid, PCBs, Sodium hydroxide (solution), Sulfuric acid	127,362

SOURCE: EPA, "TRI Explorer," accessible at http://www.epa.gov/triexplorer/facility.htm (9 July 2003).

Bacterial Contamination of Surface Waters

The bioregion's most widespread water pollutant is fecal bacteria from sewage and animal waste. More than 68 percent of all posted waterways in Tennessee are posted because of improperly treated human waste coming out of failed septic tanks or collection systems, or from improper connections to sewers or sewage-treatment plants.[50] (For a complete list of posted waterways in the Southern Appalachian Bioregion, see appendix 1.) Bacterial contamination is most severe in large urban areas, particularly Knoxville and Chattanooga, where the problem comes mostly from leaking sewer lines, and in overdeveloped rural areas—especially Blount and Sevier counties in Tennessee, where failing septic tanks are the culprits. Some smaller towns, such as Red Bank, Lake City, and White Pine, Tennessee, also have inadequate sewage treatment plants.[51] Even the relatively modern sewage facilities of the major urban areas may pour raw sewage into the rivers when they are overwhelmed by an influx of water from heavy rains. In many systems, faulty connections between sanitary sewers and storm sewers exacerbate this problem.

Agriculture is another major contributor to surface water pollution—chiefly in the form of nitrate runoff from fertilizers and manure—and to a lesser extent in the form of pesticide and herbicide contamination. But the contamination is more diffuse: agriculture occupies much more land area than urban or industrial activities—about 47 percent of the land area in Tennessee, as compared to only two percent for urban and industrial areas.[52]

State of the Reservoirs

TVA calls the waters backed up behind their dams "lakes," and the name has stuck. But this name promotes forgetfulness of the historic Southern Appalachian landscape, and so we do not use it. Southern Appalachia is too mountainous for real lakes of any considerable size. The permanent floodwaters created by the dams are more accurately termed "reservoirs."

Of the bioregion's four mainstream reservoirs, Fort Loudoun is the dirtiest. It receives a massive nutrient overload from housing developments along its banks, from sewage-contaminated streams in Blount and Sevier counties, and from the dirty waters of Knoxville's urban creeks.

The sediments of Fort Loudoun Reservoir are laced with zinc, PCBs, and other industrial wastes. Industries along the tributaries once dumped toxins openly into the waters, though regulation has generally put a stop to that. Most notorious among the historic polluters were Robertshaw Controls, which illegally discharged toxic materials as late as 1989;[53] Rohm & Haas on Third Creek; David Witherspoon, Inc., on Goose Creek (which is now a Tennessee superfund site); and Alcoa Aluminum and TVA's Singleton Ways facility on the Little River embayment. But all of these facilities are either closed or much cleaner today.

Siltation in Fort Loudoun from urban construction and eroded banks is intense. Knoxville's urban creeks disgorge muddy brown plumes far out into the river—a phenomenon readily observable from the bridges or the bluffs on the south bank. Trash is ubiquitous, petroleum slicks not uncommon. Extensive overdevelopment along both sides of the river and destruction of riparian corridors assure a steady influx of runoff from streets and parking lots and of fertilizers and other chemicals from suburban lawns. The toxics, the turbidity, and the silt combine to eliminate many forms of benthic (bottom-dwelling) life; nevertheless, the high nutrient levels support substantial populations of pollution-tolerant shad and enormous carp.[54]

Downstream from Fort Loudoun is Watts Bar, which, though less contaminated by bacteria, still harbors within its silts the deadly legacy of decades of nuclear weapons production. In Watts Bar the dirty waters of Fort Loudoun Reservoir are diluted by the cleaner Clinch and Little Tennessee rivers and many smaller streams. Yet below the dam at the Lenoir City Industrial Park, the Yale Lock Factory, Railroad Shed, and Public Works Garage all have contributed toxic wastes to the river, as have Staley, Viskase, and Kimberly Clark at Matlock Bend, ten miles downstream. Still farther downstream, at its confluence with the Clinch River, Watts Bar still receives some nuclear and chemical releases from the Oak Ridge Reservation, though these have been much reduced.[55] Moreover, toxic materials in Watts Bar's sediments continue to accumulate through the food web, producing potentially dangerous levels of contamination in some catfish, bass, sauger, buffalo, and carp.

Watts Bar Dam empties its tail waters into Chickamauga Reservoir. Here further dilution produces somewhat lower densities of pollutants, though radioactive and chemical contaminants from Oak Ridge—including PCBs, cesium 137, and cobalt 60—have settled in the Chickamauga sediments too. The Hiwassee River empties into Chickamauga Reservoir, bringing with it the "black liquor" from the Bowater paper mill upstream. Downstream, as in the Pigeon River, the sediments have been contaminated with dioxin.[56]

Chickamauga in turn empties into Nickajack, where a new load of sewage and industrial contamination pours into the water from the urban and suburban sprawl around Chattanooga. At Moccasin Bend the river receives the waters of Chattanooga Creek, which drains an industrial desert of hazardous waste sites, leaking sewers, abandoned landfills, and illegal dumps. High levels of PCBs, lead, arsenic, and chlordane contaminate its silt, and suspended sewage makes the water unfit for bodily contact. People living in the nearby low-income African American communities of Piney Woods and Alton Park have reported high rates of cancer, miscarriages, breathing problems, headaches, and eye and skin irritations.[57] Less than two miles downstream, the city adds the technically treated effluent from its sewage plant, a black liquid liberally dosed with chlorine. Still farther down, the waters of Stringers Branch discharge the inadequately treated sewage of Red Bank.

All this nutrient influx stimulates the growth of huge mats of filamentous algae and Eurasian water milfoil, which frequently clog the shallow waters of Mullins Cove and Bennett Lake. Until the early 1990s, TVA regularly combated these water plants with thousands of pounds of herbicides, adding still more toxic chemicals to the water. Although TVA no longer uses this method to control all aquatic weeds, it continues to use herbicides to control about sixty acres of plants around private docks and public and commercial recreational facilities in Marion County, Tennessee. This area includes about twenty acres in Mullins Cove and Bennett Lake area of the Nickajack Reservoir. TVA has, however, implemented the use of mechanical harvesters to deal with the remaining aquatic plants in a less noxious way.

The silts of Nickajack Reservoir are contaminated with PCBs, which are also found in potentially dangerous amounts in catfish. TVA's Sequoyah Nuclear Plant leaked PCBs into Nickajack for many years after a transformer fire in 1982,[58] but this was certainly not the only source. According to the Tennessee Department of Environment and Conservation, sediment levels at Nickajack violate EPA's standards, and its waters are "impaired"; yet, TVA continues to rate the health of Nickajack as good—indeed, as the best in the entire Tennessee River system.[59]

Suffocating Waters

In free-running creeks and rivers, the water splashes and foams, absorbing oxygen from the air. Stagnant or slowly-moving water, particularly if warmed by direct sunlight, absorbs less oxygen and is less able to support animal life. This is why many aquariums require aerators.

Moreover, the excessive influx of organic matter into a river or stream in the form of sewage, fertilizers, yard waste, and agricultural runoff reduces dissolved oxygen. Bacteria, protozoa, and fungi that decompose these nutrients draw oxygen from the water, sometimes generating toxic materials as wastes. High nutrient levels can also promote algae blooms, which turn the water green. As the algae grow, they create organic compounds from water and carbon dioxide by photosynthesis. Consequently, when they die and are decomposed, they return more organic material to the water than they absorbed from it, which may lower oxygen levels still further.

Dams contribute to oxygen depletion in several ways: by making the water sluggish and slow, by maintaining a pool of deep water far from surface exchange, and by piling the water up under great pressure. These problems are greatest at the tributary dams, which tend to be higher than dams on the Tennessee itself. During the warm summer months, water behind the high dams separates into two layers: a warm top layer, which contains most of the dissolved oxygen; and a cooler bottom layer, which contains relatively little oxygen and therefore cannot adequately support aquatic life. TVA dams release cold and oxygen-poor water from the bottom of the reservoirs. As a result, the tailwaters below the dams suffer from unnaturally cold temperatures, low dissolved oxygen, and extreme turbulence.[60]

TVA has gone to great lengths address the problem of dissolved oxygen and the related problem of inadequate flows below some dams when the turbines are not generating. At Douglas Reservoir, for example, oxygen is injected directly into the water upstream of the dam through dozens of miles of hose laid along the reservoir bottom.[61] At other locations TVA has adopted other innovations: downstream weirs (small dams that provide additional control of water flow and create artificial waterfalls to oxygenate the water), turbine venting (the use of baffles to improve the turbine's ability to draw air into the water), surface water pumps (which force warm, oxygen-rich water downward to be drawn in by the generating turbines), and air compressors (which inject air into the water flowing through the turbines).[62] These remedies appear to be improving water quality.[63] Yet, as was noted in the introduction, it is not a sign of ecological health that a river that once did an excellent job of aerating itself must now be kept on artificial respiration.

Moreover, such improvements generally affect only the tailwaters. Oxygen levels remain inadequate to support healthy aquatic communities in many of the deep tributary reservoirs—including Douglas, Norris, Boone, South Holston, Watauga, Cherokee, Nottely, and others.[64]

Hot Water

Heating a lake, river, or stream can harm its inhabitants just as much as infusing it with silt or toxic chemicals. Ecologists have dubbed this unwanted heat "thermal pollution." TVA's power plants, both coal and nuclear, are located on rivers, which they use as sources of cooling water. After running through the plant's cooling system, the heated water is discharged back into the river. At the nuclear plants and at one coal plant, the water is first passed through cooling towers (monumental structures that dominate the landscape for miles around) before being discharged into the river. Still the water may be quite hot—often exceeding a hundred degrees Fahrenheit. Some TVA coal plants that do not have cooling towers must cut back generation at certain times of the year in order to meet EPA temperature limits.[65] In the winter frozen cooling towers may be bypassed altogether. Heated water flowing directly into a much cooler river disrupts the local ecology, perhaps accelerating the growth of algae, and encouraging the growth of nonnative organisms (a problem discussed in chapter 4). Moreover, because, other things being equal, hot water absorbs and retains less oxygen than cold, an influx of heated water also reduces dissolved oxygen.

Urban and suburban development is another source of thermal pollution. On sunny summer afternoons, streets, parking lots, and roofs become sizzling hot. The sun bakes the soil of grassy lawns, too, making it much hotter than shaded soils. During a sudden thundershower, rain falling on these surfaces is heated, and the first flush of runoff into nearby streams or wetlands brings with it a sudden rise in water temperature, along with an influx of toxic chemicals, trash, silt, and bacteria. As the rain continues, the water temperature falls again just as suddenly. It is well known that sudden temperature changes can be fatal to aquarium fish. Similarly, these extreme fluctuations, which do not occur in naturally forested streams, may kill sensitive aquatic fauna (amphibians, fish, or insects), and so account in part for the ecological poverty of urban waterways. Thermal pollution of creeks is increased by the widespread elimination of shade trees and other vegetation along their banks. Sunlight shining directly into a creek heats the creek bottom and raises the water temperature, decreasing dissolved oxygen, and making the stream less hospitable to many species.

Flotsam

Heavy rain falling anywhere in the bioregion may, as it collects in rivulets and streams, gathering speed and force, move objects along the ground. All across the land, from the freshest of mountain streams to the filth of urban gutters and storm sewers, this water gathers carelessly discarded trash: plastic bottles, paper and plastic wrappers, rags, cigarette butts, lightbulbs, plastic shopping bags, beer bottles, steel and aluminum cans, bits and pieces of appliances, old tires, and Styrofoam. Swirled and tumbled by the rushing waters, this trash moves downstream toward the Tennessee. Because all our major rivers are pooled by dams, much of the trash eventually reaches a reservoir where the water flow is slower. There most of it sinks to the bottom, deposited in a permanent geological layer of silt, but a portion of it stays afloat, landing, when the current slackens or the water level fluctuates, on the shorelines of quiet bays.

By traveling by boat in virtually any of the bioregion's reservoirs late in a dry summer, or in the fall or winter when the water is down, one can see in these dismal bays or in the shoreline vegetation all the detritus of a throw-away culture—the plastic, paper, aluminum and broken glass; the tangled monofilament fishing line; the unidentifiable fragments of broken consumer products.

Over time, water currents carry both the sunken and floating debris downstream, where it eventually backs up behind the dams. TVA monitors the depth of the water behind the dams to determine the amount of accumulated debris. Once the accumulation becomes significant, cranes and barges are used to remove the sunken material.

To deal with this pervasive litter, communities along the river, beginning with Chattanooga in 1989, have initiated river rescues, in which hundreds of local volunteers, equipped by TVA with gloves and garbage bags, fan out along the shoreline to pick up trash. Knoxville's first river rescue, organized in 1991 by the Foundation for Global Sustainability (and later supported by other organizations), netted about five hundred tires. The items collected by the river rescuers have been varied: motorcycles, refrigerators, sunken boats, shopping carts, hypodermic needles, pesticide containers, gas tanks from cars and trucks, six-foot globs of floating asphalt, dead cattle, and barrels of toxic waste—to mention a modest selection. The bulk of the trash, however, is more mundane: paper or Styrofoam cups and wrappers, plastic or glass bottles and jars, and recyclable aluminum cans. In areas where river rescues have been active, there has been a gradual reduction of litter along the shoreline, as large items (especially tires), accumulated over decades, are removed. Yet each year much of the trash is renewed as the spring rains bring a fresh influx into the reservoirs. In 2001 volunteers working with TVA removed 2,600 cubic yards of rubbish—enough to create a foot-thick blanket stretching over two football fields. Roughly 11,500 cubic yards were collected between 1996 and 2001.[66]

River Traffic

Of people who interact with the bioregion's rivers in any significant way, most seem to value them primarily for their recreational benefits—that is, for fishing or boating.[67] Yet powerboats and jet skis degrade both the rivers themselves and the experience of those who prefer more quiet recreations—the more powerful and more numerous the boats, the worse the problems. The wakes of powerboats erode unprotected banks and pierce the silence of the river while spreading fumes—and sometimes petroleum slicks—across the water. Many vessels are ostentatiously oversized and overpowered for the Tennessee—as can be seen on any home football Saturday in Knoxville, when the "Vol Navy" docks at the municipal waterfront. Poorly driven power boats are a danger to swimmers and to one another. Many boaters toss their trash overboard; crews of larger boats sometimes illegally dump sewage.

As of yet, there have been no official studies conducted to calculate exactly how much human waste is released into the waterways by boaters in our bioregion. However, a 1999 TVA survey of Norris Lake's twenty-three marinas found that less than 40 percent of houseboats were using mandatory sanitation systems. The study estimated that boaters dumped as much as much as 400,000 gallons of untreated waste into that reservoir in one boating season. Extrapolating from TVA's figures, Assistant U.S. Attorney Guy Blackwell concludes that more than 7 million gallons of untreated human waste was being discharged annually from the 120 marinas on TVA's lakes and rivers in East Tennessee, tainting about half of Tennessee's 652-mile river system.[68]

One reason for damming the Tennessee was to provide for commercial navigation. But, though barges still ply the river, they have never provided the economic bonanza once envisioned, largely due to cheap competition from trains and trucks. And lately barges have played an increasingly important role in the destruction of Southern Applachian forests, being a primary means for transporting wood chips and whole logs down to Mobile Bay for export (see Chapters 4 and 9). Yet barges can be both an economically and environmentally efficient way to transport large amounts of materials.

Unfortunately, barge terminals also bring dangerous industrial operations close to the water. At Knoxville's Volunteer Asphalt plant on the south bank of the Tennessee just below the confluence of the Holston and the French Broad, for example, a tank rupture in the early 1990s poured many tons of tar into the river. For many years after, huge globs of tar, some six feet in length, could still be seen, melted onto downstream rocks.

Wetlands

In healthy, undisturbed ecosystems (of which none now are left in our bioregion), surface water is usually quite pure and safe even for drinking. In part, this is due to the operations of wetlands. Wetlands—"swamps" in less-pretentious jargon—are natural water filters. The water in a wetland is shallow, diffused over a large area, and relatively slow-moving. This provides time for lush aquatic vegetation to absorb nutrients, toxic metals, and other contaminants. Water leaving a wetland is often quite pure— which is why constructed wetlands have been proposed, for example, as a means of cleaning up Knoxville's Second Creek.

In heavy rains, wetlands also serve as retention areas, absorbing a great influx of water and releasing it slowly, which helps to prevent downstream flooding. They are, moreover, crucial habitat or breeding grounds for many species of birds, fish, reptiles, insects, and amphibians.

Unfortunately, these functions have until recently been poorly understood, and wetlands have been viewed as noxious impediments to progress. The farmers who began to till the Tennessee valley about two centuries ago observed two things about wetlands: 1) they were a source of pesky and sometimes dangerous mosquitoes, and 2) when drained, they yielded a rich soil, ideal for crops. So, industrious farmers drained them. As a result, Tennessee has already lost more than 60 percent of its original wetland area over the past century, and about 7 percent of the remaining wetland acres are threatened by pollution and loss of hydrologic function.[69] Because there are few low-lying plains in Southern Appalachia, there have never been many wetlands here. In this century they have become quite rare, and their unique ecosystems are disappearing. The few that remain harbor at least ninety threatened and endangered species.[70] Forested wetlands are especially important to biodiversity. Over 90 percent of the bird species living in eastern North America use forested wetlands at some time in their life cycle.[71]

Because agriculture has been diminishing in Southern Appalachia for some time, the destruction of wetlands has gradually decreased. (The wetlands that have already been lost, of course, are generally not being restored.) Current federal policy mandates "no net loss" of wetlands. If a wetland must be destroyed, the loss must be compensated for by artificially constructing a wetland of equal or greater acreage than the one removed, or by enhancing or preserving an existing wetland.[72] While this has helped to stabilize the quantity of wetland acreage, it has not prevented loss of quality. The compensatory wetlands are often constructed far from the natural wetland that was destroyed and generally prove less suitable as habitat. Thus, despite the "no net loss" policy, wetland degradation continues.

A significant case in point is the Turkey Creek Wetland off Lovell Road in west Knoxville, the largest remaining wetland in Knox County and a habitat of the rare flame chub, dozens of aquatic birds, and many other sensitive aquatic species. In the late 1990s, the city of Knoxville and Turkey Creek Land Partners bulldozed an extension of Parkside Boulevard through this wetland and filled in other parts of it for commercial development, allegedly compensating for this destruction by creating new wetlands upstream. The newly constructed wetlands are so fragmented and so tightly hemmed in by intense commercial development that their value as habitat will, in the final reckoning, almost certainly be far less than the value of the original.

· Groundwater

Much of the Tennessee valley is characterized by karst (limestone cave) topography and faulted, fractured, and folded bedrock. This allows for rapid, voluminous, and largely unpredictable movement of groundwater. Sinkholes abound, providing inlets to often extensive cave systems, underground watercourses, and sometimes even underground rivers or lakes. Water entering the sinkholes flushes whatever has been deposited near them into the underground water system, often with surprising force. During a heavy summer thunderstorm, a cave that was dry only an hour before can rapidly fill with roiling water—a peril borne in mind by all experienced spelunkers. Because of this water flow, cavers sometimes find the same sorts of trash that are so common along riverbanks and streams, including items as large as truck tires and shopping carts, deposited deep underground on high ledges in a cave that is, for the moment at least, dry.

Because rainwater can plunge directly underground in a karst system, it is not filtered or purified as it would be if it seeped through many layers of loam, clay, or sand. Some people in rural areas, apparently unaware that the water that circulates through the caves beneath their land is also the water they drink from their wells, use sinkholes as garbage dumps or disposal pits for dead livestock. Some service station owners in Tennessee have contaminated the groundwater by dumping waste oil into sinkholes. But dumping any polluting substance into a sinkhole is illegal.[73] Astonishingly, the practice remains fairly common. In one particularly egregious case, the Burnett Demolition & Salvage Co., a contractor for the city of Knoxville, was charged in 2002 by the Tennessee Department of Environment and Conservation with dumping approximately eight hundred truckloads of hazardous debris into a sinkhole in south Knox County. The debris that filled the sinkhole came from the old Norfolk Southern Railroad Coster Shop remediation project and contained arsenic, lead, polychlorinated biphenyls (PCBs), diesel fuel, and other contaminants above regulatory limits.

Sinkholes and caves play an important role in Tennessee Valley drainage. In the First Creek and Ten-Mile Creek drainage basins in Knoxville, for example, 10 to 15 percent of the runoff drains into sinkholes.[74] Development that disrupts these natural storm sewers or causes caves to collapse may produce flooding in areas not previously flood prone.

Not all of the bioregion has a karst topography, however, and where clay soils and more solid bedrock prevail, groundwater movement is slow.

Moreover, because of the hilly terrain there are no large, continuous aquifers; groundwater contamination therefore tends to be localized.

Sources of groundwater contamination in our bioregion include wastewater ponds (often associated with mines, industrial sites, and coal-fired power plants); contaminated industrial or hazardous waste sites (such as the David Witherspoon sites in the Vestal community of south Knoxville); landfills (of which there are hundreds in our bioregion); leaking underground storage tanks; agricultural herbicides, pesticides, and fertilizers; suburban lawn-care products; underground injection of wastes (such as occurred until 1984 at Oak Ridge National Laboratory); and failing sewers and septic tanks. The three most common groundwater contaminants in our bioregion are petrochemicals, fecal bacteria, and nitrates from fertilizers and sewage.

Petrochemical contamination occurs most frequently in the form of gasoline, diesel fuel, fuel oil, or motor oil. It is often, though not always, associated with leaking underground storage tanks. Petroleum contamination of groundwater is also common beneath storage facilities, such as the tank farm just off of Middlebrook Pike in west Knoxville. Frequent spills and runoff from that facility have been known to be a significant source of pollution contaminating both Third Creek, which runs through the site, and the groundwater beneath it.

Septic wastes from failing septic tanks also contribute significantly to contamination by nitrates and fecal bacteria. In the human body, some nitrates are converted to nitrites, which reduce the blood's ability to transport oxygen, a condition especially dangerous to fetuses and young children. They may also react to form carcinogenic N-nitroso compounds.[75] Wells that are most susceptible to nitrate and bacterial contamination are those that are located close to animal feeding operations and municipal or personal septic systems that are either improperly maintained or incorrectly installed.

Leaking landfills may contaminate wells with hazardous waste, a problem suffered for many years by residents of the Tennessee community of Witt in Hamblen County. Often in these cases no cleanup is possible, and the only solution for residents is to move or obtain their drinking water from elsewhere.[76]

Another important problem associated with groundwater is that water from private wells is often contaminated by lead from pipes, not from the groundwater itself. In about 90 percent of the cases, these wells have pumps made before 1995 that contain lead alloy parts. Much of our groundwater is soft and acidic, which makes it corrosive enough to leach lead from the pumps. Lead in drinking water can produce irritability, anemia, blood enzyme changes, hypertension, miscarriages, and learning disabilities.[77] Fortunately, for this problem there is in most cases an easy solution: let the tap run for a few seconds to flush the system before drawing water for human consumption.

The bioregion's most contaminated groundwater underlies the U.S. Department of Energy's Oak Ridge reservation. There mercury and radioactive fission products—mainly tritium and strontium 90—are found in the groundwater down gradient from nuclear waste burial grounds at Oak Ridge National Laboratory. The Department of Energy is attempting to reduce off-site discharge of these contaminants by capping the burial grounds with clay and implementing other containment measures.

Between 1951 and 1984, four seepage pits, known as the S-3 ponds, were used by the Y-12 nuclear weapons plant to dispose of over twenty-seven million gallons of various liquid wastes, including concentrated acids, caustic solutions, mop waters, and by-products from uranium recovery processes, including uranium and other heavy metals. These unlined pits were designed to allow the liquid either to evaporate or percolate into the ground. They are now capped, paved, and used as a parking lot, but the groundwater beneath them is still heavily contaminated with nitrates, cadmium, and uranium.

Likewise, at the K-25 gaseous diffusion plant, numerous burial grounds consisting of unlined trenches, sludge ponds, and leaking underground storage tanks have produced complex and extensive groundwater contamination by toxic and radioactive pollutants.

Most of the groundwater contamination on the Oak Ridge Reservation appears still to be confined to the watershed in which it originated; however, the contamination is migrating (at various rates for various sites), and there is no known way to eliminate it.[78] At Oak Ridge National Laboratory, for example, a plume of water about fifty feet under ground was recently discovered to be carrying radioactive strontium 90 into the Clinch River. This one plume, from an unknown source, was estimated to be responsible for at least 10 percent of the strontium 90 contamination in the Clinch.[79]

Despite the enormous scope and variety of groundwater contamination, water from most of the wells in our bioregion is still safe to drink. If you drink well water, however, it is a reasonable precaution to have it tested regularly—especially if you live near potential sources of contamination.

Municipal Water Supplies

According to the Tennessee Department of Environment and Conservation, "Drinking water criteria insure that water supplies contain no substances that might cause a public health threat, after conventional water treatment."[80] Modern water treatment facilities do remove most pollutants effectively, so that even if they draw from polluted waters, the resulting tap water is usually safe to drink. Pollutants can, however, enter the water after treatment. With old plumbing systems, lead may leach into the water between the treatment plant and the faucet—though, fortunately, this is becoming rarer.

It should be noted, however, that the chlorine used at the water treatment plant to kill bacteria is itself a pollutant that reacts with other chemicals to form carcinogenic byproducts, especially chloroform. Recent studies suggest that drinking chlorinated water may increase the risk of cancer of the gastrointestinal and urinary tracts, but no conclusive evidence has yet been found.[81] Of course, water untreated with chlorine presents greater and more immediate dangers of infectious disease such as typhoid fever, cholera, and dysentery. Yet, to put the matter in a broader perspective, the necessity for chlorine treatment itself arises largely from human degradation of the environment. Before the arrival of Europeans, an abundance of small springs and clear streams supplied the inhabitants of the Southern Appalachians with pure, unchlorinated drinking water. Now many of the springs and all the streams are unsafe for drinking, and drinking water must be extracted by complex industrial processes from reservoirs, rivers, and groundwater that are fouled with sewage and toxic waste.

Meeting Future Water Needs

We normally associate water shortages and water conflicts with the western states, and certainly Southern Appalachia has, by comparison, an abundance of fresh water. Yet even here supplies are finite, and there are worries of potential conflict ahead. Officials in water-poor Atlanta have expressed an interest in tapping the Tennessee River to slake their city's voracious and growing thirst; already a Chattanooga water utility, the East Side Development District, has been piping water from the Tennessee River to Fort Oglethorpe, Georia, just across the state line.[82] This diversion has heightened fears that neighboring states may increasingly look to the Tennessee River to meet their water supply needs, creating water shortages even in Southern Appalalachia.

Some East Tennessee communities have already experienced water shortages. During the hot summer of 2000, for example, public officials in Pikeville closed schools and car washes, urged residents to boil drinking water, and requested that local industries curtail their water consumption due to supply failures.[83] Pikeville depends for its water on underground wells that, in this summer, came hazardously close to drying up due to the combination of growing demand and drought.

Currently, about nine billion gallons of water per day is withdrawn from the Tennessee River basin. Fortunately, most of it (about 96 percent) is returned for downstream users, but as development escalates, withdrawals are constantly increasing.[84] Big new municipal withdrawals, such as those contemplated by Atlanta, would remove much more water without returning it and could, especially during drought years, imperil East Tennessee's water supply.[85]

Chapter 4

BIOTa

Stan Guffey and John Nolt

The ecological health of a bioregion is largely dependent on the status of its native biota—the species and the ecological communities they comprise. This chapter examines the biota of Southern Appalachia.

Humans and the Land

Human influences on living and nonliving nature in Southern Appalachia can be roughly divided into five temporal phases, each differing in human population density, economy, and technology, and human perceptions of their role in nature.

Twelve thousand years of Native American life in Southern Appalachia, from the first settlers to the complex societies at the time of Euro-American contact, represent the first phase. Native American culture during this period, as was noted in chapter 1, was far from homogeneous and unchanging. In their evolving adaptations to local environments and increasing population densities over time, Native Americans shaped the landscape and affected its nonhuman inhabitants, as all societies do. Early hunters of the Paleo-Indian period may have contributed to the demise of the Pleistocene megafauna,[1] but there is no evidence of other extinctions caused by Native Americans. The Native Americans differed in population density and economy from the Euro-Americans who displaced them, but the most salient differences involved their relations to nature. Native Americans lived in the landscape. The Euro-Americans who followed lived more on the landscape than in it.

The second phase of the human ecology of the Southern Appalachians is the period of Native American contact with the Euro-Americans. Soon

111

after Europeans arrived, their diseases decimated Native American populations and disrupted their cultures. Continued conflict and trade with Euro-Americans and catastrophic population decline led to substantial reconfiguring of Native American societies and their increasing participation in Euro-American economies. Many villages and agricultural fields of the Native American landscape were abandoned, and for a time the area of forest increased. Eastern woodlands bison increased in numbers in the Southern Appalachians as the numbers of Native American hunters declined, but only for a few decades. Soon, instead of hunting merely for subsistence, Native Americans began hunting to supply the Euro-Americans' demand for animal hides, which they traded for iron goods, firearms, and other items of European manufacture. Native American and Euro-American market hunters in the seventeenth and eighteenth centuries decimated white-tailed deer populations, extirpated beaver from the Southern Appalachians, and hunted the eastern woodlands bison to extinction.

Euro-American settlement, the third phase of human habitation, began in earnest in the 1750s and was inexorable by the 1780s. By the 1830s, with the removal of all but a tiny group of Cherokees, the period of Euro-American settlement and consolidation was largely over. The primary concern of initial arrivals was subsistence—obtaining food from the land. Hunting of still-abundant game was essential and remained so in economically isolated areas into the twentieth century. But the settlers were primarily farmers who began by clearing land for house sites and agricultural fields. Their concept of a settled, civilized landscape was of open fields and pastures interspersed with small woodlots, with forests confined to the distant margins and ridgetops. The abundance of timber and, for a time, wild game supported their notion that resources were inexhaustible.

The Euro-Americans were also conscious participants in a wider trade and money economy. Surplus agricultural production and natural products obtained from land and forest were fungibles that could be exchanged, by barter or sale, for products not available locally and for manufactured goods. Local processing and manufacturing followed closely on the heels of initial arrivals. Gristmills—and later powder, saw, and hammer mills—were established on streams and smaller rivers throughout the region; and iron smelting, making use of the abundant timber needed for charcoal, developed in areas where suitable ores could be found.

Settlement and consolidation changed the structure and composition of ecological communities. Valley forests were rapidly cleared, both as a source of raw materials, and because they were an obstacle to the establishment of fields and pastures. Most of the land in the Tennessee valley was cleared by 1840. Bison and elk disappeared during this period, and large predators declined in abundance. Many smaller mammals were probably lost from valley communities and populations of others declined, but there is no record of extinctions. Passenger pigeons and Carolina parakeets were still relatively abundant, at least regionally, around 1840, but their population decline had most certainly begun. Although there are no records of plant extinctions during this period, populations of many forest species were undoubtedly reduced, and it is conceivable that at least a few narrow endemics passed into extinction without notice. Aquatic fauna, particularly mussels, were probably negatively affected by erosion from cleared lands, but there are no known extinctions. The devastating effects of pollution and engineering on aquatic communities would come later.

The fourth phase of human habitation of the upper Tennessee region, from about 1840 to about 1890, was a period of population growth, expansion of urban areas and transportation networks, and increased integration into regional and national economies. Fields and pastures continued to displace forests, and with the expansion of steam-powered water and rail transport, lumber became a commodity for export. Mineral resources began to be exploited on a large scale to feed the demands of national industry. Mining and processing of copper and other metals (including gold) began in the Copper Basin of southeastern Tennessee and northwestern Georgia in the 1840s. As was noted in chapter 1, the direct and indirect environmental effects (and social costs) of these enterprises were enormous. Mining in the coal deposits of the Cumberland highlands began about mid-century and increased after the Civil War, although most of the extensive environmental and human degradation of coal extraction would come in the twentieth century.

The second half of the nineteenth century saw the demise of the passenger pigeon and the Carolina parakeet. Although extinction of both species would not finally come until the early twentieth century, by 1880 they had become rare and ecologically extinct.[2] The first mussel extinctions[3] and the extinction of the harelip sucker[4] probably occurred around this time as well, resulting from pollution, runoff from cleared lands, and the beginnings of navigational improvements on the Tennessee River. Wolves and cougars were relentlessly hunted and continued to decline in the second half of the nineteenth century, but would persist, in low numbers, in the high mountains of the Blue Ridge into the twentieth.

The fifth (but not the final) phase of human habitation of the region is the phase of power and control. Twentieth-century environmental and economic history was a continuation of nineteenth-century patterns, expanding the scale and scope of landscape modifications and consequent effects on native biota. Three significant "events" mark the environmental history of the twentieth century in our region: the industrial-scale logging of the Blue Ridge forests at the beginning of the century and the return of industrial forestry near the end of the century; the near-total conversion of the Tennessee River system into a series of impoundments and tailwaters; and the chestnut blight, which all but eliminated the American chestnut from the region's forests. These three events are emblematic of the effects of human activities on native biotic communities: injudicious exploitation; habitat destruction and fragmentation; and the spread of invasive non-native species. We begin our discussion of the state of the region's biota in this fifth phase with a look at its most conspicuous habitats, our forests.

Industrial Forestry

Picture the North American continent as a vast lawn. Then imagine a huge lawn mower moving slowly from east to west. The mower begins in the east about the middle of the nineteenth century. By the end of the twentieth, it has mowed virtually everything from the Atlantic Coast to the Pacific Northwest. This mower is industrial forestry. In the nineteenth and early twentieth centuries, industrial forestry stripped away virtually all the forests east of the Mississippi, leaving bare earth, depleted soils, erosion, and raging floods in its wake. Eastern forests having been thoroughly exploited by the 1930s, the mower moved on to the pine forests of the Rockies and the redwood forests of California, and north to the vast rainy woods of Washington and Oregon. (Cutting began in the West, of course, with the arrival of white settlers and continued in the East as it began to devastate the West. Moreover, trees were cut not in wide swaths across the landscape, but in a patchwork. Still, the general trend was from east to west.) The northwestern corner of the lawn is now nearly mowed. But the eastern—and particularly the southeastern—end has been growing for many decades. Southern Appalachian hills, which less than a century ago were left treeless and brown with mud, are, many of them, once again forested, green, and ripe for logging. So the mower has returned and the cutting frenzy has begun again.

Today the southeastern states are once again producing most of America's industrial wood, and their output is rapidly increasing. In 1952 the 13 southern states were responsible for 41 percent of the country's wood fiber output. By 1997 this figure had risen to 58 percent. During the same period, the Pacific region's share of the nation's production dropped from 24.8 to 16 percent. By 2002 the South was producing nearly 60 percent of the U.S. total—more wood fiber, in fact, than any other *country* in the world.[5]

Industrial forestry came to the region around the beginning of the twentieth century with the logging of the forests of the Blue Ridge. Valley forests had been cleared in the nineteenth century to convert the land into an agricultural landscape. Timber from the valley forests was also used locally at a prodigious rate for construction and fuel, and some lumber, primarily from the remaining forests of the ridge and valley, was shipped out of the region, especially after the Civil War. By 1890 the valleys and ridges were largely cleared and in agricultural use. The only large tracts of old-growth forest that remained in the region were in the Blue Ridge, and these became the focus of the region's first exploitation by industrial-scale forestry.

Industrial forestry is total-extraction forestry, characterized by clear-cutting large areas over relatively short periods of time to supply distant markets with whole logs, sawn lumber, or wood chips. Beginning in the first decade of the twentieth century, lumber companies began to penetrate the mountains, clear-cutting and extending railroads as they went. By the early 1930s all but the most inaccessible areas of the southern Blue Ridge had been logged, mostly by clear-cut, and the boom was over. During that span of less than thirty years, 60 percent of the area of Great Smoky Mountains National Park, about 300,000 acres, had been clear-cut by corporate loggers, yielding an estimated (and astonishing) two billion board feet of lumber.[6] Similar levels of cutting occurred in the mountains outside the area of the future park.

The landscape still bears the scars of the clearing of the Valley and Ridge forests in the nineteenth century for agricultural land, lumber, and fuel and of the cutting of the Blue Ridge forests in the early twentieth century to supply lumber to national markets. Deforestation and farming of ridges increased water runoff, plaguing the Tennessee River system with heavy siltation and devastating floods. Among the several reasons for the establishment of the Tennessee Valley Authority in 1933—particularly its dam construction and forest restoration—was to tame a wild river that had become wilder through the loss of forests. Industrial logging in the Blue Ridge had similar effects on hydrology, and among the reasons given for the establishment of national forests and Great Smoky Mountains National Park was protection of watersheds. The forests of the Blue Ridge have largely returned, and with the decline of subsistence farming, large areas of the Ridge and Valley have again become forested, setting the stage for the current cycle of logging.

Still, despite these relatively recent—and perilously temporary—gains, we have lost much and are on the verge of losing much more. Across the South, natural forests have declined to just over half their original extent, from an original 356 million acres to about 182 million acres.[7] Well over 98 percent of old-growth deciduous forest is gone.[8] Much national forestland and potentially all of the privately owned (especially corporate) forests are "available" for "harvest." Only about 2 percent of the landscape has protected (i.e., wilderness) status, and most of this is in the Blue Ridge section. Still more troubling is the relatively small size of protected areas and the lack of connectivity between protected units. Only the 465,000 acres managed as wilderness in Great Smoky Mountains National Park are of sufficient size to maintain ecological integrity and function in the face of large-scale natural disturbances.[9]

Impacts of Logging

The forest-products industry argues that there is room for more logging here, since overall in recent decades tree growth has exceeded cutting. But that depends on how and where you look. It is not true for high-quality saw timber, which has long been logged much faster than it grows.[10] Nor is it true for softwoods; between 1990 and 2002 removals of softwoods across the South surpassed growth by about 10 percent.[11] The growth of southern hardwoods has outrun cutting, but not for long. According to U.S. Forest Service projections, hardwoods across the South will, on average, be falling faster than they can regrow by 2025. Cutting of hardwoods on *private* land (which comprises the majority of southern forests) will exceed growth around 2010.

Moreover, the picture varies considerably from state to state. In Tennessee, where hardwood forests are still extensive, overall cutting of hardwoods may not exceed growth until about 2035, but in North Carolina it will occur before 2015 and in Georgia before 2010. From this point on, as far forward as anyone can predict, natural *hardwood* forests, too, will diminish. Regionwide, the outlook is unrelenting. County-by-county projections show cutting of all forest types on private land significantly exceeding growth between 2000 and 2050 for virtually every county in our bioregion.[12] We are losing the forests.

This picture is complicated by the fact that the *softwood* growth rate has been increasing and will soon surpass the cutting rate—but only because softwoods are ever more frequently grown on pine plantations. Natural forests are shrinking as the pine plantations expand.[13] Moreover, within a few decades even the pine plantations themselves will be cut faster than they can regrow. Cutting of softwoods and hardwoods is forecast to increase across the South by 56 and 47 percent respectively between 1995 and 2040; even allowing for the rapid growth of softwoods on pine plantations, softwood removal will exceed growth on private lands by 2045.[14]

Forest health will suffer proportionately. Most forms of logging require extensive systems of forest roads and skid trails, where the forest soil is bulldozed and laid bare. Heavy logging equipment compacts the soil. Soil is washed off into streams, contaminating them with silt and nutrients.[15] The forest system does, however, quickly begin to replace lost nutrients and biomass. If the system is left undisturbed long enough, eventually the lost biomass and nutrients are restored. In northern hardwood forests, this takes sixty to eighty years.[16] In Southern Appalachian forests, the process varies widely with forest and soil type, but seems generally to be quicker, because of the longer growing seasons here.[17] It would be wrong, however, to conclude that in the long run there will be no net loss when a forest is logged. An undisturbed forest not only maintains itself, but actively builds soil, enriching the foundation from which all land-based life springs. If succeeding "harvests" occur as soon as the nutrients and biomass return to pre-cut levels, this enrichment is halted and the system no longer grows richer. If such harvests occur earlier, it is reversed and the system becomes poorer. In any case, removing the trees always leaves the forest system and forest land poorer in nutrients and biomass than if it had been left alone.

Nutrients and total biomass are only part of the story. It takes as much as a century or more for a Southern Appalachian forest to mature and become old growth. Old-growth communities have a characteristic biota that differs from the species composition of earlier successional (or recovery) communities, and many characteristic species are lost, at least locally, in heavily logged areas.

Among the species lost are herbaceous perennials that grow on the forest floor. A study of Southern Appalachian forests by Duffey and Meier compared forest plots that had never been logged to plots that had been clear-cut fifty to eighty-five years before. Average understory plant density was significantly higher in unlogged plots than in those previously logged. They suggested that during the first fifty to eighty-five years after clear-cutting, species richness and cover may actually decrease. Duffey and Meier offered several hypotheses for this continued biotic loss, but were unable to confirm any of them. They did, however, reach this conclusion: "The data presented here strongly suggest that recovery requires at least several centuries, longer than the present logging cycles of 40–150 years for Appalachian cove hardwoods." This conclusion is still controversial but has received support from subsequent work.[18]

Similar results, and similar heated debate, have come from studies of the effects of clear-cuts on terrestrial salamander populations in the Southern Appalachians. Studies by Ash and Petranka et al. demonstrate the near total disappearance of salamanders from clear-cuts in Appalachian forests.[19] Petranka estimated a loss of about 14 million salamanders per year due to clear-cutting in National Forests in western North Carolina. While this number represents only about 0.3 percent of the total estimated population of salamanders in this area, Petranka estimates that, with continued clear-cutting at current levels and population recovery times of fifty

Fig. 7. *A ridgetop clear-cut in East Tennessee. Copyright by Mignon Naegeli.*

years, populations would be chronically reduced by about 8.5 percent, or about 267 million animals. Recovery time, as in discussions of the effects of clear-cutting on forest-floor herbaceous vegetation, is at the heart of the controversy among academic researchers. Ash estimates that salamander populations return to pre-clear-cut levels in 20–24 years, while Petranka estimates 50–70 years.[20]

What can we make of these conflicting results? There is no disagreement that clear-cuts result in at least short-term local extinctions among forest-floor perennials and salamanders. How long it takes for populations to recover (if ever) is a matter of ongoing research. Part of the difficulty in arriving at comprehensive answers is the heterogeneity of Southern Appalachian landscapes and forests, and the lack of suitable long-term baseline data for comparison. However, what the conflicting results do show is that recovery takes time, be it measured in two to three decades or in centuries. And, as Petranka's analysis demonstrates, continued clear-cutting results in chronic biotic loss, the magnitude of which is a function of clear-cutting rate and biotic "recovery" time (whatever that might be). Petranka's estimate of chronic loss of salamanders assumed continuation of clear-cutting at 1981–90 rates, but an increase in the rate and scale of logging, such as we are likely to see across Southern Appalachia in the next few decades, would increase the loss and recovery times nonlinearly by reducing recolonization from forested areas.

Logging does sometimes increase the species richness of some groups and the densities of some species, but these increases are usually only temporary as forest succession proceeds (or until monotypic stands are planted). Populations of small mammals (mice, voles, and shrews) and their predators may increase in the years after clear-cutting,[21] and increased browse may benefit deer and bear populations. Herbaceous species richness in clear-cuts is usually higher in clear-cuts than in mature forests, but is dominated by pioneer and weedy species, including many non-native and potentially invasive species.[22] Clear-cutting may also favor the growth of commercially valuable trees—oaks, for example—by reducing competition from other species. But from an ecological point of view, and from the point of view of native species conservation, such benefits are generally outweighed by the resulting fragmentation and destruction of habitat and degradation of soil and water.

The forest industry argues that clear-cutting and other even-aged management practices are superior to the alternative of high-grading, the practice of cutting only large, healthy, commercially valuable trees, leaving those that are less healthy and less marketable. It is undeniable that high-grading has seriously damaged forests by eliminating the fittest trees, leaving the weakest and most disease-prone. High-graded forests are now dominated by species of little commercial worth and by damaged or unhealthy trees. Such forests are characterized by the forest industry as "low grade." This judgment, however, reflects the commercial value of the trees, not their ecological value. Trees that are low grade commercially may still serve important ecological functions, such as shading and enriching the soil, improving water quality, preventing erosion, and providing habitat for understory plants and for many species of animals.

Given the region's history of high-grading, the most sustainable logging practice is probably to cull non-native, genetically inferior, and diseased trees and leave the healthy ones—much as the traditional farmer saved the best corn for seed. In some areas, this is still best accomplished by old-fashioned horse-logging, which avoids the soil compaction associated with the use of heavy machinery. Such painstaking practices may not have much appeal in a competitive market economy, but market economies are designed to maximize short-term productivity, not long-term health. Health is a more elusive goal, demanding not blind obedience to the law of supply and demand, but effort, choice, self-discipline, and personal responsibility.

Pine Plantations and Chip Mills

We have so far considered only damage that occurs when a logged forest is allowed to regrow naturally. But often nowadays clear-cut land is poisoned with herbicides, fertilized with petrochemicals, and then replanted with neat, geometrical rows of pine trees (typically fast-growing loblolly pines) to create a "pine plantation" or "tree farm." The pines are planted very close together (sometimes on a six-by-six-foot grid), protected from fire (which is essential for natural biodiversity), and, worst of all, recut on rotations as short as ten years, so that nothing like a natural forest ever has a chance to become established. Only the pines, some weeds, a small variety of understory plants, and a smattering of hardy, adaptable fauna live in these sad mockeries of forests. The resultant loss of biodiversity is catastrophic.[23]

Pine plantations are transforming the land on an almost unimaginable scale. Between 1953 and 1999 planted pine across the South increased from 2 million to 32 million acres—an increase of 1,600 percent. During this same period, natural pine forest declined from 72 million acres to 34 million acres.[24] The trend is expected to continue far into the future. According to the best available projection, pine plantations in the 13 southern states will expand to 54 million acres by 2040—approximately the size of North and South Carolina combined. In Tennessee the area of pine plantations is expected to increase by an additional 120 percent by 2040—the largest percentage increase among the 13 southern states.[25]

Unlike natural forests, pine plantations are intensively managed—often with petrochemicals. Nearly 1.6 million acres of planted pine were fertilized in 1999—an increase from 1990 of nearly 800 percent. Since 1969, nearly ten million acres of forest have been artificially fertilized in the South, exceeding the forest fertilization area for the entire rest of the world. Comparable data on herbicide application are not available, but a reasonable estimate is about two million acres annually.[26]

Fig. 8. *A stand of pines damaged by the southern pine beetle. The deciduous trees in the foreground are healthy, but all the pines in the background are dead. Turkey vultures perch in the pines. Copyright by Mignon Naegeli.*

Chemical and biological experiments on such a grand scale are not without risk. In 1999, after pine plantations had been expanding rapidly for several decades, populations of the southern pine beetle suddenly exploded, devastating pine trees across the South. The infestation peaked in 2000, but it continues, and stands of dead pines are evident nearly everywhere. At Big South Fork and in the Daniel Boone National Forest in Kentucky, dead trees are so numerous that rangers have posted warning signs altering hikers to the dangers of their falling limbs. From the air, huge stands of denuded pines are visible in the pine plantations themselves, along with frantic, widespread clear-cutting to salvage pulpwood from the dead trees before they rot. The results of the clear-cutting are also plainly visible: bare, baking soils and severe erosion.[27]

Periodic outbreaks of the southern pine beetle, which is native to the region, have occurred before, but none have been so virulent or widespread as this. Though a series of mild winters (possibly the result of global climate change) undoubtedly played a role, it is evident that monocrop pine plantations, which consist almost solely of trees favored by the beetle, especially loblolly and shortleaf pines, serve as incubators for such epidemics.[28]

Wood from pine plantations is sometimes used for saw logs, but an increasingly large portion is ground into chips for the manufacture of paper and construction materials, such as oriented strand board. (Hardwoods are also increasingly used for this purpose.) Because, unlike saw mills, chip mills can use very small logs, forests can be cut when the trees are very young—sometimes as little as ten years old, and this encourages ever more rapid cutting and recutting. Huge regional chip mills, fed by streams of logging trucks (often hundreds per day), debark the logs and grind them into chips. To supply a big chip mill, which may run around the clock, loggers must strip and haul away dozens of acres of forest each day from a sourcing radius of about seventy-five miles.[29] Mills in our bioregion include the International Paper mill near Caryville, Tennessee, the Blue Ridge Paper mill in Canton, North Carolina (both formerly owned by Champion International); the Bowater mill near Chattanooga; the J. M. Huber mill in Spring City, Tennessee; and the Weyerhauser mills at Kingsport, Tennessee, and Union Mills, North Carolina.[30] The chips are mostly used in regional or national markets, but some are transported by barge down the Tennessee–Tombigbee waterway to Mobile Bay, where they are exported internationally.

Though chip mills produce chips for a variety of purposes, by far the largest single use is for paper products, chiefly packaging. Recycling rates have improved over the past decade, but packaging is still typically used just once, then thrown into the trash, whence it becomes a waste-disposal problem (see chapter 8). This is a less than optimal use for hard-pressed forests.

Logging and the Law

The forest-products industry claims to be wiser now than in earlier years; it claims to curb forest abuses and promises to follow a code of "best management practices" (BMPs), which would reduce erosion and some of the other destructive effects of logging. These promises, however, do not ensure environmental protection. These impacts may be reduced, but they are not eliminated even by the strictest adherence to BMPs. In their environmental impact statement on the chip mills, TVA, the U.S. Army Corps of Engineers, and the U.S. Fish and Wildlife Service noted: "BMPs, as currently designed, do not protect all possibly impacted resources. Intensified timber harvesting and uncertain BMP compliance at least pose risks of adverse impacts on a localized or site-specific basis in the sourcing area to wildlife, karst [limestone cave] features, water quality, aquatic species, endangered and threatened species, aesthetics, and archaeological sites. Hard mast [acorn and nut] production [needed by many forms of wildlife] could be adversely affected regionally."[31]

As to the uncertain compliance, the forest industry responds that its compliance record is improving and that it is policing itself under the new federal Sustainable Forest Initiative. But the industries touting the initiative are among those cutting most rapidly. And since BMPs are voluntary, not legal requirements, they are widely ignored. ForestWatch, a regional citizens' watchdog group, has documented extensive violations of BMPs.[32] These violations are likely to increase as the cutting frenzy intensifies and good stands of trees become rarer and more expensive. It defies credence that in such circumstances multinational corporations under intense competitive pressures will steadfastly honor merely voluntary guidelines. TVA, the EPA, the U.S. Department of the Interior, the Alabama Department of Environmental Management, and the Tennessee Wildlife Resources Agency have all expressed doubts about the adequacy of voluntary BMP compliance.[33] Moreover, since 1997 attempts to develop comprehensive, binding laws in Tennessee (e.g., the Tennessee Forest Resources Protection Act) have, due largely to industry lobbying, failed to make it out of legislative committee.

Even if BMPs were passed into law, and even if the law provided strict penalties, and even if these penalties were enforced (three big ifs), the result would still be a continual and generalized deterioration of the land. The future will be marked by more logging roads and machinery in the forest, followed by more human intrusions, more trash, more erosion, invasions of non-native organisms, damaged soils, and diminished wildlife habitat—not to mention the loss of the trees themselves.

Southern Appalachia *can* support a sustainable logging industry (a massive, wood-derived paper industry is more problematic), but this requires some sacrifice of immediate gain by the majority of foresters and landholders and stiffly enforced legal sanctions against the few who lack scruples.

Conserving Forests

Because individual landholders control most of the region's forestland, the work of conserving the forests falls primarily upon individuals, rather than upon industry or the government. Private owners or investors hold 73.2 percent of the region's timberland; industry possesses 6 percent; the U.S. Forest Service has 17 percent; and other government agencies (including the National Park Service, which operates the Great Smoky Mountains National Park) retain 3.8 percent.[34] Moreover, most of the Forest Service land is or will eventually be available to private corporations for logging or mining. Of the 632,348 acres of the Cherokee National Forest, for example, only 66,469 acres (10.6 percent) is permanently protected.[35] Further protection of public forestlands is thus an urgent priority.

Since so much of the bioregion's forestland is in private hands, the health of forests also hinges on the development of an environmental ethic among individual landholders, many of whom allow their land to be logged by less-than-scrupulous operators. There are, however, some hopeful signs. The September 1995 issue of the *Journal of Forestry,* for example, heralded the appearance of a new constituency among forest landholders. An increasing number of landholders, said the journal's contributors, were managing their lands in accord with environmental values. One article urged professional foresters to regain lost leadership and prestige by moving ahead with an environmental ethic of their own.[36]

Unfortunately, forest owners who manage their land primarily for conservation or aesthetics tend not to own much. According to a recent survey, landowners in the South who intend never to log their land control only 12 percent of the total private timberland acreage. The largest tracts are generally the most important for conservation purposes; yet, the fewer than 1 percent of nonindustry owners who hold more than 500 acres control 65 percent of the private timberland intended for eventual "harvest."[37] It is largely these big landowners who will determine the fate of the forests.

The Wounds of Development

Logging is not the only threat to Southern Appalachian forests. Sprawling suburbs and other forms of development also take their toll. The 13 southern states will likely lose 12 million acres of forest (about 8 percent of the total forestland) to development between 1992 and 2020 and an additional 19 million acres by 2040. The total—31 million acres—represents a loss much larger than the entire state of Tennessee.[38]

Many other forms of development—most important, mining, road building, and the damming of rivers—have fragmented, scarred, smothered, inundated, and depleted the land and will continue to do so.

Mining, particularly strip mining, has a long and grim history in Central Appalachia, but we are relatively fortunate in that coal mining in our bioregion is confined to its western edge on the Cumberland Plateau. There, however, large amounts of coal are extracted with a wide range of consequences for human environments and biodiversity. We also have other minerals of worth, and the extraction of these, too, has caused the land to suffer. In chapter 3 we mentioned the devastation wrought on the Nolichucky River by the kaolin, feldspar, and mica mines near Spruce Pine, North Carolina, and on the Ocoee by the mines of Copper Basin. Copper Basin is, of course, the bioregion's most extreme example of mine damage, though now, after nearly a century, its wounds have begun to heal. Today mines are better regulated, but the regulation is only as effective as vigilance and public pressure force it to be. Vigilance must be constant, even (in fact, especially) when the land at issue is part of our national forests. Southern Appalachian national forests are under increasing pressure to become privatized and are subject to various forms of exploitation—including mining, from which they are not legally protected.

Perhaps more destructive even than mining is the vast network of roads that provide automobiles access to everything. Aside from the interior of the tourist-besieged Great Smoky Mountains National Park, few places in our bioregion are more than a mile or two from a road. Roads fragment habitat, preventing the movement and migration of wildlife. Road construction creates huge wounds of open earth, from which soil washes into streams, silting and sometimes acidifying them. As roads grow and widen, they bring more traffic and fouler air, and development flows out along them, radiating across the countryside.[39] Where roads go, development inevitably follows: subdivisions, gas stations, convenience marts, shopping malls, golf courses, and industrial areas multiply, devouring rural land and rural life. In 1984 urbanization passed agriculture as the primary cause of loss of forestland in the Southeast. Although the southeastern region as a whole has seen an increase in forest acreage since 1990, the

Blue Ridge (especially in western North Carolina) and southern Ridge and Valley are already seeing a decrease that is projected to continue.[40]

Like road construction and the development it brings, road maintenance also degrades natural vegetation. To maintain roads—and also railways and utility rights-of-way—nearby plants are often sprayed with herbicides. The herbicides may run off into streams, killing aquatic life, or harm rare and sensitive plants. Bunched arrowhead, a small flowering plant that grows only in Henderson County, North Carolina, and Greenville County, South Carolina, has been nearly lost to road, railroad, and utility maintenance and to the clearing and development of the wooded swamps in which it grows.[41] Salt used to melt snow and ice may also damage streams or roadside vegetation.

Dams, too, have radically altered the land. The reservoirs they have created are huge, permanent floods that have drowned many thousands of acres of prime forest and farmland. This land is already lost. But the constant fluctuation of the water level continues to damage vegetation along the shores. To balance the demands of recreation and flood control, TVA raises reservoir levels in the late spring and lowers them in the fall, a pattern not synchronous with the natural cycles of Southern Appalachian rivers, which run high in the winter and low in the summer. As a result, the reservoirs are bordered by mud flats of a kind that did not originally occur in this bioregion. Few native plants and animals are specifically adapted to living there, and they are ecologically unproductive.[42]

Fires

Prior to the last half of the twentieth century, fire was a significant factor in Southern Appalachian ecosystems. Because of the high rainfall, lightning-caused fires are uncommon in our region, but they do occur, particularly on dry, south-facing ridges. Even more important has been the cultural use of fire to shape the landscape. Native American use of fire probably dates from the most ancient times. Euro-American settlers in the region adopted and modified the practice, employing periodic burnings into the twentieth century.[43] Early in the century a national ethos of fire suppression developed, with disastrous consequences for southeastern longleaf pine forests and the conifer forests in the western United States.[44]

Well-intentioned efforts to prevent forest fires can reduce biodiversity in our region as well. The table mountain pine, *Pinus pungens,* for example, needs fire to reproduce. Its cones do not open automatically in the fall, but remain closed, protecting the seed inside, until opened by the intense heat of a fire. Fires also prepare the seed bed for germination and reduce competition by eliminating fire-sensitive tree species. Before the development of fire-suppression policies with the establishment of the Great Smoky Mountains National Park and the surrounding national forests, fire was a common and natural occurrence in the tree's native habitat. In Swain County, North Carolina, for example, 30–50 percent of the land burned every year, and every acre burned within five years, enabling the table mountain pine to reproduce regularly. But for most of the past century, fires have regularly been extinguished, and as a result the table mountain pine was dying out.[45] Over the past decade, however, more enlightened managers in the park and national forests have allowed some fires in table mountain pine communities to burn and have even initiated controlled burns.[46]

Oaks, particularly those species in communities on drier ridges, are less fire sensitive than competing beeches and maples, and they tend to grow best where there are occasional fires. Where burning has been suppressed, oaks are crowded out by the beeches and maples, which can tolerate more shade. Paleoecological evidence indicates that many of the oak-dominated communities in the region may have developed as a consequence of Native American use of fire.[47] Fire suppression gradually shifts the dominant forest composition from oak to beech and maple.[48]

Mammals

The history of human exploitation, and sometimes elimination, of wildlife in the upper Tennessee valley is long. The earliest known human inhabitants, the Paleo-Indians, may have overexploited their big-game food sources. The mastodon, the horse, the camel, and a prehistoric bison, all of which inhabited large portions of North America until 10,000 B.C., were extinct by 7,000 B.C., victims perhaps of climate change, likely hastened by overhunting.[49] Later inhabitants seem to have understood the immorality and ultimate destructiveness of such behavior. The Cherokees had legends of a distant time when overhunting threatened animals with extinction—though the great time lapse makes it unlikely that these legends sprang from Paleo-Indian sources. The hard-pressed animals, the legends relate, retaliated by inventing diseases, which they inflicted upon humans by magic. To avoid disease, the Cherokees believed, the hunter must apologize to the animals he kills and take certain ritual precautions that in practice limited the extent of the kill. Implicit in this mythos was a concept of human forbearance in preserving a natural balance.[50]

Yet during the eighteenth century, even the Cherokees, in concert with the incoming settlers, hunted the white-tailed deer almost to extinction in a frenzy comparable to the slaughter of the Great Plains bison a century later. This fact, it has been argued, shows that the Cherokees were just as exploitative as the settlers, though ineffective until they acquired guns. That conclusion, however, is hasty. By the time the deer were being slaughtered, war, disease, and whiskey were already dissolving the bonds of Cherokee culture. To survive, the remnants of the tribe needed guns to defend themselves against the Europeans and iron tools to work the marginal land to which they had been forced to retreat. These could be obtained only through trade, and the only marketable commodity the Cherokees could provide was deerskin. (A "buck," our slang word for a dollar, originally meant a deerskin used in trade.) Whether the Cherokees would have overhunted the deer if their own survival had not been at stake remains uncertain.[51]

The slaughter of the deer may have a more intriguing, alternate explanation. Diseases brought by European explorers had been decimating Native American tribes at least since the early sixteenth century. In 1738 a smallpox epidemic killed nearly half the Cherokee people.[52] Given their account of the origin of disease, it would have been natural for the Cherokees to conclude that the animals had invented new diseases and were engaged in magical aggression against them. Such reasoning could easily have eroded the original reluctance to overhunt and perhaps even sparked retaliation against the deer. Thus, the very ideology that supported sustainable hunting before the European invasion may have served to rationalize overhunting as the plight of the Cherokees became desperate.[53]

By the middle of the twentieth century, few deer remained, hiding warily in mountain fastnesses. In the 1950s, state wildlife officials began to reintroduce white-tailed deer, using individuals from as far away as Wisconsin. The reintroduction effort, in concert with the creation of large no-hunting preserves, such as the Great Smoky Mountains National Park, has restored deer populations, and since the 1950s herds have been increasing throughout the bioregion.

In fact these efforts have succeeded too well. Statewide, the deer now number over 800,000,[54] well in excess of the population prior to European settlement.[55] As the deer increase and range more widely over the humanly redesigned landscape in their search for food, new problems arise. Deer are now common in many suburbs, where they raid gardens and eat pesticide-poisoned vegetation. Many collide with automobiles.

Because the deer's predators (wolves and mountain lions) have been eliminated, there is no natural check on deer populations. Consequently, human predation in the form of hunting has become necessary to prevent the deer from ravaging crops, wildflowers, and forest plants and reducing the understory cover needed by songbirds.[56] Deer are already damaging endangered and threatened plants.[57] Unfortunately, while nonhuman predators tended to cull the sick and weak and so promote the herd's long-term health, hunters prefer to shoot the biggest and strongest—with what long-term evolutionary effect remains to be seen.

The white-tailed deer ultimately survived, but the bison did not. Bison were not abundant in our region in the late prehistoric and early historic periods, in part due to Native American hunters. But with the decline of Native American populations, bison numbers increased. Then came the influx of Euro-American settlers. By 1800 woodland bison had become rare, even in the areas of higher density in the central basin region of Tennessee and Kentucky. Within a decade or two they were extinct.[58]

Eastern elk, too, once inhabited the forests of Southern Appalachia. They were gone from the region by the start of the Civil War.[59] In 2000 the Tennessee Wildlife Resources Agency began its first efforts to reintroduce elk into the Cumberland Mountains, with the goal of creating a hunting resource. In 2001 Great Smoky Mountains National Park introduced elk in Cataloochee Valley in an experimental program to evaluate the possibility of restoring this lost biotic component to the mountains. However, these reintroductions, and others in other eastern states, are using the Rocky Mountain elk, *Cervus elaphus nelsoni.* The eastern elk subspecies, *C. elaphhus Canadensis,* once native to most of eastern North America, is presumed extinct.[60]

Deer, as Aldo Leopold so eloquently notes, are enemies of mountains, and the defense that mountains have evolved against them is the wolf:

> I have lived to see state after state extirpate its wolves. I have
> watched the face of many a newly wolfless mountain, and seen
> the south-facing slopes wrinkle with a maze of new deer trails.
> I have seen every edible bush and seedling browsed, first to
> anemic desuetude, and then to death. . . . I now suspect that
> just as a deer herd lives in mortal fear of its wolves, so does a
> mountain live in mortal fear of its deer. And perhaps for better
> cause, for while a buck pulled down by wolves can be replaced
> in two or three years, a range pulled down by too many deer
> can fail of replacement in as many decades.[61]

Leopold was writing of the mountains of the Southwest, but (though vegetation recovers more quickly here) the principle is the same in the Southern Appalachians. Our mountains, too, once defended themselves with
wolves. But now, though the deer are recovering and non-native wild hogs
ravage the land, the mountains' long-evolved defense is gone.

Wolves hunted deer, bison, and elk. As the settlers eliminated their
prey, wolves turned increasingly to farm animals instead. The result was
an escalating war of extermination by humans against wolves that lasted
until about 1920, when, somewhere deep in the Smokies, the last wolf cry
was silenced.[62] (The Southern Appalachians may have been home to two
kinds of wolves: the red and the gray. However, whether the red wolf is a
distinct species is controversial; genetic studies suggest that it is a mix of
wolf and coyote.)[63]

Though driven from the East, gray wolves, like elk and bison, maintained refuges in the western United States and Canada. But the smaller
red wolf, whose range never extended beyond the Southeast, was hunted
almost to extinction and is now, despite the taxonomic controversy, listed
and protected as an endangered species. By the mid-1970s, there were
as few as one hundred red wolves left in the entire world, mostly in
Louisiana and Texas. Many of these could not find mates and had begun
breeding with coyotes. A captive breeding program was established, using
seventeen captured red wolves in 1976. Early in 1991, wolves from this
program were released in Cades Cove in the Great Smoky Mountains National Park. Five pups were born that spring. At least seven of the wolves
released in the first few years were killed by disease (canid parvo virus),
by fights with other wolves and coyotes, or by their drinking antifreeze
that leaked from cars.[64] By 1998 continuing disease problems, and the

failure of the animals to establish home ranges in the park with resultant conflicts with livestock on nearby private lands, had reduced the number to four, and the Fish and Wildlife Service decided to remove the remaining animals and terminate the project.[65]

Red wolves are solitary canid predators, and that ecological role is now filled by coyotes, which had entered the park by at least the early 1980s, and probably earlier. Since that time coyote populations throughout the region have increased.[66] Perhaps the greatest failure of the Smokies red wolf reintroduction effort is the public-relations damage it may have done to other large predator reintroductions. It is likely that the idea of reintroducing pack-hunting gray wolves and cougars will receive less serious consideration.

Nobody knows when (or even if) big cats—variously called cougar, mountain lion, or "painter"—disappeared from the Southern Appalachians, but, like wolves, they were relentlessly persecuted. The last reported killing of a cougar in the area of Great Smoky Mountains National Park was in 1920. Gainer reported in 1928 that cougars were extinct in Tennessee, with the possible exception of a few surviving individuals in the Great Smokies.[67] There have been numerous reported sightings since the 1920s. The possibility remains that these wary hunters have persisted in some of the more remote areas of the Blue Ridge, but many wildlife officials believe that if there are any big cats here, they are released or escaped western cougars, not the native eastern variety.[68]

The black bear was nearly eliminated early in the twentieth century by hunting and the extensive logging that destroyed much of its habitat. After the creation of the Great Smoky Mountains National Park in 1934, the bear population increased, but it dropped suddenly again as a result of the chestnut blight, which eliminated one of the bears' main food sources. Eventually the bears turned to acorns and began to rebound, but drought and the decline of oak trees brought the bears to the brink of starvation in 1992, and many came out of the mountains to forage on the edges of suburbs. There are now about eighteen hundred bears in the Smokies,[69] but these are still harried by the relentless development that is hemming the park in from all sides and by poachers, who kill as many as eighty bears per year.[70]

River otters were gone from Southern Appalachia by the middle of the twentieth century, primarily as a result of hunting, trapping, and habitat loss. A reintroduction effort began in 1984. Between then and 1994, 487 otters were released in Middle and East Tennessee, including 140 in Great Smoky Mountains National Park. Populations are established, and otter pups have been spotted in the wild.[71]

The large, sociable Carolina northern flying squirrel and its cousin, the Virginia flying squirrel, may still be found in the high spruce-fir forests of the Southern Appalachians, the southernmost extent of their ranges. But not for long. As the spruce-fir forests die and the climate warms, their ranges will contract northward. Being endangered species, these squirrels are likely never to return.[72]

Fourteen kinds of bats occur in Tennessee. Nine of these live in caves and mine shafts; the others live in trees. Two of the cave dwellers, the gray bat and the Indiana bat, are endangered. Both have been hit hard by pesticides, which make the insects they eat poisonous. Many of the caves in which these bats are found have been fitted with gates to prevent human intrusion and disturbance while enabling the bats to come and go freely. But the gates have killed some bats by altering cave temperature. The gray bat occurs in two caves in our bioregion: Pearson Cave in Hawkins County, Tennessee, and Nickajack Cave, just north of Chattanooga. Their populations seem to have stabilized, or even increased somewhat, but the Indiana bat, which lives in New Mammoth Cave in Campbell County, continues to decline. (Both bats also inhabit other places outside our bioregion.) Some researchers think the decline of the Indiana bat is due in part to logging, which destroys the big trees in which the bats forage and often roost.[73]

In addition to white-tailed deer, some other mammals have increased in abundance in the human-dominated landscape. Opossums, raccoons, skunks, rabbits, red foxes, coyotes, and feral dogs and cats all do well in rural, suburban, and even urban areas. These "garbage raiders" are often a nuisance, but the greatest concern is their predation on more sensitive species of smaller mammals and birds.[74]

Birds

Prompted by citizen action, government intervention, and regulations—specifically hunting restrictions and the banning of the pesticide DDT—have saved many native birds from extinction. However, two remarkable

Southern Appalachian birds, the passenger pigeon and the Carolina parakeet, are gone forever, both having been hunted to extinction early in the 1900s, before the age of nongame wildlife protection.

The Carolina parakeet, the only member of the parrot family native to the eastern United States, was a large and flamboyantly colored bird, about thirteen inches long, mostly bright green, but with a yellow head and neck and orange cheeks and forehead. These birds flew and roosted in large flocks, but people shot them because they raided orchards, because ladies prized their feathers for hats, and because their bright colors made them useful for target practice. Their unwariness undoubtedly hastened their extinction. When one was wounded, its distress call caused the rest of the flock to circle around until all were shot.[75] The last known member of the species, a male named "Incas," died in the Cincinnati Zoo in 1918.[76]

Nearly everyone knows the story of the passenger pigeon, a bird once so numerous that its flocks used to darken the sky for hours as they passed overhead. Ornithologists estimate their pre-colonial population at two to three billion, making them at that time probably the most abundant bird species on earth.[77] Their extensive winter roosting grounds along river-bottoms in the Southern Appalachians are still recalled in the names of two rivers—the Pigeon and the Little Pigeon—and in the name of the city of Pigeon Forge. Though the pigeons tended to mate farther north, they wintered in Southern Appalachia in great numbers.

Despite their profusion, passenger pigeons were quickly exterminated. The initial blow was habitat destruction. As forests were cleared in the nineteenth century, especially in the upper Mississippi valley, the beech nuts and acorns the pigeons ate became too scarce to support the huge flocks, and populations declined. Then, in the last half of the nineteenth century, market hunters across the eastern United States blasted hundreds of millions from their roosts or from the sky. The meat was shipped by rail to New York and Chicago, where it had become fashionable to dine on squab (young pigeon). The business was lucrative, and the big city cash did much, no doubt, to soothe consciences.

The pigeons' last precipitous decline surprised even conservationists. As their numbers fell, the pigeons ceased to mate.[78] Apparently their mating instinct was triggered—in a way that we will now never understand—by something in the presence of the great flock. The last known passenger pigeon, a female named Martha, died in the Cincinnati Zoo in 1914, the very place where the last known Carolina parakeet would pass four years later.[79]

Bachman's warbler, which once inhabited the dense undergrowth of river swamps across the Southeast, persisted longer than the Passenger Pigeon and the Carolina Parakeet. It may have occurred sporadically in our bioregion; if so, it was driven away long ago by lumbering and the draining and filling of wetlands. For a while, it clung to life along the Carolina coast, but it has not been spotted since the 1960s. It wintered in Cuban forests, and the conversion of these forests to sugar-cane fields may have been the fatal blow. The last of these birds that anyone saw or heard were adult males. They sang for weeks, apparently in a futile effort to attract mates.[80]

One bird that was saved by prohibitions on hunting birds of prey, and by the ban on DDT in the early 1970s, was the bald eagle, the symbol of the American Republic. The bald eagle was gone from Southern Appalachia by 1961 and was in danger across North America. But in 1980, with environmental DDT levels falling, the state of Tennessee began a reintroduction program. Successful nests have become fairly numerous in western Tennessee, and the birds have begun to return to Southern Appalachia. Golden eagles have not recovered as well, perhaps in part due to the greater attention devoted to bald eagles. Tennessee began its golden eagle reintroduction effort in 1981. There are now a few nests in the western part of the state and occasional sightings in the eastern part of the state.

Other large predatory and fishing birds rescued by the DDT ban were the osprey, peregrine falcon, and great blue heron. Osprey were never completely eliminated from the bioregion, but by 1980 there were only about five remaining nests, all on Watts Bar Reservoir. In the succeeding years, they enlarged their territory, and by 1995 the number of active nests had risen to over sixty.[81] Today, osprey are found throughout the region and populations continue to recover from past declines. Peregrine falcons were gone from Southern Appalachia by about the mid-twentieth century. They are still endangered, but restoration efforts have begun to bring them back. Peregrines like to nest on skyscrapers and eat urban pigeons, and at least one has even taken up residence in downtown Knoxville.[82] They have also returned to the Great Smoky Mountains National Park.

Great blue herons have recovered dramatically. Originally rare in the upper Tennessee valley, they have become more common, in part because of the building of the dams, which created suitable habitat. Their numbers, however, declined precipitously nationwide in the 1950s as their prey became contaminated with DDT. By 1973 only 135 active nests could be found in the entire Tennessee valley. Today there are thousands. The reduction of toxic industrial discharges has aided the herons' recovery, as has the protection of wetlands, but the most important factor was probably the banning of DDT.[83] Black-crowned night herons seem to be making a similar recovery.

Several game birds have been saved from near extinction by restrictions on hunting. Wild turkeys were common throughout Southern Appalachia until the 1930s, when overhunting, poultry diseases, and the chestnut blight (which eliminated one of their main food sources) caused a sudden population decline. By 1952 there were only about a thousand wild turkeys in Tennessee, mostly in inaccessible mountain areas. Restocking and conservation programs directed by the Tennessee Wildlife Resources Agency have brought the turkeys back, and their statewide population had reached 75,000 by 1993.[84]

Giant Canada geese, a subspecies of Canada geese, were hunted relentlessly throughout the nineteenth century. By 1920, they were presumed extinct, but a small population was found in the Midwest in 1960. Carefully managed, Giant Canada geese have recovered, and flocks of these large, beautiful waterfowl are once again plentiful on the Tennessee River, sometimes attaining nuisance status.[85]

Wood ducks were also almost lost to hunting and to the logging of the old trees in whose cavities they nested. Hunting restrictions and the installation of millions of nesting boxes have begun to restore this colorful and distinctive bird.[86] Another game species, ruffed grouse, has seen population declines in recent decades, but this may be a return to a more natural pattern, as the forests mature. Grouse require young, successional forests (saplings), so their habitat tends to increase after logging.[87]

Unfortunately, these successes among raptors and game species occur against the backdrop of increased threats to other groups. Today, more birds are jeopardized than ever before—not by hunters, but by roads, suburban development, and clear-cutting. Some of the worst losses are among songbirds—a fact that anyone with a long memory and an eye or ear for birds knows well. Worldwide, two-thirds of all songbird species are declining, and those in Southern Appalachia are no exception.[88] Many Southern

Appalachian songbirds migrate in the winter to the tropics, where the destruction of the rain forests has created intense competition for habitat.[89] In the tropics they also encounter DDT, which is still used by Central and South American nations to control mosquitoes. Radar studies over Louisiana indicate that the number of migrating songbirds fell 50 percent between the mid-1960s and the late 1980s.[90] Of the 76 neotropical migrant species known to occur in the southern United States, only 12 showed population increases between 1966 and 1996, while 31 declined, the remainder being relatively stable.[91]

If, despite the perils in the South, the songbirds manage to return, they find their northern, breeding habitat also under attack, as roads, suburbs, industrial parks, and clear-cuts fragment the forest.[92] Many songbirds must nest in large, uninterrupted forest stands, for only there are their eggs and young safe from predators. Nest predators, such as blue jays, crows, grackles, chipmunks, weasels, possums, cats, and raccoons, live mostly at the forest edges. So do brown-headed cowbirds, which lay their eggs parasitically in the nests of other birds. Deep in the forest these enemies are rare. As forests are fragmented, deep forest habitats disappear and fewer birds survive.[93] This effect is measurable and profound. One researcher simulated songbird nests by setting out artificial nests containing quail eggs in small, medium, and large forest tracks. In some of the smaller tracts, nearly all the nests were raided by predators. But in the largest only one in fifty were. This largest tract was the Great Smoky Mountains National Park.[94]

Populations of many Southern Appalachian songbirds have dropped in recent years. The cerulean warbler, Swainson's warbler, Bachman's sparrow, grasshopper sparrow, Bewick's wren, yellow-billed cuckoo, least flycatcher, Baltimore oriole, worm-eating warbler, and blackburnian warbler are all in decline. Bachman's sparrow and Bewick's wren may already be gone from the bioregion.[95] Spring is not yet silent, but the music is fading.

For some birds, the crucial factor in survival is not only the size of the forest habitat, but its age and composition. The red-cockaded woodpecker nests only in cavities in living pine trees that are infected with a fungus, and these are found primarily in old-growth forests in the Southeast, most of which have been eliminated by logging. As a result, only a few thousand red-cockaded woodpeckers still exist. Although this endangered species was never abundant in our region, which is on the margins of its range, other birds that nest in tree cavities—bluebirds, red-headed woodpeckers, and barn owls, for example—have also diminished because clear-cutting has eliminated many of the old, dead, or dying trees that provide the habitat they need.[96]

Many birds are declining because of the destruction of wetlands. The American bittern, least bittern, woodcock, king rail, and common gallinule are fading into memory as the swamps in which they live, nest, or breed are hemmed in or bulldozed for agriculture or development.[97] Ducks and other waterfowl have declined nationwide as a result of the draining of the prairie wetlands in which they breed.[98]

The demise of the high-elevation spruce-fir forest of the Blue Ridge—from imported insects (see below) and air pollution (see chapter 2)—is also taking its toll on birds. Populations of black-capped chickadees, solitary vireos, red-breasted nuthatches, American robins, black-throated green warblers, and golden-crowned kinglets at Mount Collins in the Great Smoky Mountains all dropped by more than half as the trees died. Other species declined, too, but not as dramatically.[99] The small, yellow-eyed saw-whet owl ranges from Canada south to Guatemala, but almost all the recorded sightings in our bioregion have occurred in the spruce-fir forests. With the loss of these forests, the owl may not return. The rare olive-sided flycatcher has also used spruce-fir forest for breeding and nesting ground and is likely to vanish with the forest.[100]

Not all bird species, of course, are in decline. Urbanization and the fragmentation and destruction of forests seem to have benefited mockingbirds, pigeons, mourning doves, robins, cardinals, blue jays, English (house) sparrows, grackles, and starlings. Many birds have learned to feast on garbage or the leavings of a littering fast-food culture. Crows enjoy cornfields and the bountiful roadkill.

Amphibians

Amphibians—frogs, toads, salamanders, and newts—inhabit worlds smaller than ours and move in a different sort of time. Often furtive and silent, they easily escape notice. And Southern Appalachia is a world center of salamander diversity; they are a characteristic and unique component of our native biota. Yet, out of sight, out of mind, amphibians suffer many depredations. Development, clear-cutting, siltation, acid rain, toxic water pollution, and the destruction of wetlands all take their toll. But something else is at work that scientists do not yet fully understand. Amphibians everywhere—in Australia, in Central and South America, in Europe, and in the western United States—even in remote and relatively undisturbed habitats—are declining in unprecedented numbers. Many species have abruptly disappeared.[101]

The causes are still debated. Initial speculation centered on pesticides, herbicides, and other pollutants, but ozone depletion and the resultant increased exposure of eggs, zygotes, and larvae to ultraviolet-B radiation have also been implicated.[102] More recently, fungal and viral infections have been identified as causes of amphibian population declines in the United States and Canada.[103] However, demonstrated sensitivity of amphibians to water-quality declines and of eggs and larvae to UV radiation suggest that synergistic effects may be operating: environmental stressors may increase amphibian sensitivity to pathogens. A major difficulty in deciphering the causes of observed amphibian declines is the lack of long-term studies of amphibian population dynamics in our region and worldwide.[104]

Because research is scanty, nobody is quite sure what is happening to the amphibians of Southern Appalachia. The Great Smoky Mountains National Park is one of the world's most important centers of salamander diversity, boasting 27 species.[105] Somewhat surprisingly, these are concentrated at the higher elevations. But whether their populations, or the populations of other regional amphibian species, are declining is not known.[106] Global climate change is likely to reduce or eliminate some Southern Appalachian salamanders, especially those endemic to the cool higher elevations. As the peaks heat up and the spruce-fir forests fall, these salamanders will have nowhere to go.[107]

One thing, however, is certain: clear-cutting devastates salamander populations—and, as was noted above, recovery times are long at best. Because the salamander fauna is so rich, and at the same time so incompletely understood, large clear-cuts, developments, and even road construction have the potential to wipe out unrecognized narrow endemic species or evolutionarily significant units.[108]

Principal among the narrow endemics are the cave salamanders. Although they are not federally listed as endangered or threatened, two cave-dwelling species, *Gyrinophilus palleucus,* the Tennessee cave salamander (found in Tennessee, Georgia, and Alabama), and *Eurycea luciguga* (found in southwest Virginia), are state-listed. An additional thirteen species and one subspecies of salamanders and one species of frogs are state-listed in one or more of the states of the upper Tennessee valley.

Rarest and most endangered among native cave salamanders is the Berry Cave salamander *(Gyrinophilus gulolineatus)*. Originally classified as a subspecies of the Tennessee cave salamander, which, ironically, is the state amphibian of Tennessee, the Berry Cave salamander is now generally regarded as a distinct species.[109] Only four populations are known, all of them in East Tennessee. All but one are threatened or already extirpated. The salamander's common name refers to Berry Cave in Roane County, where the only apparently healthy and unthreatened population still exists. It has also been found in Mud Flats Cave in west Knox County, but this habitat was severely degraded several years ago by silt from the Gettysvue housing development. Subsequent efforts to find the salamander there have failed, and this population may already be lost; in any case, continuing pollution and siltation from this development render its survival there precarious. Several specimens were found in a ditch in Athens, Tennessee, in 1953. The exact location is now unknown (the collector has died and left no record), but the fact that it was within the city limits indicates that this population, if it still exists, is also threatened. The only other place where the salamander has been found is the Meades Quarry/ Cruze Cave complex in south Knoxville. This habitat would be seriously degraded and perhaps destroyed by construction of the proposed James White Parkway extension (see chapter 9).[110]

The Berry Cave salamander is thus an endangered species, though it is not so listed at either the state or federal level. One of the authors (Nolt) has petitioned the U.S. Fish and Wildlife Service to list it, but for many years now the Fish and Wildlife Service has refused to act on such petitions on the grounds that nearly all of its funds for listing threatened and endangered species have been tied up in court-ordered actions. These actions are the result of lawsuits filed, ironically, by environmental groups attempting to force designation of critical habitat for still other threatened and endangered species.[111]

The gridlock in listing arises ultimately from political and economic pressure. Since listing can limit the activities of private landowners and developers, recent Republican-dominated congresses have been less than generous in funding it. But because the plight of many species is dire, legal squabbles arise as to how best to use what little money is available, further impeding the process. The upshot is that many threatened and endangered species remain unlisted—and thus have little or no legal protection.

Fish

It is the sport fish that come first to mind when we think of fish in the Tennessee valley: bass, sauger, crappie, pike, sunfish, walleye, catfish, carp—and, in the mountains, brown, brook, and rainbow trout. Many of the most prominent game fish—including the brown trout, rainbow trout, carp, striped bass, yellow perch, and some varieties of walleye and muskellunge—are not native, having been introduced mainly for the benefit of fishermen. But the greatest species diversity is in the small, largely unnoticed fish of the creeks and streams: darters, madtom catfish, daces, chubs, and other minnows. In fact, because of the profusion of these little fish, Tennessee is home to about three hundred species—among the greatest diversity of freshwater fish in the nation.

Reservoirs provide ample habitat for most of the sport fish, which generally remain plentiful. However, four large and distinctive species became extremely scarce or disappeared from the upper Tennessee soon after the building of the dams: the huge lake sturgeon, the shovelnose sturgeon, the paddlefish, and the muskellunge. (The muskellunge occasionally caught at Norris Reservoir and elsewhere are not native fish, but stocked individuals of a northern variety that are apparently incapable of reproducing in Tennessee waters.)[112] None of these three species are yet extinct, but all are extirpated from much of their original ranges and in decline elsewhere. Recently there have been efforts to reintroduce the lake sturgeon into the Clinch and French Broad rivers, but it is too soon to know whether these will succeed.[113] Old-timers say that big paddlefish (spoonbills) used to be caught regularly up on the Holston and French Broad before the building of Fort Loudoun Dam. They are scarce or nonexistent there now, though they still inhabit the bigger reservoirs of the Tennessee.

It is the smaller, less-noticed fishes, however, that have been most threatened by human intrusions. Many of these can feed or reproduce only on clean, gravely or rocky streambottoms and thus are very sensitive to silt. The harelip (or rabbit-mouth) sucker, which once ranged widely in the upper Tennessee valley, was probably the first fish in our bioregion to fall victim to human negligence. This silvery fish, which needed extremely clear water, vanished quickly in the late 1800s as the streams became turbid with silt from clear-cutting and agriculture. It was last seen in 1893 and is now almost certainly extinct.[114]

The slender chub, an inhabitant of gravel shoals, may have already disappeared as a result of habitat degradation. One important cause appears to be coal-washing operations along the Powell River in Virginia. The fish is listed by the federal government as "endangered," but despite repeated searches, it has been seen only once in the last fifteen years: a single specimen was found in the Clinch River in the fall of 1996.[115] The western sand darter has not been found in the Clinch since 1980 and in the Powell since 1987, and it may now be gone from our bioregion, though it persists elsewhere. A gravel quarry on U.S. 25E is thought to have destroyed its Powell River habitat. The yellowfin madtom, a three-inch catfish with venomous stinging spines, originally ranged through much of the Tennessee valley. Dams and pollution degraded most of the streams and rivers it once inhabited, confining it to three widely separated locations: Copper Creek (a tributary of the Clinch River in southwestern Virginia), a small section of the Powell River, and Citico Creek in Monroe County, Tennessee. These populations are all small and precarious.[116]

The blotchside logperch—one of the largest darters, attaining lengths of up to eight inches—is missing from seven of its fifteen known locations. Its isolated populations are separated by reservoirs, and more may disappear soon, threatening the species with extinction. The channel darter disappeared from the Tennessee River below Knoxville shortly after the building of the dams. For a while it remained fairly abundant in the Clinch and the Powell, but recent investigations have failed to detect it, though it still survives in relatively large numbers in the Big South Fork of the Cumberland River. Two of six known populations of the duskytail darter have vanished, and at least one more (in the Little River) is unstable.[117] The longhead darter is gone from half of its known locations, and other populations are declining. Seven of sixteen known populations of the ashy darter have probably been extirpated, though it remains common at Big South Fork and has apparently expanded its range into the New River.[118] The palezone shiner inhabited only one place in our bioregion: Cove Creek in Campbell County, Tennessee. Norris Dam and mining-related pollution extirpated it from that habitat half a century ago, but it still exists in some locations in Kentucky and in the Paint Rock River system in northern Alabama. The spotfin chub is also gone from most of its former range, but persists in North Carolina above Fontana Dam and at several other locations.[119]

Much is unknown about these and other small, unobtrusive fishes, some of which have not even been classified biologically. Because there is little funding to study them, it may well be that some long-unobserved species still exist and that some that are not yet classified as extinct are, in fact, already gone. Yet it is clear that many of these small fish species are declining and that the main causes are siltation, the damming of rivers, and pollution.

Until the late 1950s, some of the rarest of these fish inhabited the relatively pristine Abrams Creek, which meanders though Cades Cove in the Great Smoky Mountains Park and into the Little Tennessee River. But in June 1957 the creek was intentionally poisoned with rotenone for fourteen miles below Abrams Falls in an effort to exterminate "rough" fish and enhance populations of non-native rainbow trout for sport fishing. The poisoning was a joint effort by the National Park Service, the U.S. Fish and Wildlife Service, TVA, and the Tennessee Wildlife Resources Agency (then called the Tennessee Game and Fish Commission). A number of specimens collected as their dead bodies floated to the surface after the poisoning were sent to the Smithsonian Museum, and it was among these specimens that the Smoky madtom was first discovered many years afterward. At first presumed extinct, it was later found to be clinging to life in nearby Citico Creek.[120]

One small denizen of Tennessee waters has made international news. The snail darter was catapulted into prominence after a population was discovered in the Little Tennessee River in 1973, during construction of Tellico Dam. In one of the first and most controversial applications of the newly passed Endangered Species Act, a court decision halted construction of the dam. But in the waning hours of the 1978 session, the U.S. Congress hastily passed a rider exempting the dam from the Endangered Species Act and mandating its completion—to the delight of real-estate investors, who would soon reap a financial bonanza by selling "lakeside" housing units. In November 1979 the gates closed and the dam flooded the valley, creating Tellico Reservoir and obliterating the snail darter's entire breeding habitat on the Little Tennessee. The snail darter has since been found elsewhere, and some have been captured and introduced into the lower Hiwassee and the Holston River tailwaters below Cherokee Dam, where they appear to have established healthy breeding populations that have expanded into the French Broad.[121]

One of the most abundant species in the Tennessee River, particularly in turbid Fort Loudoun Reservoir, is the common carp. Carp are not native to Southern Appalachia. They were imported from Eurasia in the nineteenth century. But they are tolerant of silt and of low dissolved oxygen levels and so have a competitive advantage in degraded waters. Canoeists, who can move quietly enough to observe carp in the shallows, are familiar with the long, thick clouds of mud they stir up as they dart along the bottom. These mud clouds decrease light penetration, impairing aquatic plants and discouraging foraging waterfowl. The sediment may also settle on the eggs of more desirable fish, depriving them of the oxygen needed for normal development.[122] Large carp occur in extraordinary numbers in Fort Loudoun Reservoir, and they are prominent along the Knoxville waterfront, where some visitors enjoy exciting them into a slurping, frothing frenzy by dropping food scraps into the murky water.

Southern Appalachia's only native species of trout, the brook trout, began declining about 1900, but since the 1970s the decline has slowed or been halted.[123] Brook trout are smaller than the non-native rainbow and brown trout, which have been introduced into many clear mountain streams. They are, however, exceedingly beautiful; the dark-green to slate-grey breeding male has a strikingly spotted dorsal fin and an orange or scarlet belly with spots of orange or scarlet along the side. Much of their initial decline was probably due to heating and siltation of streams resulting from clear-cutting early in the twentieth century. Overfishing and competition from the non-native rainbows and browns accelerated the losses. Slight acidity in streams gives the brook trout an advantage over its competitors, so that they have increasingly retreated to the more acidic higher-elevation streams.[124] However, increases in acid runoff are making some of these streams so acidic that they may soon no longer support brook trout either.[125] Then brook trout may be caught in a squeeze between competitors moving up from below and acidity moving down from above. Because they are confined to cold mountain waters and are at the southern extent of their native range, global climate change may eventually lead to the demise of the Southern Appalachian brook trout as well as other mountain species.[126]

Given the degree to which aquatic communities of the Tennessee River system have been modified through impoundments, pollution, and siltation, it is surprising that more species have not been lost or endangered. It is at least conceivable that undescribed fish species of restricted range in the Tennessee River system or its numerous tributary streams have passed from the scene unknown to ichthyologists.[127]

Mussels and Snails

The Tennessee River Valley still holds one of the richest freshwater mussel (or clam) faunas in the world, but probably not for long. Native mussels are nearly all declining, smothered by silt, poisoned by toxics, and deprived of fresh, rapidly flowing water by the dams that have turned almost the entire river system into a series of impoundments from headwaters to mouth. Silt interferes with the filter-feeding of mussels and may prevent reproduction. Some of the remaining mussel populations consist entirely of fifty- to seventy-year-old individuals, which have been incapable of reproducing since the dams were built. Mussels have also been injured or killed by low dissolved oxygen, acid mine drainage (which erodes, and sometimes eats holes in, their shells), fertilizers, pesticides, and other toxic pollutants. They are more vulnerable than many other species to local disturbance, since they are stationary and cannot move to cleaner water.

The numbers are humbling. Tennessee was once home to 130 mussel species.[128] Sixteen species are extinct, or presumed extinct.[129] Of those that remain, 42 in the Tennessee river system statewide and 27 in the upper Tennessee are on the federal endangered species list.[130] Many species that were once widespread are now clinging to life only in isolated and endangered communities. Some of these communities consist entirely of aged individuals that have not been able to reproduce for decades and will almost certainly pass into extinction. A freshwater mussel sanctuary at Chattanooga that harbored nearly 100 species in 1939 had 44 in 1969 and 11 in 1978.[131] Because of the dams, losses have been greatest on the main stem of the Tennessee, but they are high even in the tributaries. The species count for the lower section of the North Fork of the Holston River dropped from 32 to 9 between 1918 and 1978. Similar losses have occurred in other tributaries. The headwaters of the Clinch River contained 18 mussel species in 1918; a 1978 survey found only 5. The headwaters of the Powell harbored

21 species in 1918; this has dropped to 7.[132] Establishment of a mussel sanctuary on the upper Powell in Tennessee and Virginia,[133] and efforts at captive propagation at Virginia Polytechnic University, may curtail the rate of extinction, but the list of the lost will almost certainly continue to grow.

To the uninitated the names of these endangered or threatened creatures are poetic, quaint, and sometimes downright odd: Appalachian elktoe, rough pigtoe, fine-rayed pigtoe, rough rabbitsfoot, birdwing pearlymussel, Dromedary pearlymussel, tuberculed blossom pearlymussel, cracking pearlymussel, little wing pearlymussel, shiny pigtoe pearlymussel, white wartyback, orange-foot pimpleback, clubshell, fanshell, Cumberland combshell, tan riffleshell, Cumberland monkeyface, Appalachian monkeyface, oyster mussel, pink mucket, ring pink, purple bean, Cumberland bean, winged mapleleaf, pale lilliput . . . and on and on. Some produce freshwater pearls. Some are strikingly beautiful. All are declining. The turgid blossom pearlymussel, green blossom pearlymussel, big-river yellow blossom pearlymussel, acornshell, Cumberland leafshell, sugarspoon, angled riffelshell, narrow catspaw, forkshell, round combshell, and Tennessee riffelshell, all former natives of our region, are probably already gone.

Several native species of snail are also in trouble. The brownish-yellow noonday snail (noonday globe), small and dome-shaped, has been found only at Blowing Springs, Cliff Ridges, and Nantahala Gorge in Swain County, North Carolina. U.S. Highway 19 through Nantahala Gorge has disrupted its forest habitat.[134] The three other federally listed gastropods in our region are aquatic or seep species with limited ranges. Anthony's river snail and royal marstonia are found in the Sequatchie Valley, and the Knotty elimia in the Hiwassee Valley. Five other snail species are listed as endangered or threatened by one or more of the states of the upper Tennessee region. There are likely more species with declining populations.

Probably the most abundant freshwater clam in the upper Tennessee, and certainly the most widespread nationwide, is the non-native Asian clam, *Corbicula fluminea*. Zebra mussels, a more recent import from the Baltic via northern Europe and the Great Lakes, may eventually become at least as abundant as the Asian clam. When the zebra mussel was found in the Tennessee River in 1991, there was widespread concern that their population would explode as it has in the Great Lakes and parts of the Ohio and Mississippi rivers. So far this has not happened, but if it does, they may eliminate many of the native mussel species already in peril and other species of aquatic life as well.

Zebra mussels can reproduce at an astonishing rate, blanketing the muddy bottoms of rivers or lakes and coating any hard object they can find. In Lake Erie, less than a decade after their introduction, researchers were finding as many as 100,000 per square meter of lake bottom. Zebra mussels clog water-intake pipes and the cooling systems of power plants, forcing expensive cleaning and prevention measures. They also build colonies on top of other mussels, competing for their food and smothering them. Thus, they may virtually eliminate other mussels from the Tennessee River.

Perhaps their most dramatic effect is to clarify the water. Zebra mussels consume microscopic algae that color the water green. They also filter out brown particles of silt. When present in large numbers, zebra mussels can make the water visibly clearer, as they have in Lake Erie. This has several important consequences. Because sunlight penetrates deeper in clear water, aquatic plants will be able to grow deeper than before and will likely colonize new areas. Portions of the river now suffering from nutrient overload might be cleaned up as zebra mussels remove algae and nutrients from the water. Because algae are eaten by insects and microscopic animals, which in turn are eaten by fish and aquatic birds, effects will ripple up the food chain as zebra mussels deplete the algae. Many fish and bird species are likely to decline; others—particularly bottom-feeders such as catfish and drum that could learn to eat zebra mussels—may increase. Fish that spawn in gravel, such as smallmouth bass, may decrease as zebra mussels crowd gravel beds.[135]

The spread of zebra mussels is an instance of a larger pattern: unique native plants and animals are increasingly being pushed aside and replaced or eliminated through predation or disease by imports from around the world. This is happening not only in the water but on land and not only in Southern Appalachia. It is happening everywhere.

The Homogenization of Nature

A surprising number of the most prominent plants and animals in the Southern Appalachian landscape are non-native—and they are often the same plants and animals that now occur in climatologically similar regions around the world.

Two of the bioregion's most common bird species, English sparrows and starlings, are European imports. In the nineteenth century the sparrows expanded with the British empire around the world. Starlings were first introduced to North America at the performance of a Shakespeare play in New York in the early 1900s. From there they swept across the continent in a matter of decades, becoming one of the most numerous of birds.[136] Vast flocks now darken Tennessee valley skies in the fall. The homogenization of nature extends also into the region's wild places. The wild hogs of the Southern Appalachian mountains, which root up and destroy native forest vegetation, are European in origin.

Many Tennessee valley fish, including the rough and undesirable carp, are non-native, as was noted above. Eurasian water milfoil, an imported ornamental aquatic plant that transforms aquatic ecosystems, grows so thickly in some eutrophic reservoirs that it fills swimming holes and tangles outboard motors.

Red clover, timothy, and bluegrass were brought to Southern Appalachia from Great Britain by settlers whose livestock (also imported species) did not thrive on the Native American grasses.[137]

Pink-blossoming mimosa, foul-smelling ailanthus (tree of heaven), and rough, lilac-flowered paulownia (princess) trees, all three of which are prominent in disturbed landscapes across our bioregion, are imports from Asia. Some unattended urban lots are dominated by these species and appear more Asian than American. Princess trees are especially troublesome on rocky outcrops in the mountains, where they can overcome native flora.

Common privet, a bushy semi-evergreen European hedge plant, has become naturalized throughout Southern Appalachia and now dominates the understory of many regenerating forests. English ivy, escaped from suburban plantings, also grows aggressively on the forest floor, smothering native plants. Japanese grass and the invasive periwinkle do likewise.[138] Multiflora rose, a native of East Asia with numerous small, white, sweet-smelling blossoms, chokes pastures with its briars and weedlike growth. Shrub honeysuckle and especially the fragrant Japanese honeysuckle, with its tough and tenacious vines, can easily strangle or overgrow many native species. Both were imported from the Orient.

The champion strangler and over-grower, of course, is another Asian import, kudzu. Kudzu was introduced into the United States at the 1876 World's Fair in Philadelphia. A legume and nitrogen-fixer, it was used effectively during much of the twentieth century to fertilize and loosen soil that had been worn out and compacted by overfarming; it was also

Fig. 9. *Kudzu overgrowing trees by Route 95 on the Department of Energy's Oak Ridge, Tennessee, reservation. Copyright by Mignon Naegeli.*

introduced to control erosion, though for the latter purpose it has proved ineffective. Now naturalized, it grows rapidly on disturbed land, smothering all other plants, from the tiniest mosses to the tallest trees.

Such invasive aliens leave their natural biological controls (such as insects and fungi) behind, and so acquire a competitive advantage over native species. Some hybridize with the natives, threatening their genetic distinctiveness.

More damaging still are the imported insects, fungi, and microorganisms that attack native plants. The Japanese beetle, a prominent lawn and garden pest (but damaging also to forest trees), is a familiar example. Southern Appalachian forests—already fragmented by roads, logging, and development and degraded by acid rain and ozone pollution—are also facing a mounting accumulation of biological assaults. Road building and other forms of human disturbance facilitate the spread of disease and invasive organisms by opening routes of infection into the heart of the forest.

Perhaps the most destructive disease outbreak so far—some would call it the greatest ecological disaster in North America[139]—was the chestnut blight, the result of a fungus on chestnut trees from Asia imported early in the twentieth century. Though at the time of European settlement chestnuts may have constituted as much as a quarter of all the standing timber in eastern forests, by the late 1930s the chestnut blight had killed virtually all the mature chestnut trees of Southern Appalachia, depriving many forest creatures, including black bears, of the nuts that were until then a staple of their diet. The rotting stumps of the huge old chestnuts are still visible here and there in the Smokies. Chestnut sprouts still emerge from old root systems, but generally die of the blight within a decade. The Allegheny chinquapin, a relative of the chestnut, has also been decimated by the chestnut blight.[140]

But the Chestnut blight was not the first imported organism to kill chestnut trees and chinquapins. Both began to decline about two hundred years ago as a result of infections by the soil-born algal fungus, *Phytophthora cinnamomi*. By 1877 this fungus had become well established in North Carolina, killing many chinquapins and chestnuts in the Piedmont region.[141]

Since 1930, efforts have been underway to breed a blight-resistant chestnut. But these are likely to be hampered by yet another invasive organism: the chestnut gall wasp. Illegally imported on smuggled budwood, this pest first became established in a chestnut orchard in southern Georgia in

1974. The wasp lays its eggs in buds and flowers on which the larvae feed, producing a characteristic gall. Severe infestations can be lethal.[142]

For over a century now, white pines have been falling to white pine blister rust, a disease that attacks five-needle pine species, causing galls that eventually girdle branches and stems. White pines are tall trees that grow mostly in the cool, wet climate of the mountains. A European native, the pathogen arrived in North America as early as 1892 and spread quickly, especially in areas containing gooseberries and currants, its alternative hosts. Programs to eradicate these alternative hosts have helped to slow the pathogen—though such programs may not be without their own ecological effects.[143]

Many dogwood trees have been killed by dogwood anthracnose, a fungus thought to have been introduced with imported Chinese dogwood trees in the late 1970s. The infection begins as leaf spots that may eventually enlarge and kill the entire leaf. Stem and branch cankers can form, causing girdling and subsequent death. Mature trees often die within two or three years as a result of repeated defoliation. Air pollution also plays a role in dogwood anthracnose disease, probably by weakening the trees.[144] Losses among the dogwoods are foreboding, since they are prime soil builders and important sources of high-protein fruit for migratory birds.[145]

Beech trees are ailing from a combined attack by an exotic fungus and an exotic scale insect. The insect, which was introduced into Nova Scotia around 1890 and progressed southward, reaching the Smokies in 1993, feeds on bark, leaving tiny holes that provide entry for the *Nectria coccinea* fungus.[146] Stands of dead beech are now common in the park.

The European mountain ash sawfly, probably introduced from Europe into Canada early in the twentieth century, defoliates American mountain ash, which grows at high elevations along the crest of the Smokies. These trees have been dying at unusually high rates, and the sawfly is a likely culprit.

Many native elms have fallen to Dutch elm disease, which was introduced into the United States in 1930. The disease has hit urban trees the hardest, but it is slowly felling forest elms as well. An outbreak of the disease in the 1980s killed hundreds of American elms in the Little River area.[147]

Butternut (white walnut) trees have been decimated by the butternut canker, a disease that first appeared in southern states about seventy years ago. It has since killed 90 percent of the butternut trees in the Southern Appalachians.[148]

After the chestnuts died, oaks took their place. Now the oaks are declining, too. Part of the problem is that fire suppression has reduced the fire-resistant oaks' competitive advantage over other trees. But the oaks are also suffering from invasive pests. The Asiatic oak weevil is present in large numbers and is a voracious feeder on leaves, reducing regeneration and vigor. Moreover, oaks in the Southern Appalachians are likely soon to face the devastating attack of the gypsy moths.[149]

Introduced from Europe into Massachusetts sometime between 1867 and 1869, gypsy moths have advanced gradually down the Appalachian Mountains, defoliating forests in their wake. They are especially fond of oak leaves. Infected forests literally crawl with their caterpillars, which at times, along with their spoor, drop from the trees like rain. By the time the caterpillars spin their cocoons, few leaves remain. A few gypsy moths have already reached the Smokies, but there have been no permanent infestations yet. When the main "front" moves in, probably within the next two decades, yet another major stress will be added to the bioregion's already beleaguered forests.[150]

Such invasions present forest stewards with painful dilemmas. There are, for example, over eight hundred species of arthropods known in the oak forests of the Great Smoky Mountains National Park, including many delicate moths and butterflies. If chemical or biological controls are used to fight the gypsy moth there, many native creatures may be killed as well. If such controls are not used, the gypsy moths may kill them anyway by destroying their habitat.[151]

In 1993, a related invader, the Asiatic gypsy moth, was inadvertently introduced into North Carolina by a munitions ship docked near Wilmington. More voracious and faster spreading than its European counterpart, this moth could have done even greater damage,[152] but fortunately that invasion was eradicated.[153]

Sometimes an invasion has ramifying effects across whole ecosystems. The balsam woolly adelgid, an insect carried here on plants from Europe in 1908, has almost eliminated mature Fraser fir trees from the high Southern Appalachian Mountains, which are the trees' only habitat. By volume 91 percent of the mature Fraser firs are dead. Only a few small stands are left, and these are infested. Infested trees usually die within seven years.[154] Though the adelgid prefers mature trees, so that in many places juvenile Fraser firs still grow thickly beneath the bleached trunks of the dead adults, the loss of mature trees threatens the species' ability to reproduce.

Because of the complex interdependence of other species with the Fraser firs, their demise has initiated a cascade of further declines. The most prominent side effect is the blow-down of red spruce trees, some over two centuries old, which used to be shielded from the wind by the Fraser firs. We have already mentioned the birds that are disappearing and will likely disappear. But there are also more subtle effects. At least eight specialized species of moss and liverworts grow mainly or only on the bark of Fraser firs. As the firs die, these tiny plants die with them. Moreover, the dead trees have allowed sunlight to penetrate the forest canopy, drying out the moss mats that once dappled the moist forest floor. And the loss of the moss in turn threatens other species, including the spruce-fir moss spider.[155] Where this cascade of effects will end remains to be seen.

The balsam woolly adelgid has recently been joined by the hemlock woolly adelgid, an import from Asia so destructive that it has the potential to kill every eastern and Carolina hemlock tree that is not actively protected by pesticides. (But spraying hemlocks, which often grow in most shady thickets near streams, could contaminate the streams, and nobody is seriously considering widespread application of pesticides in the forests.) The loss of the hemlocks would initiate further effects, since the eastern hemlock is crucial habitat for already stressed neotropical migrant birds and an important component of stream ecosystems, providing cooling shade and vital nutrients to the flowing waters.[156] The first hemlock adelgids in our region were found in forests south of Great Smoky Mountains National Park in 1999. The following year they were identified in the park itself. The hemlock woolly adelgid invasion has begun.

These are merely examples. The problem is global in scope, and the influx of new invasive organisms, often termed "biological pollution," is continual. It is as if all the world's species had been thrown into a blender, mixing competitors, diseases, parasites, and predators from each unique ecosystem with competitors, diseases, parasites, and predators from all the others. Many delicate, unique, and rare species are too fragile to survive this treatment and are being lost. And everywhere tough, aggressive, and weedy species are moving in to replace them. We are blurring the outlines of Creation, leaving a homogenized, standardized, scrambled, and depleted world in its place.

Chapter 5

population and urbanization

Mary R. English and Sean T. Huss

There's no getting around it. People alter the places in which they live, work, and play. In the early years of European settlement of the so-called New World, settlers regarded domestication of the wilderness as both necessary and virtuous.[1] Over the past 150 years, however, this dominant view slowly has altered. People have become aware that domestication of the natural environment is not unequivocally good.

In our bioregion, as elsewhere, it has meant an increase in pollution and a decrease in natural assets. Can these costs be avoided? Not completely, if people wish to use and inhabit the region. But the environmental costs to the bioregion do not have a one-to-one correlation with the *number* of people living here. Instead, the kind and degree of these costs depends much more on *patterns* of development and on people's *practices and values*.

In many ways, this region is ideally suited to habitation. Compared with the arid West, it has plentiful water resources. Compared with the frigid North and the torrid Deep South, it has a temperate climate, requiring little artificial heating and cooling. Compared with the flat plains of the Midwest, it has forested hills into which development can be nestled.

The bioregion also has rare and significant environmental qualities. These need to be protected. They can be, even as the region accommodates more people, but only if wasteful land-use practices change radically. These changes will not occur easily. They will require a widespread change in values.

Values can be categorized as *core values* and *instrumental values*. Core values are deeply held and concern ends in themselves, such as privacy, freedom, security, love, and friendship. Instrumental values also may be deeply held, but they concern the means to achieve ends: for example, owning a house on a large tract of land to achieve privacy, or owning an all-terrain vehicle to achieve a sense of freedom.

Instrumental values can change, but they usually change slowly and with difficulty. Most of this chapter illustrates how far we have to go. The conclusion of the chapter suggests that we may, nevertheless, have some cause for hope.

Population and Urbanization Trends

Patchwork population growth. The bioregion's population more than tripled during the twentieth century, growing from about one million to over three million people. (During the same period, the U.S. population as a whole nearly quadrupled, growing from about 76 million to about 281 million.) As shown in figure 10, more than two-thirds of the region's population were living in Tennessee by the end of the 1900s. Population growth was especially pronounced in the last decade in all but the Virginia portion of the bioregion, which still lags behind its historic 1950 high of approximately 262,000 people.

Given the fixed area of the bioregion, its trend in average density has followed the same trend as its population. As shown in figure 11, density increased from about 40 to about 110 people per square mile during the 1900s. The trends for the individual portions of the bioregion were somewhat different. Of the bioregion's 24,076 square miles, Tennessee accounts for 54 percent; North Carolina for 25 percent; Virginia for 14 percent; and Georgia for 7 percent.[2] The average density in the Tennessee portion surged ahead of the region's average, especially in the last half of the twentieth century, with North Carolina and Georgia showing major density increases in the last three decades. In contrast, Virginia portion's density was slightly lower than it was at mid-century.

Dilute urbanization. Much of the population growth has occurred in or near cities—a main reason why Tennessee has had the greatest population gains. As of 2000, the bioregion had eight "urbanized areas": Asheville, Bristol, Chattanooga, Cleveland, Johnson City, Kingsport, Knoxville, and Morristown. All but Asheville are located primarily in Tennessee. Together, these urban areas accounted for 43 percent of its population as of 2000.

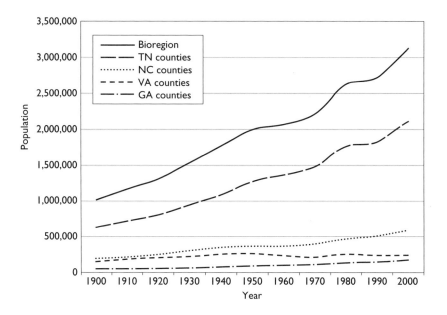

Fig. 10. *Population in bioregion by year (1900–2000).*

Urbanized areas generally consist of a core place and contiguous areas with densities of at least 1,000 people per square mile; to qualify as an urbanized area, the area's total population must be at least 50,000.[3] Data on these areas provide better "snapshots" of population density than do data on metropolitan areas, which, except in New England, are composed of whole counties.[4]

Compiling data for urbanized areas is more difficult than for metropolitan areas, because their boundaries do not follow jurisdictional boundaries. Once compiled, however, the data can be revealing. Census information on the bioregion's urbanized areas shows that, while their populations increased somewhat during the 1990s, their areas increased more. This resulted in a slight decrease in the average density of the urbanized areas. The Knoxville and Asheville areas are the most dramatic examples of this change. The Knoxville area's population increased by 38 percent, but its area increased by 55 percent; the Asheville area's population increased by 100 percent, but its area increased by 118 percent. What do these trends indicate? In a word, *sprawl:* more people occupying a lot more space for residential, commercial, and industrial purposes.[5]

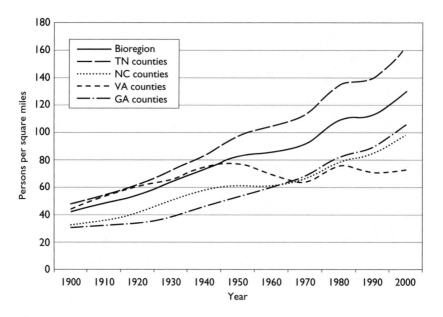

Fig. 11. *Population per square mile (1900–2000).*

Because a dilute form of urbanization—sometimes called "suburbanization"—is spreading across the bioregion's landscape, more of its people are counted as "urban" than as "rural." According to census data for the bioregion's 61 counties, 54 percent of region's total population was considered to be urban in 2000, up from 46 percent in 1990. (In the United States as a whole, 79 percent of the population was urban in 2000.) The bioregion has passed a demographic watershed: its population is no longer predominantly rural.

More wealth, more newcomers, more seniors. One change correlated with the trend toward urbanization and suburbanization is a trend toward greater wealth. As shown in figure 12, the bioregion's median household income rose by nearly 11 percent between 1979 and 1999, although in Virginia it dropped by nearly 10 percent.

Statistics can be deceiving, however. The real picture of household income may be obscured by the influx of wealth that has occurred with the migration of prosperous newcomers into the bioregion. According to a separate study done on the state of Tennessee, 86 percent of the population growth in the state during the 1990s was due to net in-migration rather than natural increase.[6] Some of the migration into the bioregion is due to industry and jobs; some is due to the region's low cost of living, temperate climate, and outdoor recreational opportunities, all of which attract "near-retirees" (aged 55–64) and seniors (aged 65 and over).[7] In the bioregion in 2000, nearly 15 percent of the population was age 65 and over; up from less than 12 percent in 1980.[8]

The hidden poor. With prosperous newcomers, young and old, settling here, growing inequalities of wealth can easily be overlooked.[9] Granted, boom times mean more jobs in construction, tourism, retail, and services. This can and has meant a decline in poverty, at least according to federal standards. In 1979, 16.1 percent of the region's people were living in poverty, according to the U.S. Census Bureau; by 1999, this percentage had declined to 13.6 percent.[10] Unfortunately, many of these jobs are poor-paying and seasonal, and many are not located in remote rural areas or in blighted downtowns. Some people in the region are riding high on the tide of population growth and dilute urbanization, while others are struggling to stay afloat.

Effects on Agriculture

Agriculture is not an unmitigated good. Agriculture consumes 70 percent of the water used in this nation. In contrast, industries use 20 percent and residences use 10 percent.[11] Agriculture is also a major source of water pollution from fertilizers, herbicides, pesticides, soil erosion, and animal wastes—and of air pollution from herbicides, pesticides, and dust.

Pollution problems can be minimized, and sensible, low-irrigation practices can be followed. Nevertheless, agriculture inescapably alters the natural landscape. Putting land into agriculture has adversely affected natural areas and wildlife habitats in the Southern Appalachians, as elsewhere.[12]

Agriculture is essential, however. It feeds us. While we may be content eating apples from the state of Washington and tomatoes from Mexico, we eat most efficiently when we eat locally, avoiding long-distance hauling. In doing so, we eat fresher food, and we support the region's economy. But eating locally is only possible if local lands remain in farming. In our bioregion, many do not.

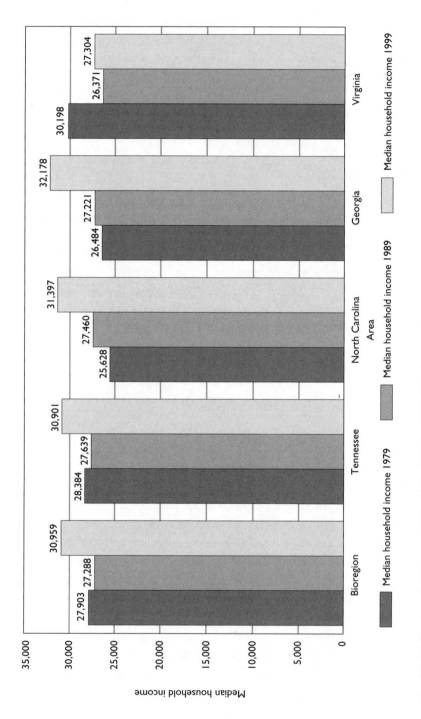

Fig. 12. *Median household income, 1979, 1989, 1999 (in 1999 dollars).*

In the bioregion, the number of farms declined 10 percent between 1987 and 1997, from nearly 47,000 to about 42,000. Similarly, the number of acres in farming declined 10 percent between 1987 and 1997, from nearly 4.5 million to about 4 million.[13]

Many factors can influence a farmer's decision to sell. Some farmers are driven out by neighbors' complaints and lawsuits about noise, dust, and odors from farm operations, as surrounding land becomes suburbanized. This possibility, however, is alleviated by "right to farm" laws.[14] More often, farmers are not driven out; they are worn out or lured out.

Farmers may be ready to retire (the average age of a full-time farmer in Tennessee was well over fifty-five as of 1997),[15] or they may be exhausted from trying to make a living in the face of rising expenses and declining revenues. Lacking pension plans, their land is often their retirement "nest egg." The egg usually will be much larger if the land is sold to a developer rather than to another farmer, and typically far more developers are knocking on their doors.

Should we care if land in the bioregion that traditionally was used for farming is being converted to other uses? In some cases, we should not. Some of this farmland was marginal at best, with thin soils and steep slopes. But some of it was prime farmland—land that is flat or gently rolling, often in the rich bottomland of a nearby river or stream.[16] This land is well-suited for many kinds of agriculture. Unfortunately, because of its topography and location, such land also is well-suited for many kinds of development. Statewide in Georgia, North Carolina, Tennessee, and Virginia, a total of about 2.1 million acres of open land was converted to development between 1992 and 1997; of this total, nearly 28 percent was prime farmland.[17]

Land that is uniquely suited to growing food should be used for just that, not for bricks, mortar, and asphalt. We may not care now, but we will if food becomes more scarce due to burgeoning populations, declining transport fuels, or national security concerns. A few years ago, worrying about the long-term availability and healthiness of imported food would have seemed paranoid. Today it does not. Local agriculture matters and the pressures of urbanization are taking their toll.

Effects on Water and Air

Traditionally, large "point sources" such as industries and sewage plants were the main water-pollution culprits. Today, agriculture has become a major source, as noted above. But agriculture is not the only problem. Just as important are a host of dispersed problems often associated with urbanization or suburbanization, including the following:

Runoff from roads, driveways, and parking lots. As more surfaces are paved to accommodate more cars and trucks, more motor oil, antifreeze, and other chemical products run off into culverts and streams.

Runoff from roofs, patios, and plazas. All impermeable surfaces cause rapid water runoff, taxing stormwater systems and preventing water from gradually percolating into the ground to recharge underground aquifers.

Runoff from construction. Who has not encountered a mudslide from a construction site? Mud is the most obvious evidence of construction damage to our water supplies, but even if it is not present, construction poses other, more insidious problems. As well-established vegetation is removed, water is no longer gradually absorbed into the terrestrial environment. Instead, it ends up as silt-laden water heading down the Tennessee River toward the Mississippi Delta.

Contamination from landscape management and chemically treated greenery. Our American heritage contributes to a longing for large private green spaces. Not content with low-maintenance natural landscaping, we strive for perfect lawns, gardens, and shrubbery. Out come the lawn mowers, the weed whackers, the leaf blowers. On go the herbicides, the pesticides, the plant food.

Water pollution is only part of the picture. All of these sources produce air pollution as well. As with water pollution, point sources of air pollution from power plants, incinerators, and factories historically have been the main targets of concern. Point source air pollution remains a concern, especially as our energy demands escalate and as we continue to rely on coal-fired power plants.

Major point sources of air pollution are under intense scrutiny, however. More pervasive, and a continuing problem, is non-point-source air pollution. Collectively, these other sources are called "area sources." They include small, stationary sources such as residences, but they also include on-highway and off-highway sources.

Most notable is the air pollution from cars and trucks—key sources of nitrogen oxide, sulfur dioxide, and fine particulate matter. Although improved pollution controls have helped mitigate the effects of cars and trucks, the number of vehicle miles traveled is soaring. In Tennessee, for example, this number increased 31 percent between 1992 and 2000.[18]

Off-highway vehicles are also a significant and growing source of air pollution. Most of the machinery used for road building, new construction, and landscape maintenance is powered by combustion engines, often with less-stringent controls and longer idling periods than highway vehicles. All of these engines cause air pollution, even as they work their land-transforming magic.

Between 1990 and 1999, the amount of area source nitrogen-oxide emissions in the bioregion increased 19 percent, going from about 157,000 tons to nearly 187,000 tons. As a percentage of all sources of nitrogen-oxide emissions, this increase was equally striking. Area sources accounted for 50 percent of all nitrogen-oxide emissions in the bioregion in 1990; they accounted for 59 percent in 1999.

Nevertheless, there will always be plenty of air, even though it may be polluted. The same cannot be said of water. About 97 percent of the earth's water is too saline to be directly usable.[19] Our fresh water supplies are dwindling. We are using them up faster than they can replenish themselves. In the United States, this problem is most apparent in arid areas west of the Mississippi, but it is a problem in our bioregion as well. In Sevier County, Tennessee, for example, wells for mountaintop residences are being dug ever deeper and sometimes run dry.

As we use water profligately to keep our landscaping green and our vehicles clean, we contribute to water-shortage problems. Water conservation is an idea foreign to many people in our bioregion, where water appears to be plentiful. But it soon may not be. Our practices, together with the unpredictable effects of global climate change, mean that we can no longer count on water as an abundant resource.

Effects on Flora, Fauna, and Natural Areas

Human beings matter, but so do other species. What are we doing to plants (flora) and animals (fauna) through our ever-expanding patterns of economic activity, settlement, and recreation? What are we doing to the places we cherish for their wild beauty? A recent study by Ken Cordell and Christine Overdevest makes clear that the wildlife and wildlands of Southern Appalachia are at risk.[20]

Cordell and Overdevest, two U.S. Forest Service researchers, analyzed county-level data on six types of natural lands and environmental resources across the nation: 1) undeveloped natural land (public and private), 2) public natural land (federal, state, and local), 3) wilderness, 4) forests, 5) terrestrial wildlife habitat, and 6) water/wetlands.

These categories overlap, but each one targets a particular natural asset. Cordell and Overdevest then mapped these assets against maps of human presence and activity. The maps of human pressures take into account current (2000) and anticipated (2020) pressures. They also take into account pressures arising locally (intra-county) as well as pressures arising from surrounding counties (called "ambient" pressures). Counties where high levels of natural assets coincide with high levels of human pressure are called "hotspots." Hotspots are especially valuable, and they are especially at risk.

Our bioregion is rife with hotspots defined in this way. The region has one of the richest arrays of plants and animals in the nation, but it also has some of the most intense pressures. These pressures come from industries that plunder the region's natural resources—for example, from clear-cutting, where large swaths of mixed hardwood forests are cut and replaced with a single, fast-growing species, usually pine. Pressures also come from our scattered residential patterns, especially with the advent of highways that make previously remote valleys and ridges accessible and of new technologies that drill deeper for water and that treat sewage with small package plants. And pressures come from our greater wealth and leisure, which make lakeside and hillside retirement communities affordable, owning not one but two homes commonplace, and six-hour day trips a whim of the moment.

The maps that accompany *Footprints on the Land* depict hotspots at the county level.[21] Many of our bioregion's sixty-one counties show up as hotspots, because of pressures on one or more of the six categories of natural assets. Moreover, virtually all of the bioregion's counties not labeled as hotspots are still at risk: they show up as having either moderate or moderately heavy population/land interaction indices. In other words, much of the bioregion has high natural land endowments as well as significant pressures from population and economic development.

Although Cordell and Overdevest's study was national in scope, they had several comments targeted at our bioregion. They noted the prevalence of public lands in the Appalachians (unlike the East and Midwest) and commented that the effects of population pressures on these lands are particularly evident in the Southern Appalachians. They also commented that forests in the South are in demand for homes, because of both their scenery and cooler temperatures, and that emerging hotspots can be expected in forested counties, especially in East Tennessee and western North Carolina. They also predicted a significant increase in recreational demand in the Southern Appalachian highlands. These trends, which lead to fragmented landownership and land-use patterns, affect wildlife habitat as well.

As Cordell and Overdevest imply, our bioregion is special: we have much to lose and much to save.

Effects on Shorelines

The recent story of the bioregion's environment is entwined with the story of the Tennessee Valley Authority. Created by "New Deal" federal legislation in 1933, TVA was directed to control flooding, develop hydroelectric power, improve navigability along the Tennessee River and its tributaries, encourage soil conservation, and conduct economic planning in the TVA region. The region is defined by the Tennessee River watershed. It includes parts of seven states: Alabama, Kentucky, and Mississippi, as well as Georgia, North Carolina, Tennessee, and Virginia.

To achieve its river-related purposes, TVA built a massive system of dams and lakes over nearly fifty years, from the 1930s through the 1970s.

The net social benefits of TVA have been questioned,[22] but arguably the early projects served the public good. This argument is more difficult to make for some of TVA's later projects. The Tellico Dam and Reservoir is a case in point.

Tellico Dam was TVA's twenty-fifth dam. It dams the Little Tennessee River at its mouth, where it joins the Tennessee River after flowing westward out of the mountains of Tennessee, North Carolina, and northern Georgia. First proposed in 1939 to spur economic development through navigation and recreation, the Tellico project was revived twenty years later. In 1963 TVA decided to go ahead with it, estimating that nearly 60 percent of its benefit would derive from promoting flat-water recreation and from developing the reservoir's shoreline.[23]

Holding a government agency's right of eminent domain, TVA purchased about 37,000 acres for the Tellico project, including Chota, the ancestral capital of the Cherokees, and other sacred Cherokee sites. as well as 340 productive bottomland farms.[24] Of this acquired land, less than half would be flooded. Much of the remainder would be used for redevelopment, purportedly for a planned new town to be called "Timberlake" and for industrial development.[25] In 1982 TVA sold more than 11,000 acres of the acquired land to the Tellico Reservoir Development Agency, a Tennessee state agency created to carry out economic development around Tellico Lake. The TRDA is run like a private corporation, with an executive director and a nine-member board of directors, and it has broad powers to sell, lease, manage, and regulate its land.[26] TVA retained approximately 13,000 acres of shoreline property for public use and/or use in TVA projects.[27]

Beginning in the early 1970s, the Little Tennessee Landowners Association tried to block the Tellico Dam project. With the passage of the 1973 Endangered Species Act and the discovery of the snail darter in the Little Tennessee, it appeared that the project would be stopped, even though it was 95 percent complete. In 1980, however, a rider to the Energy and Water Appropriations Act exempted the Tellico project from all federal laws and mandated its completion.

The Tellico Reservoir has a 373-mile shoreline. The Timberlake planned community idea was soon abandoned, and to date little industrial development has occurred. Instead, Tellico Lake has become home to privately developed upscale communities. The first was Tellico Village, which was established in 1985 and now has 5,800 property owners, 40 miles of shoreline, and three golf courses. Four similar communities since have been established.[28] Rarity Pointe Community Development is the most recent one to be proposed.

Under environmental review as of late 2002, Rarity Pointe would include 120 acres of TVA-managed federal land on the eastern shore of the reservoir, 212 acres of land purchased from TRDA, and 327 acres of private land. Rarity Pointe would have "residential homes, a marina and lighthouse, a lodge and spa, rental cabins, par 3 and 18-hole championship golf courses, and retail shops."[29] It also has been described as "a gated community that would include 1,200 homes costing $500,000 or more apiece."[30]

Farmland and sacred Native American sites are now being used for gated communities, thanks to TVA. Developers are making a handsome profit on these communities. So are residents and businesses who acquire and then sell their property. Long forgotten are the earlier landowners who received modest recompense for this public gift to affluent America, or the Cherokee whose lands were taken long ago. Land transfers of this sort have been found legal under the Fifth Amendment to the U.S. Constitution ("nor shall private property be taken for public use, without just compensation"), but they are morally dubious.

Concerns about social injustice are not the only problem with projects such as Tellico Lake. By encouraging residential development around Tellico Lake, TVA is encouraging sprawl. Thousands of people now live in an area that once was sparsely populated. In luring residents and businesses to the area, TRDA uses the Tellico Lake's proximity to Knoxville and its airport as a selling point. Proximity, however, means thirty miles to downtown Knoxville.

What does the development of Tellico Lake mean for the environment? It not only means the obvious changes—the loss of farmland, of a coldwater river, of forests and natural habitat—it also means more air pollution from automobiles, boats, lawn-care equipment, and construction equipment; more non-point-source water pollution from impervious surface runoff; more herbicides, pesticides, and water to maintain manicured landscaping; more fragmentation of natural habitat; more opportunity for invasive exotic plants to proliferate. Long forgotten are the rich environmental assets of the area, including but not limited to the much-discussed snail darter. Around Tellico Lake, "environment" means lakeside amenities for those who can afford it.

Fig. 13. *Shoreline development on Melton Hill Reservoir. Copyright by Mignon Naegeli.*

On paper TVA's approach to shoreline management looks good. In 1998 TVA issued a final environmental impact statement on its "Shoreline Management Initiative: An Assessment of Residential Shoreline Development Impacts in the Tennessee Valley." In its summary, TVA notes, "Residential shoreline development was found [in the Environmental Impact Statement] to generally decrease forested area, wildlife habitat, aquatic habitat, water quality, and informal recreational opportunities and increase income, employment, and property values and taxes."[31]

The summary notes that about 13 percent of the total shoreline miles along TVA's reservoirs currently is developed for residential use, with lakefront property owners having access to an additional 25 percent of the undeveloped shoreline. It concludes that TVA intends to "review existing permitting practices and establish a policy that better protects shoreline and aquatic resources, while accommodating reasonable access to the water by adjacent residents." This policy became effective in November 1999.

That's the story on paper. In practice, TVA does not look as good, despite a recent move to open up its resource-related decision making.[32] Kathryn Jackson, the TVA executive vice president of river systems operations and environment, has since been quoted as writing in a 12 November 2002 letter, "TVA's current practice is to develop land-use plans on a reservoir-by-reservoir basis and to review proposals for uses of TVA land on a case-by-case basis."[33] It's business as usual at TVA.

Anticipating and Managing Patterns of Development

The bioregion's human population will probably continue to grow for the foreseeable future, increasing by 7 percent between 2000 and 2010 (see figure 14). Some of this growth will occur because of natural increases (more births than deaths); much of it will occur because of net in-migration (more people moving in than out). Strict controls on reproduction or migration

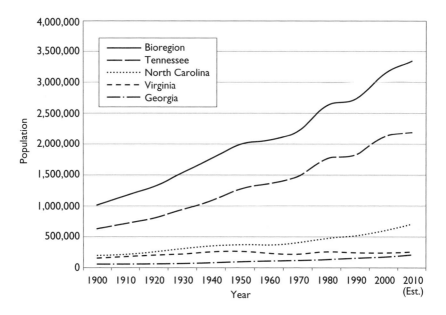

Fig. 14. *Population in bioregion by year (est. 2000–2010).*

are unlikely in the near future, and arguably they are undesirable. The question is not *whether* population growth will occur, but *where* and *how.* In the recent past, more people has meant more sprawling development, more air and water pollution, less habitat for wildlife, less prime farmland, and less available water. But this does not have to be the case. Patterns and practices matter more than sheer numbers.

To minimize the impact of more people on the bioregion's environmental health, foresight and proactive measures are needed. Foresight is needed to anticipate not only the number but also the composition of the bioregion's population. Will the proportion of people over age 65 continue to increase, as now appears to be the case? If so, they will have special housing and service needs. Will the economic divide between the richest and the poorest continue to deepen? If so, special measures will be needed to bring jobs with decent incomes to rural as well as urban areas. Will people's appetites for easy access to goods and services continue? If so, policies may be needed to promote more compact development, more public transit, and more Internet-based services.

Foresight will be required of regional, state, and local governments and public agencies. Hampered by bureaucracy and political special interests, public entities often do not serve the long-term common good. But sometimes they do. There are promising signs of governmental change in the bioregion. Some of this change reflects a quickened awareness of the need to preserve open lands. Georgia recently enacted a law establishing a Greenspace Program to help eligible, rapidly developing counties (including several in the bioregion) acquire fee simple or lesser interests in land to protect green space. North Carolina enacted a 1998 law establishing a Farmland Preservation Trust Fund; it also has a Natural Heritage Trust Fund and a Parks and Recreation Trust Fund, paid for by a tax on real-estate deed transfers. Virginia enacted a 2002 Parks and Natural Areas Bond Act authorizing the state to issue $119 million in bonds to create or expand natural preserves and state parks. Some of this governmental change demonstrates an awareness of the need to plan for growth. Tennessee, for example, enacted a Growth Policy Act in 1998 that mandated counties to consult with their municipalities and develop growth plans with urban growth boundaries. While the growth boundaries do not absolutely constrain growth, they raise consciousness of the need to manage growth. And some of the change reflects an expanded awareness of the need to deal with the environmental problems created by sprawling development: for example, "best management practices" to deal with runoff from construction and tightened controls on ozone emissions, including vehicular emissions.

Government cannot do it alone, however. Foresight also will be required of private organizations, both non-profit and for-profit, and of individuals making individual choices. One outstanding example is the Foothills Land Conservancy and its contributors. Based in Maryville, Tennessee, the non-profit, member-based Foothills Land Conservancy has, since its inception in 1985, protected more than 8,200 acres through acquisition of conservation easements or of land in fee simple. The conservancy recently acquired the development rights to a working dairy farm in Blount County, through the generosity of its owner. By donating a conservation easement to the conservancy, the owner has protected this 1,000-acre tract from development. The owner maintains possession of the farm, but if sold, it must remain in agriculture. The owner, an orthopedic surgeon, is better able than most farmers to afford this donation. It is, nevertheless, an important one: the number of dairy farms in Tennessee has decreased by more than 50 percent in the past decade.[34]

Each of us contributes in large and small ways to patterns of urbanization. We "vote with our feet" when we live in the ex-urbs, when we buy a second home in the mountains, or when we drive fifty miles to an outlet mall. Can these patterns be changed? Ultimately, the choice is up to us.

Chapter 6

FOOD

Walter Riker and John Nolt

How Are We Fed?

Day and night, seven days a week, in any weather and in all seasons, streams of smoke-spewing eighteen-wheelers rush noisily along the interstates to feed us. They roll in from California, from Mexico, from the Pacific Northwest, from Florida, and via Gulf ports from Brazil, Guatemala, Costa Rica, and points south, carrying food to a land once peopled by some of the most rugged, independent, and self-sufficient farmers on earth. But the current inhabitants of this land are beholden to the trucks, for they can, for the most part, no longer be bothered to feed themselves.

The trucks, of course, keep the warehouses, grocery stores, and restaurants well stocked. But fresh food is not everywhere to be found. Supermarkets and even small neighborhood markets have abandoned many inner-city communities. The only food available in these neighborhoods is at the convenience stores, often attached to gas stations. This food is mostly junk: highly processed, fattening, low in quality, of little nutritional value, and expensive. Thus it helps reinforce the familiar cycles of poverty, disease, and dependence.

Even in the suburbs, most of the available produce has been transported long distances and much of it is treated with chemical sprays, waxes, and colorings to preserve "freshness"—or at least the illusion of it. And even where high-quality produce is available, many people—ignorant, befuddled by advertising, or demoralized beyond caring—still choose junk, as can be confirmed by a few minutes' observation at any supermarket checkout line. The results are apparent in our diet-related illnesses (obesity, cancer, diabetes, heart disease), which are, without exception, problems of excess, not of want.

Yet many people do strive for health, and so have begun a venture that may take them beyond themselves; for personal health requires healthful food, and healthful food requires a healthy land.

Food and People

We no longer have a healthy, integral relationship with our food. We do not know where it comes from, how it is grown or raised, and, in all too many cases, even how to prepare it. Instead, we have sought convenience, low cost, and freedom from the apron and the hearth—with money to burn. Food producers have responded to our apparent interests, trading efficiency for health, quantity for quality, and taste for substance. But our apparent overriding interest in freedom from the kitchen and field has cost us in many unsettling and often unrecognized ways. We are getting what we *pay* for. The foods most readily available to us today, designed as they are with efficiency and profit foremost in mind, do not help us to live better lives.

Most of us are too fat, and we are getting fatter. Three out of five people in the United States are overweight, and the rate is going up.[1] The causes of this excess weight are complex, but there can be little doubt that processed, high-sugar, and high-fat foods contribute to the problem. We drink too many sugary sodas and flavored coffees and eat too many processed pastries and deep-fried foods and too much high-fat meat. Although we have started to eat more grains, fruits, and vegetables,[2] we are still not eating enough of them.

Excess weight is linked to increased occurrence of many health problems, including heart disease, high blood pressure, diabetes, several cancers, sleep apnea, and arthritis.[3] Consequently, overweight individuals tend to visit physicians and hospitals more frequently and spend more on health care than those who weigh less. It would be a mistake to think that all excess weight is just a result of bad food choices, but much of it clearly is.

We eat too much meat. We have made some improvement on this front—we eat less red meat and more poultry and fish than we did thirty years ago[4]—but we still eat over 190 pounds per capita annually. In fact, today we get most of our protein from meat. About a hundred years ago we got most of our protein from the grain products we ate, consuming half again as much grain (300 pounds per capita) as we do today.[5] As a result, we do not get enough fiber, we eat too much fat, and we suffer from heart disease. This is unnecessary. We eat meat because we like it and because we are used to eating it, not because we need it to be healthy. Our high meat consumption is unfortunate and wasteful, because meat production is a less-efficient way to use the farmland we have left than is grain production, returning less per resource input. Further, as currently practiced in the United States, meat production is the source of many preventable harms to communities, to individuals in and around the meat industry, and to animals. We suffer a host of ills and cause extensive harms to satisfy a mere taste.

Antibiotic resistance is another reason to be concerned about meat consumption. Meat animal producers in the United States typically raise animals in crowded, unsanitary and stressful conditions. These conditions promote disease, so meat animal producers often use large quantities of antibiotics. But not all of the animals fed the antibiotics are sick. The Union of Concerned Scientists estimates that 70 percent of the antibiotics used in animal agriculture are fed to healthy pigs, cows, and chickens.[6] It is just easier to feed antibiotics to all of the animals, and, some argue, the antibiotics fed to healthy animals can serve as "preventive" medicine. Yet the extra antibiotics given to the animals leave troublesome antibiotic residues in their meat.

These antibiotic residues contribute to a number of human health problems. They cause allergic or toxic reactions in some people. They also contribute to the development of antibiotic-resistant organisms in animals, on the farm, and even in the human gut. One analysis of store-bought chicken revealed the presence of harmful salmonella and campylobacter bacteria in about half of the samples.[7] Most of the campylobacter and one-third of the salmonella also showed resistance to common antibiotics used to treat people. The World Health Organization recently stated that accumulating evidence provides documentation that these and other resistant bacteria have significant health consequences for humans.[8] Resistant bacteria are responsible for infections that would not have otherwise occurred, increased frequency of treatment failures, and increased severity of infections. They result in prolonged duration of illnesses, increased frequency of bloodstream infections, increased hospitalization, and increased mortality.

These health consequences can be avoided, because this indiscriminate use of antibiotics is unnecessary. Several European countries have banned the use of antibiotics in healthy animals without suffering any reduction in animal-production levels.[9] The incidence of antibiotic resistant organisms often decreases—though it does not always do so—after indiscriminate antibiotic use is stopped.[10] In Finland a campaign to reduce erythromycin use for sore throats and related infections in children led to a reduction in the rate of erythromycin-resistant forms of the causal bacteria, Group A Streptococcus. In this case it appears that the need for antibiotics was increased by the use of antibiotics. Danish studies showed levels of other resistant bacteria decreased when bans were placed on the feeding of certain antibiotics to healthy animals.

Many of the foods most readily available to us contain harmful trans fatty acids.[11] These acids, often called trans fats, are made by a process known as hydrogenation. Hydrogenation turns liquid oils into semisolids, and this is done to increase the shelf life and flavor stability of foods made with the oils. Hydrogenation is not necessary to the production of food, but it has been widely used in the United States since the 1940s because it makes foods easier for producers to ship and store. Trans fats are harmful because they increase the level of "bad" LDL cholesterol in the body and decrease the level of "good" LDL cholesterol, increasing the risk of heart disease.[12] Scientists think that as few as two or three grams of trans fat a day can increase the health risks. Since our bodies require no trans fats for healthy function, we should eat as little trans fat as possible. This is not easy to do. A single glazed donut typically has four grams of trans fat.

After years of debate, the FDA has at last decided to make food processors indicate the amount of trans fats in foods on nutrition labels, but the new requirement does not go into effect until 2006.[13] Nutrition labeling is still not required in restaurants. You can, however, find out if the foods you buy at the store contain trans fats by looking in the list of ingredients for "hydrogenated" or "partially hydrogenated" oils. These ingredients often indicate the presence of trans fats. The results of such sleuthing are often surprising. For instance, large grocery stores in Southern Appalachia offer hundreds of varieties of bread for sale, but few, if any, that do not contain trans fats. Other sources of trans fats are harder to discover. Many fast-food restaurants cook foods in oils that contain trans fats. For instance, McDonald's announced in 2002 that it would stop cooking french fries in oil containing trans fats and start using an oil without trans fats, but the company has never actually made the switch.[14]

Fortunately, we have started to eat more fruits and vegetables, but until there are more stringent production standards, we need to be aware of the potential threat posed by pesticide residues. The benefits of increased fruit and vegetable consumption generally outweigh the risks presented by pesticide residues, but there is still reason to be concerned. An analysis of data collected by the USDA showed that some fruits and vegetables contain levels of pesticide chemicals in one serving that are higher than the daily limits considered safe, presenting real risks to children, pregnant women, and pets and other animals.[15] The highest levels of toxic residues were found on winter squash, peaches, apples, pears, grapes, green beans, spinach, strawberries, and cantaloupe, while the lowest levels were found on bananas, broccoli, canned peaches, canned peas, and canned corn. In general, U.S. produce is more likely to have higher residues than imported produce, and fresh produce is more likely to have higher residues than canned and frozen produce. Pesticides banned in the 1970s—like dieldrin and DDT—still show up in fruits and vegetables today. Dieldrin, a carcinogen which remains in soil even twenty years after exposure, is found today in residue tests on squash, cantaloupe, soybeans, sweet potatoes, and spinach. Dieldrin cannot be washed off.[16]

Food and the Land

Our current food system damages the land, not only here, but wherever the food is grown, and everywhere in between. Because our food comes from around the world, our eating habits in Southern Appalachia promote environmental damage almost everywhere. Rain forests in Costa Rica may be felled to provide grazing land for cattle that become the fast-food hamburger we eat in Chattanooga. A hillside in the Philippines may be deforested and soaked with herbicides and pesticides to grow the pineapple in the sundae we have for dessert. The lettuce on the hamburger may be grown in Southern California on an irrigated desert, where the shrinking water supply will dry up within a decade or two. The bread for the bun may be baked in Cincinnati from flour ground in Minneapolis that is made from grains grown on Iowa farms that douse their crops with pesticides and chemical fertilizers. The onion may be raised in Texas on eroding land that is thin and compacted by heavy farm machinery. In one way or another, all these forms of agriculture degrade the land that produces the crops that feed us.

To that damage we must add the further degradation involved in getting the food here. The exhausts of the ships and trucks that transport our food pollute the air with carbon monoxide, particulates, nitrogen oxides (NO_x), volatile organic compounds, and ozone, the effects of which are discussed in chapter 2. As local agriculture declines, population grows, and tastes become more cosmopolitan, the transportation system also grows, crowding the interstates with trucks and the landscape with the truck stops, fast-food joints, and service stations that support the transportation system.

All this moving from place to place requires enormous amounts of fuel—as, in most cases, does the growing and processing of the food itself. This fuel is made from crude oil. Since our domestic supply of crude oil is nearly used up, much of the crude must be shipped from oil-producing nations on supertankers (which themselves use still more fuel). When it reaches the United States, typically on the Gulf Coast of Texas or Louisiana, the oil is refined by polluting processes, then pumped or trucked (again at great expense of energy and fuel and with considerable pollution of the air) to truck stops and gas stations all across the country, where the diesel trucks that bring the food receive it.

Those same Gulf Coast chemical plants may supply the plastics, styrofoam, coloring, and inks in which the food is packaged—unless it is packaged in paper or cardboard, in which case forests are cut and chipped to supply the paper mills. The packaging is generally used once, after which it becomes a waste-disposal problem. Much of our food is refrigerated over long times and distances. The refrigeration requires more fuel—or electricity generated chiefly by the burning of strip-mined coal.

Most of the food bought in Southern Appalachia is processed. Precise recent percentages are unavailable, but a good estimate may be obtained by considering the stock of any regional supermarket. The unprocessed foods are the fresh fruits and vegetables, fresh meats, eggs, and some dairy products. Compare the floor area in the grocery store devoted to these items with the area devoted to such products as soft drinks, processed meat, beer, canned and frozen foods, specialty foods, mixes, snacks, sugary breakfast cereals, and candy. Since most people buy most of their food from grocery stores, this proportion is a good estimate of the preponderance of processed food in our diet. All these processed items pass through at least one and often several industrial operations, each of which requires additional truck transportation and energy. The resulting food is almost invariably less nutritious than produce fresh from the garden or farm.

To these energy and transportation costs we must add the fuel burned by our automobiles as we drive to the grocery store or restaurant to buy the food and the fuel required to bring the fuel for our automobiles to the gas station where we buy it, and the energy and pollution required to refine that fuel—and so on. All these things are now integral components of our food-supply system, and all degrade the land.

Not only do most people in Southern Appalachia no longer grow their own food; many seldom even prepare it. Thus we endure the seemingly endless proliferation of strip malls teeming with restaurant upon fast-food restaurant. Many of these go out of business almost as soon as they open; yet, they continue relentlessly to expand across the land, leaving boarded-up buildings and desolate parking lots behind. This is said to be a side effect of the "efficiency" of our economic system.

Agriculture: A Long, Steep Decline

As the strip malls, roads, and housing developments expand, farms die, and more and more of our food comes from somewhere far away. This is a disturbing transformation for a land that was once self-sufficient and agriculturally independent. Given the exponential growth of world population, worldwide loss of croplands, competition for fresh water, and global climate change, it is prudent to have a reliable bioregional food supply.

The statistics are not encouraging (see table 4). Southern Appalachian agriculture has suffered a long, steep decline over the last century.

Since 1910 Southern Appalachia has lost 6,893,283 acres of farmland, much of it to sprawling development. That amount of land, if placed contiguously, would more than fill a 100-by-100-mile square. Most of what we have lost is wooded farmland and pasture, but since 1925 (while population has climbed), we have lost nearly a quarter of our available cropland, about 674,000 acres.[17] Though much of this land has reverted to forest, and so is healing and growing richer, and some was so steep that it should never have been farmed in the first place, much of the best agricultural land is being damaged or made barren by development.

Advocates of development sometimes herald this loss as a good thing, signaling the transformation of an agricultural economy into a modern,

Table 4

TRENDS IN FARM NUMBERS AND ACREAGE IN THE
SOUTHERN APPALACHIAN BIOREGION

Year	No. Farms	Total Farm Acres	Average Acres Per Farm
1910	124,862	11,019,016	88.2
1925	126,162	9,400,400	74.5
1935	152,489	9,568,747	62.8
1945	140,777	8,606,619	61.1
1954	118,707	8,008,922	67.5
1964	77,236	6,434,184	83.3
1978	49,048	4,758,353	97.0
1987	46,718	4,466,336	95.6
1997	41,969	4,125,733	98.3

SOURCE: U.S. Bureau of the Census, Census of Agriculture. Farm data are for the counties included in the bioregion. Farm data are not collected for every year. Since 1920, the Census of Agriculture has been taken at approximately five-year intervals, with some variation. We have attempted to present available farm data in roughly ten-year intervals.)

diversified economy. King Midas may celebrate his power to turn everything into gold, but if he turns his farms and fields into gold, his stomach will eventually inform him that he has erred.

It is astonishing how little care we have taken, for example, to insure a fresh supply of locally grown vegetables. The soils and climate of the Tennessee valley are quite good for the cultivation of many vegetables, including asparagus, broccoli, cauliflower, carrots, lettuce, onions, green peas, and many others. Some of these can be grown in multiple crops per season. Certain highly nutritious greens—cabbage, kale, collards, and brussels sprouts, for example—can be cultivated into the late fall and even the winter. But we do not commonly grow any of these vegetables commercially.[18] Some of the vegetables that are raised here—tomatoes and eggplants, for example—are often sold outside the bioregion, while we import (mostly tasteless) tomatoes and eggplants from elsewhere.[19] A large portion of our fruits and vegetables come from the irrigated deserts of California.

Regional farm losses have been partially offset, of course, by increases in productivity. Production per acre of virtually all crops has increased with the development of new crop varieties and new agricultural chemicals, and Tennessee's total production of corn, soybeans, and tobacco (though fluctuating widely from year to year with shifts in rainfall and weather) has generally held steady or increased. Yet the production of many vegetable crops has steadily declined, either in absolute quantities or per capita.[20] This indicates a statewide movement away from agricultural diversity and toward monoculture—the reliance on a single kind of crop, or at most a small group of crops. Monoculture is economically efficient in the same way that factory mass-production is economically efficient, but it is also risky. A single crop can be destroyed by a single kind of insect, a single disease, or a shift in climate or precipitation. In agriculture, as in nature, diversity makes it probable that much will survive. Monoculture makes it probable that, under certain conditions, very little will survive.

Agricultural plagues remain possible, despite the advances of modern chemistry. Many weeds, fungi, and insects have evolved rapidly in response to the pervasive use of herbicides, fungicides, and pesticides, and they are now immune to much of the contemporary chemical arsenal. Crops are also under stress from changing climate and weather patterns, increasing ultraviolet radiation due to ozone depletion, and ground-level ozone (see chapter 2). Putting all our agricultural eggs into one basket, then, is imprudent. Yet the farms of Southern Appalachia are mostly monocultural, their chief products—apart from hay—being cattle, sorghum, soybeans, corn, and tobacco.

An especially unsettling aspect of the agricultural decline has been the disappearance of small farms. Table 4 shows that average farm acreage has increased since World War II, as small farmers have been driven out of business by low crop prices and stringent competition. Small farms are important—strategically, ecologically, and culturally—for they are mainstays of local food-supply systems, havens of crop diversity, and molders of sturdy, independent character. It is worth being reminded that Thomas Jefferson was emphatic in insisting that the small farmer is indispensable to preservation of democracy.

But many today regard Jefferson's view as outdated. When the farm statistics were announced in the summer of 1996, showing yet further declines in Tennessee agriculture and in the small farm, a University of Tennessee agricultural economist was quoted in the *Knoxville News-Sentinel* as saying, "The smaller farms are not inherently less profitable, but they may have to change if the market conditions require that they get bigger, and I think they have done that."[21] It seems that the structure of twenty-first-century agriculture is to be determined not by Jeffersonian wisdom, but by the dictates of the inexorable god Market Conditions, a being whose demands we must obey, though they undermine much that we once thought good.

The demise of the small farm is a nationwide phenomenon, more prominent in many other places than it is here. In portions of the Midwest and Southern California, much of the agricultural land has been acquired by huge corporations—Tenneco, Goodyear, Exxon, Prudential Insurance, Bank of America, and others—that have rooted out and exterminated the culture of the family farm. Not so in Southern Appalachia. In this respect, as in many others, we have been protected by the rugged hills, which naturally divide the land into small holdings and have so far made large-scale corporate farming impractical. As a result, Southern Appalachian farms (those that remain) are still on average much smaller than farms nationwide. Most are still in private hands, and remnants of the traditions of the family farm survive here.

But a close look at the remaining family farms reveals a disturbing trend. The very nature of family farming is changing, and not for the better. The small farms in our area are becoming increasingly one-dimensional, focusing more often than not on the production of beef cattle to the exclusion of all else (see table 5).

In 1925, almost half of East Tennessee's 72,246 farms were producing sweet potatoes, while today less than one-tenth of 1 percent grow them. Back then over three-quarters of our farms grew apples. Today less than 2 percent grow them. Almost all of the apples in our grocery stores come from Washington state. Farms in our bioregion are still relatively small and are still owned by private citizens, but they are not very much like traditional family farms anymore. Though remnants of the family farm remain, this tradition is dying here, as it is everywhere else. While overall population has grown exponentially, farm population is now much less than a tenth of what it was in 1920. The skills and experience of food production are vanishing, along with the frugal character and solid spirit that once grew from the land. Correspondingly, there is a rise of dispiritedness, improvidence, and dependence among those who live in detachment from the soil that gives them life.

Table 5

COMPARISON OF NUMBER AND PERCENTAGE OF FARMS
PRODUCING SELECTED FARM ITEMS IN TENNESSEE
BIOREGION COUNTIES FOR YEARS 1925 AND 1997

Farm Products	No. of Farms (1925)	% of Farms (1925)	No. of Farms (1997)	% Farms (1997)
Soybeans	12,890	17.8	206	0.8
Cowpeas	7599	10.5	0	0.0
Potatoes	33,994	47.1	222	0.8
Sweet Potato	18,296	25.3	6	0.0
Cabbage	916	1.3	29	0.1
Cantaloupe	734	1.0	24	0.1
Tomato	5,533	7.7	198	0.7
Watermelon	3,203	4.4	60	0.2
Apples	55,543	76.9	356	1.3
Peaches	40,073	55.5	174	0.6
Pears	26,485	36.7	69	0.3
Plum/Prune	19,676	27.2	0	0.0
Beef cows	35,311	48.9	16,414	59.9
Dairy cows	43,757	60.6	939	3.4
Milked cows	64,068	88.7		
Swine	38,376	53.1	489	1.8
Chicken	65,765	91.0	1,217	4.4
Total	**72,246**		**27,409**	

The Squeeze on Farmers

Why have we lost so much of our agriculture? The answers, of course, are primarily economic, but the direction of the economy is determined by the choices of individuals. The ultimate answer lies with each of us.

One cause has been increasing urbanization and the rising land costs and property taxes that it brings. As suburbs swallow up the countryside, available land becomes scarce and land values increase. Rising values attract speculators who buy up large tracts, wait for the value to increase still further, and then sell them to real-estate developers. This further reduces the available land, driving costs still higher. Nearby farmland also increases in value, raising property taxes and often erasing the farmer's thin profit margin. The farmer then has no choice but to sell. Those who would like to start new farms are deterred both by the taxes and the cost of the land itself. This is due in large part to population growth, which was discussed in chapter 5.

A second cause of the agricultural decline is the "corporatization" of food production. The activities involved in producing food may be divided into three categories: farming, the production and sale of inputs to farming (chemicals, seeds, and equipment purchased by the farmer), and marketing. The nature of this corporatization can be understood by considering the amount of money generated by activities in each of these three categories since 1910. Table 6 gives the essentials. In constant 1984 dollars, farming was a thirty-billion-dollar enterprise in 1910 and accounted for about 40 percent of the money generated by food production. In absolute dollars, this figure has remained stagnant, but the money generated by marketing and inputs has increased dramatically. The result is that farming currently accounts only for about 8 percent of the wealth produced by agribusiness. The wealth generated by America's farmland is, in other words, no greater than it was in 1910—despite the fact that farmers feed a much larger population and are much more productive.[22] The money has instead flowed primarily to the corporations that control inputs and marketing.

Corporate dominance of marketing is largely the result of choices of individual consumers. It is we who have acquiesced to the advertising that teaches us to prefer convenient, heavily packaged, far-transported, highly processed food to nutritious produce direct from local farms. Thus many of the local farms have withered and died from lack of consumer support.

The corporations, of course, have planned and promoted this process. With the rise of television advertising in the 1950s, many people were con-

Table 6

WEALTH GENERATED BY THE SECTORS OF
U.S. AGRIBUSINESS, 1910 AND 1990

Sector	% of American Agribusiness (1910)	% of American Agribusiness (1990)
Farming	40	8
Input	14	25
Marketing	46	67

ditioned to esteem corporate food products as symbols of modern afflu-ence. Tasteless, textureless, highly processed Wonder Bread replaced the fresher and heartier breads that had been baked locally from locally milled flour. As corporate marketers of farm products inserted themselves between farmers and consumers, many of them—Del Monte, General Mills, Ralston Purina, Kraft, and Tyson, for example—obtained monopolies or virtual monopolies on various areas of distribution, undercutting and destroying local distribution systems, forcing farmers to sell through them, and siph-oning away the profit.[23]

The squeeze came from the input side as well. As mechanization and competition intensified, farmers were forced to rely less on their own labor and ingenuity and more on mechanical and chemical inputs. Among the suppliers of these inputs, as well as among the marketers of farm outputs, there was all too often a conspicuous lack of competition. Tires, petroleum products, chemicals, and rail transportation were controlled by a few large corporations. There were many farming communities where feed had to be bought from Ralston Purina or not at all.[24] Three decades ago, Texas agricultural advocate Jim Hightower noted: "Before the first sprout breaks ground, American farm families are over their heads in debt to such corpo-rate powers as Bank of America (production loans), Upjohn Company (seeds), The Williams Companies (fertilizer), International Minerals & Chemical (pesticides), Ford Motor Company (machinery), Firestone (tires), Ralston Purina (feeder pigs), Merck & Company (poultry stock), Cargill (feed), Dow Chemical (cartons and wrappings), Eli Lilly (animal drugs), Exxon (farm fuels), and Burlington Northern (rail transportation)."[25] The situation is little different today, except that those farmers whose debts were most precarious are no longer farming.

So the squeeze on farmers continues. The corporations that control inputs and marketing have huge reserves of capital and can withstand wide fluctuations in markets and weather, leaving the financial risks to fall almost entirely on the farmers. Money, not land, must therefore be the farmers' cen-tral concern. In a national survey in 1994, 83 percent of farmers interviewed said that their biggest challenge was simply making a profit.[26] Squeezed between the profit-taking corporations that supply inputs, on the one hand, and those that market the produce, on the other, the farmer is forced to adopt the corporate values of specialization, efficiency, mechanization, and growth simply to survive. Many do not, and have given up in disgust.

The squeeze exists because the rise of mechanized and chemical agriculture and industrially processed foods has diminished the productive role of the farmer relative to the food industry as a whole, transferring more and more of the responsibility (and hence more and more of the profits) to the suppliers of inputs and the marketers. The farmer is thus crunched between two jaws of a vice—the corporations that control inputs on one side, and the corporations that control marketing on the other. One obvious way to loosen the vice is to return to the farm or farming community some of the productive functions (and hence the profits) that have been usurped by these corporations.

There are many strategies for doing this: using cover crops, compost, or manure produced on the farm rather than chemical fertilizers; bartering among farm households; using draft animals or human labor instead of petroleum-powered machinery wherever possible; making greater use of wind power and solar energy; avoiding routine spraying of pesticides and herbicides (using them only as needed); marketing more directly to local consumers rather than to corporate food processors; organizing marketing co-ops or schemes of community-supported agriculture (discussed below); experimenting with small-scale processing on the farm of such items as cheese, preserves, or cured meats; and minimizing debt.[27]

We will probably never return to a healthy agriculture, however, if we expect the initiative to come from hard-pressed farmers (the few who are left). The crucial element is an educated public that appreciates the advantages of a sound, healthful local food supply and is prepared to pay somewhat higher prices (at least initially) or take some financial risk to secure it.

One way to accomplish this is through community-supported agriculture—an arrangement whereby a number of consumers pay to the farmer in advance a fixed price for a year's supply of certain specified foods. If the harvest is good, they get a bounty and a bargain; if it is poor, they accept the loss along with the farmer (who also has less to eat). This guarantees the farmer a livable income and spreads the risk normally borne by the farmer among the consumers, each of whom carries only a small portion of it. Such an arrangement enables small farmers to avoid debt and to hold their own against competition cushioned by greater capital. It also knits communities together and increases their independence, cuts out middlemen, provides consumers with nutritious produce direct from the farm, lowers the cost and environmental impact of food transportation, and increases everyone's care for the land. Several farms in our bioregion have promising community-supported agricultural projects already underway, and with greater consumer demand there could be more; for many are the people—both old and young, both experienced and inexperienced—who would love to farm, if only they could make a living at it.

Food Systems from Farm to Plate

A food system is the set of pathways that food travels as it makes its way from producers to consumers. In the simplest food systems, consumers eat what they themselves produce. Most of us get our food through a much more complicated set of pathways. In these more complex systems, food travels from producers through various processors, distributors, sellers, and preparers before it reaches consumers. These complex food systems are supposed to have several advantages—greater variety and lower price being chief among them—but they also have costs.

Among the costs are those associated with food miles—the distance traveled by food as it makes its way from where it is grown or raised to where it is consumed. In 1969, the U.S. Department of Energy estimated that food in the United States traveled an average of 1,346 miles.[28] Another study, using data from 1980, determined that fresh produce in the United States traveled an average of 1,500 miles.[29] An analysis of the 1997 arrival data for the terminal market in Jessup, Maryland, found that fruit distributed there traveled an average of 2,146 miles, while vegetables traveled 1,596 miles.[30] In a study of the Chicago terminal market, researchers calculated that produce arriving by truck traveled an average of 1,518 miles in 1998.[31] This is a 22-percent increase in the average distance traveled since 1981, when the average for Chicago was only 1,245 miles.

As the distances reported in these studies are relatively consistent, no attempt is made here to determine the exact distances food travels to the Southern Appalachian bioregion. The following sections, where the supply channels for some major foods are discussed, provide ample evidence that our food travels just about as far as food anywhere else. California is by far the largest producer of fresh fruits and vegetables in the country. Each year over 485,000 truckloads of fresh produce leave California and make their way around the country.[32] Each year we consume 83.9 million pounds of bananas in Southern Appalachia,[33] almost all of which we import from Costa Rica (26 percent), Ecuador (21 percent), Colombia (17 percent), Honduras (13 percent), Guatemala (12 percent), and Mexico (5 percent).[34] Our food system is powered by fossil fuels and emits huge amounts of CO_2. It also encourages the building of more and bigger roads and is one of the main reasons we are surrounded by eighteen-wheel transfer trucks on the highways.

We no longer grow our own food in Southern Appalachia. Instead, we have most of it shipped in from all over the country. We even send the beef we grow away to be slaughtered for us. There is reason to doubt that this is the most efficient food system—or the most healthy.

Beef

Beef Supply Chains

The farmers in Southern Appalachia sell a lot of cattle each year (about 560,000 head in 1997),[33] and we eat a lot of beef (a little over 200 million pounds per year, or about 340,000 cattle each year),[36] but there are no major slaughterhouses in our bioregion.[37] In fact, the beef cattle we sell often travel thousands of miles away to be slaughtered, processed, and packaged. Beef is then sent back to us, through several channels, but there is no way to know whether it comes from the cattle we sold. Almost two-thirds of all cattle slaughtered for meat in the United States are killed and processed in the Great Plains region, while nearly all of the remaining one- third are slaughtered in the Corn Belt and the West.[38] This section traces the long, winding path that beef takes from farm to plate.

The cows we eventually eat begin life on many different kinds of farms.[39] (See figure 15, "Beef Supply Chains.") Many of these cattle may come from beef cow/calf farms, farms that specialize in the production of beef cattle. But these beef farms are not the sole, or even the main, source of cattle in the supply chain. Many of the cattle entering the supply chain come from dairy farms and from other kinds of farm operations. Though there are some large farms that contribute big numbers of cattle to the beef supply chain, most of the contributing producers run relatively small

Beef Supply Chain

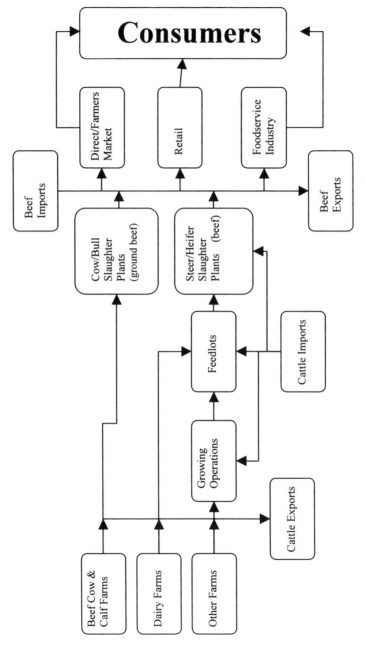

Fig. 15. *Beef supply chains.*

farm operations. In East Tennessee, for instance, where pastures are plentiful, a few beef cattle represent an easy, low-overhead source of supplemental income for many farm families. In fact, on most of the farms in our bioregion, beef cattle represent the primary farm product. So many different kinds of farmers here find it advantageous to keep a few head of cattle on the farm. In 1997, for instance, farmers in our bioregion sold an average of twenty-one cows each.[40]

The cattle that enter the beef supply chain come from all over the country.[41] Farmers from every county in our bioregion contribute cattle to the beef supply chain.[42] A significant portion of the cattle in the U.S. beef supply chain come from Canada and Mexico.[43] The U.S. cattle industry imported an average of 2.2 million cattle per year throughout the 1990s from these two countries. This figure represents roughly 7 percent of the cattle slaughtered in the United States during that same time period.[44]

There are two kinds of cattle slaughterhouses. Some specialize in steer and heifer slaughter, while others concentrate on cows and bulls.[45] In 1996 approximately 35.7 million cattle were slaughtered in U.S. cattle plants. Of these, 80 percent were steers and heifers, and the remainder were cows and bulls. There are several reasons for this division of labor.

First, the steers and heifers are smaller than cows and bulls and are shaped differently. Slaughter-line equipment must be set up according to the shape and size of the animals to be slaughtered. Many feel it is more efficient to set the equipment up in one way or the other, rather than switching it back and forth, so they specialize in either steers/heifers or cows/bulls.

Second, the different kinds of cattle produce different kinds of meat. Steers and heifers produce cuts of beef generally thought to taste better than those of cows and bulls. Cows and bulls produce a leaner, less tasty cut of meat. Cow and bull meat is often mixed with trimmings from steer and heifer carcasses and turned into ground beef.

Third, steer and heifers are fed differently than cows and bulls. Steers and heifers that are to be grown out for beef are fed a concentrated grain diet, often of corn rations. These cattle are often shipped to the major grain-producing regions of the United States to be slaughtered. Cows and bulls are fed grass and forage. Moreover, because the size of these animals makes transportation very expensive and would require large slaughterhouses to have extremely large catchment areas, cow and bull slaughterhouses tend to be smaller and more geographically dispersed than steer and heifer plants.

Cows and bulls that enter the beef supply chain most often move directly from beef and dairy farms to slaughterhouses.

The calves that will end up at steer and heifer plants are weaned from cows at six to ten months of age, when they weigh between 300 and 600 pounds. Most of these calves are sent on to "growing operations." At growing operations the calves are pastured on grass and roughage until they weigh enough to be sent on to feedlots. Many of these growing operations are integrated with the various farm operations that produce the calves. This is often the case in our bioregion. For instance, since grass grows so readily in Southern Appalachia, weaned calves often stay at home through the growing operation stage. In effect, the two stages are completed on one farm here. In other parts of the country, however, feeder calves are moved around to different growing operations, depending on the pasture and forage conditions in these other places. When the calves are big enough to be moved to feedlots—500 to 750 pounds—they are referred to as "feeder" calves.

Feedlots buy feeder cattle and "finish" them to slaughter weight, which is 900 to 1,400 pounds. This takes three to six months. The finished cattle, commonly (and somewhat perversely) called "live" cattle, are now ready for the slaughterhouse. Feedlots are generally located in states where feed, especially corn, is grown. As a result, feedlots are becoming increasingly concentrated in certain regions of the country. About 75 percent of the feeder cattle purchased by slaughter plants now come from feedlots located in five states in the Great Plains: Colorado, Nebraska, Kansas, Oklahoma, and Texas.[46]

Feedlots vary a great deal in size, from small, seasonal, family-run operations to large commercial feedlots. Some feedlots grow their own feed, but this practice is rapidly disappearing. Large commercial feedlots purchase all of their feed, hire almost all of their labor, and confine their cattle to small holding areas. These large commercial feedlots are taking an increasingly large share of sales to slaughterhouses. In 1992 the USDA surveyed the largest steer and heifer slaughter plants. These plants together accounted for roughly 87.6 percent of all steer and heifer slaughter in that year. The USDA survey found that these plants bought cattle from 19,395 sellers. Most of these sellers (89 percent) were small farmer feedlots, but these sellers accounted for only 14 percent of the cattle sold in the survey. These small operations, often part of diversified farm operations, sold an average of less than 200 cattle each, in two to three transactions per operation. The

150 largest commercial feedlots (less than 1 percent of the sellers) each sold an average of 65,000 cattle, and accounted for 43 percent of all cattle sales to the steer and heifer plants included in the survey. These largest commercial feedlots sold 9.75 million cattle to the slaughter industry, three times as many as the small farmer feedlots combined. The next 144 largest feedlots sold 3.3 million cattle—16,000 each—or roughly the same amount as the 17,000 small farmer feedlots combined.

The live cattle sold to slaughterhouses are killed and processed into wholesale cuts of beef. These wholesale cuts are sold to other processors, for instance, retailers and foodservice operators. Retail processors turn the meat into the cuts sold at retail outlets, such as grocery stores. Some of this work is done right in the grocery store. Foodservice processors turn the wholesale cuts into value-added products sold at various restaurants and fast food outfits.

The Ugly Side of U.S. Beef Production

The giant feedlots that form part of our beef production system are often ugly, stressful places for cattle to live. Cattle naturally want to graze, but at these feedlots they are packed into small dirt or concrete lots and fed concentrated diets of grain. They cannot move around much, and since the ground is compacted dirt or concrete, staying in one place can be uncomfortable if not painful. Of course, the idea is not to make the cattle comfortable, but to make them grow as quickly as possible. Moving around would not encourage growth, but would instead waste calories.

The waste and dust from these giant feedlots plagues nearby communities. The dust, which may cause respiratory problems,[47] can be reduced by spraying the dirt lots with water, but the effort is costly. One Texas feedlot—of 32,000 head capacity, and hence not one of the largest—calculated that it would need to apply 180,000 gallons of water daily in order to keep the dust down. Moreover, tens of thousands of cattle generate a lot of excrement, which is generally kept in giant lagoons or ponds until it can be moved or used in some way. It is not unusual, though, for waste products to leach into the soil and enter the groundwater, or for holding facilities to overflow and release wastes into rivers and streams during heavy rains. And then there is the smell; residents in some communities have been awakened at night by its intensity.

Animals that enter the slaughterhouse are "stunned" before they are killed. This is supposed to make the actual butchering of the animals painless. Many slaughterhouses use bolt stunning, in which an air-gun pneumatically thrusts a metal bolt against the head of the animal, destroying part of its brain and rendering it insensate. Sometimes the stunning does not work on the first try, and the animals must be stunned again. These stunned animals are hung upside-down by the leg on a chain for conveyance through the slaughterhouse. Their throats are slit, the animals are skinned, and their entrails are removed. The carcasses are then butchered into smaller pieces for further processing. Occasionally stunned animals regain consciousness while chained upside down. When this happens, the frightened animal will thrash about, dislocating and rending its own leg and ripping its own skin apart, until the workers can finally kill it.

Slaughterhouse work is dangerous for the people who perform it. The largest slaughterhouses kill about five thousand cattle each day.[48] This is a brutal pace, especially when you consider the weight and size of the animal carcasses hurrying down the line and the sharp knives the workers must use to do their work properly. The workers wear a chain-mail shirt to protect their hands, wrists, stomachs, and backs during work.[49] Even so, many of them still suffer frequent cuts, get knocked down by the carcasses, or slip on the wet, bloody floors. The turnover rate at slaughterhouses is very high, which exacerbates these problems.[50] The injury rate at slaughterhouses is incredibly high—three times higher than the average rate for other American factories.[51]

One of the most dangerous jobs at the slaughterhouse is performed by the late-night clean-up crews. Each night, after several thousand cattle are killed, someone has to clean the slaughterhouse before the next workday. The workers who do this use high-pressure hoses that spray a mixture of hot water and chlorine. This fills the slaughterhouse with fog, reducing visibility severely. Some workers ride the conveyor belts, spraying as the move through the plant, some climb up onto catwalks, and others get under the machines. Under these conditions, workers sometimes accidentally spray one another with the hot chemical water, get caught in machines, or get sick from the stench and fumes.[52]

Slaughterhouse jobs were once considered good work. The workers were protected by unions and they received recognition for the possession of a real skill. They were able to live good, middle-class American lives. All this has changed. The big meat plants have broken the unions and reduced the workers to mere cogs in their meat grinders.[53] They pay near-poverty wages and offer no incentives. American workers no longer want these jobs, so the meatpacking industry now relies on migrant workers, who come mainly from Mexico, Guatemala, and El Salvador. Some take their wages and go back home, some try to settle here, and many others simply wander from meat plant to meat plant. Many of these workers are illegal immigrants, up to one-quarter by some estimates.[54]

Meatpacking communities are suffering as well.[55] The poor, uneducated, and often transient workers who take slaughterhouse jobs are easy prey for drug dealers and other criminals. The transient population itself often includes criminals, which drives up the crime rate around many slaughterhouses. And since these poor workers often have no insurance, they drive up health-care costs. In 1991, a year after IBP opened a slaughterhouse in Lexington, Nebraska, this small town had the highest crime rate in the state. Within ten years the number of serious crimes had doubled. Gangs moved in, and the town became a major center for illegal drugs. The number of Medicaid cases also doubled.[56]

Pork

Pork Supply Chains

We eat more than 160 million pounds of pork each year in our bioregion. That's about one pound of pork per person each week for the whole year, or roughly one hog for every three people per year. It takes almost 1.2 million hogs to feed the pork consumers in our bioregion alone. But hog and pork production is not as geographically concentrated as the beef production industry,[57] so the hogs we eat do not have to travel as far as the cattle. In fact, one of the major hog-producing regions in the United States, the North Carolina–Virginia area, overlaps with our bioregion. The western Corn Belt is the other main hog-producing region in the United States. Together five states in these two regions account for roughly 60 percent of all hogs sold to slaughterhouses. This section traces the path followed by hogs from farms to the slaughterhouse, and the pork back to our plates.

Hog and pork production occurs in two distinct stages: feeder pig production and pig finishing (see figure 16).[58] Hog producers sometimes specialize in either feeder pig production or pig finishing, but most hog operations are "farrow-to-finish." Feeder pig production begins at the sow or farrowing barn, where the hogs are bred and give birth to pigs. After a

Pork Supply Chain

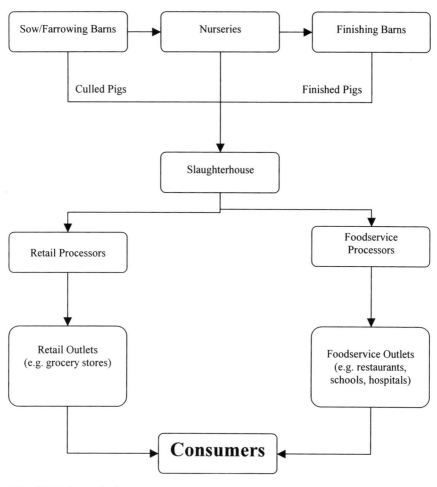

Fig. 16. Pork supply chains.

gestation period of about 114 days, the hogs farrow, or give birth to, an average of 10 pigs per litter. After three or four weeks, when the pigs weigh between 10 and 15 pounds, they are weaned. Weaned pigs usually get sent to nurseries, where they are kept until they weigh 40–60 pounds. This takes about eight weeks. At this point the pigs, now called feeder pigs, are sent to finishing farms. This is the beginning of the second stage. For hog farmers who specialize in one of the stages, this is where pigs are traded in a commercial transaction. The feeder pigs stay at finishing farms for about eighteen weeks. When the pigs reach market weight of about 250 pounds, they are sent to processors for slaughter. Sows are eventually culled from the sow barns, either because of age or because of reduced demand for pork, and sent to slaughterhouses with the other pigs.

Unlike in the beef industry, hog slaughterhouses are generally located near hog farms. Hog processors often get hogs from within 150 miles of the slaughterhouse. Most of the slaughtered hogs come from specialized hog farms. In 1997 the eighteen largest hog producers accounted for almost one-quarter of all hog sales to slaughterhouses. These eighteen producers sold over 500,000 hogs each. There are many small hog farms, but they account for only a small portion of the hogs sold for slaughter. Three-quarters of hog sellers sold fewer than 1,000 head in 1997, but these accounted for less than 5 percent of total hog sales that year. Corn Belt farms are generally smaller than the farms in the southeastern hog production area. Most of the hog farms in the Corn Belt sell between 500 and 5,000 hogs each year. In the Southeast, four out of five hog sales come from farms that sell more than 5,000 hogs each year. The largest hog producers no longer grow their own feed, preferring to purchase it instead, so they are beginning to move away from the largest corn and soybean producing areas. After the pigs are slaughtered, the carcasses are divided into wholesale cuts. This boxed pork is sold to other processors who turn it into cuts used in the retail and foodservice industries.

The Cruel Side of the Pork Industry

Pigs produced on modern industrial farms often live their whole lives indoors.[59] They move from the farrowing unit to the nursery to the growing-feeding unit to the slaughterhouse without ever getting to go outside and walk under the sun on dirt and grass. Many large pig producers practice "confinement" rearing, in which the pigs are kept confined in small cages for the duration of their lives. This is economically advantageous, because it allows one person to handle many pigs, and it keeps the pigs from expending calories by moving, but it is certainly not good for the pigs. In these confinement units pigs can only eat, sleep, stand up, and lie down. As a result, they suffer from both stress and boredom. Ammonia, released from the pigs' wastes and concentrated by the crowded indoor spaces, often harms the pigs' lungs. The slatted or solid concrete floors of the cages, which are designed not for the benefit of the pigs but to make maintenance easy, may damage the pigs' feet. Industrial pig producers often keep breeding pigs in a small iron frame—called a gestation crate or farrowing pen, and nicknamed the "iron maiden"—that prevents the animals from moving around at all. Its purpose is to keep the sows from rolling onto and crushing suckling piglets, although this could be accomplished in other ways. From all these practices, the animals suffer.

Giant hog farms are not kind to their neighbors either. Hog operators store the huge amounts of waste generated by their hogs in deep, clay-lined trenches called "lagoons." Many of these lagoons are as large as football fields. They are supposed to be leak-proof, but many are not.[60] Some present a real threat to waterways and people who live near them. In 1995 a hog waste lagoon in North Carolina collapsed, releasing 25 million gallons of hog manure into the New River. Moreover, lagoons release hydrogen sulfide and ammonia, creating a stench that sometimes sickens neighbors.[61] But even when they do not actually cause illness, these hog operations clearly reduce the quality of life of those who must live near them.

Chicken

Chicken Supply Chains

Most chicken slaughter products are produced in the Southeast.[62] Chicken producers in and around Southern Appalachia account for roughly two-thirds by weight of the chicken available to American consumers. The Central Atlantic states account for less that one-quarter of the broiler output produced in the Southeast, and the Southwest accounts for less than one-sixth. This high production has several implications for our bioregion. The most obvious is that chickens do not have to travel far to reach us, which is good, but it also means that chickens have to go farther to reach consumers in other parts of the country.

Chickens raised for meat, called broilers, are produced in seven stages.[63] These stages were once separate enterprises, but are now usually joined in one enterprise by what is known as "vertical integration." Vertical integration means that the separate entities controlling the various parts of the process are working together as a single unit. Advocates of vertical integration claim it improves efficiency and quality. Opponents see it primarily as a way for large meat concerns to control larger and larger shares of the market. In any case, most chicken operations are now vertically integrated. This section will describe the way that Tyson Foods, which handles almost a quarter of the U.S. broiler production, grows broilers for sale.

The Tyson Foods chicken supply chain begins with Cobb-Vantress, a poultry research and development company owned by Tyson Foods. Cobb-Vantress provides the grandparent breeding stock to Tyson Foods. These chickens are the grandparents of the chickens that will eventually be marketed as broilers. Grandparent chickens, which are specially selected for their desirable characteristics, lay about 130 eggs each. These eggs are incubated and hatched, producing the parent chickens, or pullets—the young female chickens that will produce the eggs that will eventually become broilers. Pullets are raised on pullet farms for about 20 weeks. Then they are moved to breeder farms when they become productive, at about 26 weeks. Their eggs are then taken to Tyson Foods' production hatcheries, where in 21 days they hatch as broiler chicks. The chicks are then sent to broiler farms, where, when they reach the desired processing weight, they are caught and sent to processing plants to be butchered. Half of Tyson Foods broilers are sold to the foodservice industry. One-third of the broilers are sold to retail outlets. About 12 percent are exported to international markets.

Cruelty in the Chicken Supply Chain

The grower houses that most chickens spend their lives in are basically giant chicken warehouses.[64] The chickens are kept in very small cages;

Tyson Foods, Inc. Poultry Supply Chain

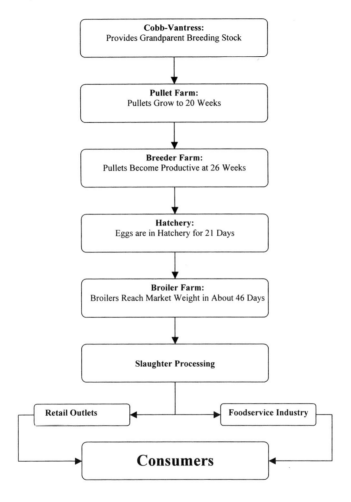

Fig. 17. *Tyson Foods, Inc., poultry supply chains.*

each chicken gets about half a square foot of space. The cages often have wire-mesh floors, which are harmful to the chickens' feet, in order to make cleaning and maintenance easier. The chickens are bred to grow twice as fast and twice as large as traditional chickens. Many suffer from heart failure before they can be slaughtered, as their hearts and lungs do not develop quickly enough to support their rapidly growing bodies. Nor do their legs develop quickly enough to support their large bodies. It is common for chickens to have trouble standing and holding themselves up.

Workers at giant chicken farms do not fare much better. Catchers, the workers who pick up the chickens and put them into containers for transportation, work in clouds of ammonia and fecal matter in hot, crowded holding pens. They get urinated on by many of the 8,000 chickens they handle each day and frequently suffer cuts, respiratory problems, and infections by salmonella or other harmful bacteria. Hangers, who work in the slaughterhouses, take chickens and hang them upside down by the feet from metal shackles. They may fasten 50 chickens a minute to the line, which amounts to more than 20,000 chickens a day. Hangers often suffer from rotator-cuff and other repetitive-motion injuries. Overall, the injury and illness rate in the poultry industry is twice the national average. Many of these workers are poor and uneducated, and many are recent immigrants. Both groups have trouble asserting their legal rights, and so are routinely exploited.

Giant chicken farms also threaten environmental health. The huge numbers of chickens generate tons of manure. Poultry manure contains nitrogen and phosphorus, both of which may harm waterways and degrade aquatic ecosystems. The EPA Chesapeake Bay Program identified poultry manure as the largest source of these chemicals in the Chesapeake.[65] Moreover, many chickens die before they even reach the slaughterhouse, and their bodies add to the huge volume of waste that is generated by chicken farms and slaughterhouses. As at the large-scale hog operations, this waste produces air and water pollution—and not infrequently an overpowering stench.

Produce

Produce Supply Chains

Produce moves about the country quickly, through a variety of channels and industries.[66] This section describes its distribution and marketing channels. (See figure 18, Produce Supply Chains.)

After produce is harvested on the farm, it is prepared and packed for shipping. Some growers pack and ship their own produce. Many lettuce growers, for example, wash and pack bulk lettuce in the field. Some large

Produce Channels

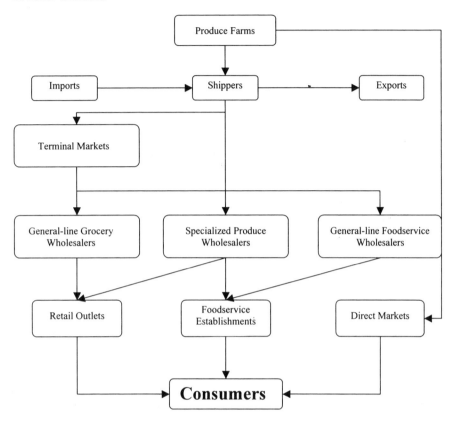

Fig. 18. *Produce supply chains.*

produce growers have packing sheds on or near the farm site. At packing sheds produce is washed, graded, packed, and often cooled for shipment. Other growers send their produce to independent packing houses, which pack it and ship it on to wholesalers, other processors, or foodservice and retail outlets.

There are several different kinds of wholesalers. In general, a wholesaler is someone who buys produce from growers and shippers in order to distribute it to customers in the retail or foodservice industries. Merchant wholesalers, the majority, take ownership of the products they distribute. Most wholesalers are merchant wholesalers. Brokers do not take ownership of the products they distribute. Rather, brokers represent either sellers or buyers in produce transactions. Jobbers are small wholesalers who buy produce from larger wholesalers.

Brokers serve buyers and sellers of produce by locating supplies and negotiating sales. Their role is analogous to that of realtors. Brokers do not own the products they handle. They represent either growers and shippers or buyers. This is a lucrative business for some. Some brokers working in the Los Angeles wholesale market make as much as $200,000 annually, with benefits.[67]

Many national supermarket chains have buying offices that perform the functions of brokers for them. These big supermarkets will obtain offices in important farming regions and have their representatives make deals with local growers and shippers. For example, Safeway, a national supermarket chain, has a buying office in Yakima, Washington.[68] This is one of the major apple-producing regions in the country. The Safeway representative in Yakima purchases apples for the supermarket chain and arranges shipment of the apples to Safeway distribution facilities around the country. The orders of the produce departments at individual Safeway stores are filled from the supplies located at these distribution centers.

General-line grocery wholesalers purchase grocery products for retailers that do not have their own buying offices, warehouses, and truck-delivery services. The general-line grocery wholesalers generally supply the individual store retailers and the small retail chains. The largest general-line grocery wholesaler, Supervalu (in Eden Prairie, Minnesota), sells a complete line of grocery and non-grocery products to 4,400 stores. Supervalu also owns and operates 431 retail stores.

General-line foodservice wholesalers sell produce and other foodservice products to businesses like restaurants, hospitals, schools, and hotels. Among the largest general-line foodservice wholesalers are Sysco and Alliant. According to Sysco, produce accounts for 6 percent of its total sales.

Specialized produce wholesalers buy and sell only fresh produce. They distribute this produce to retail stores, foodservice operators, repackers, and jobbers. Repackers buy produce in bulk and sell it to other wholesalers and retailers. Jobbers purchase produce to sell to smaller produce stores, specialty markets, small grocery stores, and restaurants not serviced by general-line foodservice wholesalers.

The fresh produce that is not processed or distributed directly from wholesaler to retailers, or distributed through direct markets, often gets distributed at either large terminal markets or supermarket chain distribution centers. These distribution centers have warehouses, refrigeration units, and docks for tractor-trailers moving the produce.

Terminal markets are large, centrally located facilities where several wholesalers, distributors and brokers are clustered together. Earlier in our history they were located near railway stations in major cities. Over the past fifty years terminal markets have played an increasingly smaller role in the food-distribution system. Most of the produce-distribution function once performed by terminal markets has been taken over by wholesalers who distribute direct to retailer and foodservice institutions. The Western North Carolina (WNC) Farmers Market is not a terminal market as such, but it performs a function very similar to the one performed by terminal markets in other cities.

The WNC Farmers Market is owned by the State of North Carolina and is run by the North Carolina Department of Agriculture and Consumer Services. The market is open seven days a week, all year. The market has five open-air truck sheds that provide 194 spaces for farmers and dealers to use to sell produce and other farms products. One truck shed is set aside specifically for direct marketing of produce from certified farmers to consumers. There are two large wholesale buildings and a small dealers building where distributors buy and sell produce for use in a variety of institutions, including grocery stores, restaurants, hospitals, schools, and roadside markets. There are two retail buildings where local and other vendors sell everything from fresh produce to handcrafted items. Between two of the truck sheds in an open area where farmers may park their trucks and sell produce, like watermelons and peaches.

Supermarket chain distribution centers are a relatively recent development. In an effort to reduce costs, some large supermarket chains have started to bypass the middlemen, and the costs of doing business with them, by purchasing produce directly from the source, and transporting and distributing it themselves. For example, Safeway's produce buying office will purchase apples and other produce in Washington and ship it directly to large distribution centers run by Safeway in different parts of the country. These regional distribution centers will then distribute the apples to individual stores as needed. This has some advantages over other distribution methods. It allows better communication between growers and the grocery stores about desirable product qualities and the timing of production and delivery. It also helps to preserve product quality by reducing the number of links in the "cold chain." Products that require refrigeration during transport may do better in this distribution system.

Some fresh produce moves directly from the grower to the consumer through a variety of direct markets. Some growers choose to market their products through farm stands and stores, roadside stands, pick-your-own operations, mail-order sales, and farmers' markets. Many of these are small, cash-based operations, so it is not easy to develop a good sense for how much produce reaches consumers in this way. And not all farmers markets report sales to the USDA. But according to the USDA's Agricultural Marketing Service, the number of farmers markets is growing rapidly. While there were only about 100 farmers' markets in the United States thirty years ago, today there are over 2,800.[69] It is estimated that over a billion dollars' worth of produce makes its way to consumers in this fashion. Unfortunately, these markets account for only a small portion, less than 2 percent, of overall produce sales to consumers in the United States.[70]

There are a large number of pick-your-own farms, roadside markets, and farmers markets in our bioregion. The Tennessee Department of Agriculture's Market Development Division runs a website called "Pick Tennessee Products" (The North Carolina Department of Agriculture and Consumer Services runs a similar site called "North Carolina Farm Fresh") These sites list farms that market products direct to consumers.

Fruit and Vegetable Production

We could grow a lot of fruits and vegetables in our bioregion, and in the past we did. But today we do not rank among the largest produce growers in the country. California is now the single largest producer of horticultural commodities. California growers currently account for over half of all U.S. production of both major fresh vegetables and fresh fruits, by value. They produce over 70 percent of each of the following fruits and vegetables: lettuce, processing tomatoes, broccoli, cauliflower, carrots, celery, strawberries, grapes, nectarines, plums, apricots, avocados, lemons, and honeydew melons. Many, if not most, of these items can be grown in our bioregion. Florida is the second largest grower of horticultural crops. Florida growers account for 14 percent of U.S. fresh vegetable production and 8 percent of fresh fruit production, by value. Florida is by far the largest producer of citrus fruits in the country, accounting for three-quarters of the U.S. citrus crop. Florida is the primary producer of processing oranges, while California is the largest producer of fresh market oranges. Florida is the largest producer of many vegetables, including fresh tomatoes, snap beans, watermelons, and cucumbers. All of these items can be grown in our bioregion.

Table 7 shows that the production of many important food crops is highly concentrated in certain parts of the country. In many cases, most of the total amount of production of the item is concentrated in five or fewer states. As a result, produce must often be shipped great distances to reach consumers. This is why our produce system requires the use of so many trucks and so much fuel. This has also led to changes in the breeding of many different kinds of fruits and vegetables. Fruits and vegetables are now bred with seemingly extraneous traits in mind. Both taste and nutritional quality are routinely sacrificed in the effort to produce fruits and vegetables that, for example, ship well, are uniform in size and ripen slowly. And although everyone knows that supermarket tomatoes pale in comparison to the home-grown ones, that mass-produced fruits and vegetables simply cannot reproduce the quality of homegrown ones, we still buy them at the supermarket.

Organic vs. Chemical Agriculture

What is nowadays known as "conventional" agriculture did not become conventional until after World War II. It began with the wide availability of cheap petrochemical fertilizers (made possible by the conversion of munitions plants to fertilizer plants) and of pesticides, such as DDT. For the greatest part of human history, all farming was organic. Farmers adopted the new technologies, not because the old methods did not work, but because the short-term efficiency of chemical agriculture generated fierce competitive pressures that forced compliance. That situation is now changing. Advances in organic agriculture (achieved mostly outside the dominant agricultural institutions) have steadily increased its efficiency, so that some farmers are now able to compete on the open market using organic methods. Other organic growers have sought niches not dominated by the competitive pressures of the marketplace, niches in which they can work slowly

Table 7

MAIN ORIGINS OF MAJOR COMMERCIAL FRUITS AND VEGETABLES

Food	1st/(%)	2nd/(%)	3rd/(%)	4th/(%)	5th/(%)	Top Producers %	U.S. Total
Apples	Washington (47.9)	New York (11.0)	Michigan (10.3%)	California (9.5)	Pennsylvania (4.6)	83.3	10,226,600 lbs
Cantaloupes	California (61.2)	Arizona (22.2)	Texas (5.8)	Georgia (5.4)	Indiana (2.0)	96.5	2,355,600,000 lbs
Grapefruit	Florida (76.3)	California (23.0)	Texas (0.5)	Arizona (0.2)		100%	70,200 boxes
Grapes	California (90.6%)	Washington (4.7%)	New York (2.0%)	Michigan (0.9%)	Pennsylvania (0.9%)	99.1%	6,836 tons
Honeydew	California (80.1%)	Arizona (12.7%)	Texas (7.2%)			100%	579,500,000 lbs
Pears	Washington (43.6%)	California (29.9%)	Oregon (24.4%)	New York (0.9%)	Michigan (0.4%)	99.1%	1,044 tons
Strawberries	California (81.2%)	Florida (10.8%)	Oregon (3.1%)	North Carolina (1.1%)	Michigan (0.6%)	97.3%	1,632,200,000 lbs
Watermelon	California (20.1%)	Florida (18.4%)	Georgia (16.0%)	Texas (15.7%)	Arizona (5.7%)	75.6%	4,073,400,000 lbs
Beans, navy	Michigan (42.3%)	North Dakota (34.2%)	Minnesota (16.2%)	Nebraska (2.1%)	Idaho (1.7%)	96.6%	548,700,000 lbs
Beans, pinto	North Dakota (41.4%)	Colorado (18.4%)	Nebraska (11.0%)	Idaho (8.2%)	Wyoming (5.7%)	84.6%	1,082,700,000 lbs
Broccoli	California (92.3%)	Arizona (7.1%)	Texas (0.5%)			100%	1,731,500,000 lbs
Cabbage	New York (23.4%)	California (14.9%)	Georgia (13.9%)	Florida (11.1%)	Texas (10.6%)	74.2%	2,739,500,000 lbs

Commodity						Total	Production
Carrots	California (76.0%)	Colorado (7.3%)	Michigan (3.9%)	Florida (3.6%)	Washington (2.7%)	93.5%	3,359,900,000 lbs
Celery	California (91.3%)	Michigan (5.7%)	Texas (2.9%)	Ohio (0.1%)		100%	1,806,200,000 lbs
Sweet corn, fresh	Florida (25.6%)	California (17.3%)	Georgia (10.9%)	New York (6.9%)	Ohio (4.7%)	65.4%	2,258,700,000 lbs
Sweet corn, processed	Minnesota (24.5%)	Washington (23.4%)	Wisconsin (21.6%)	Oregon (10.6%)	New York (7.6%)	87.7%	3,324 tons
Cucumbers	Florida (26.0%)	Georgia (20.0%)	California (16.4%)	Michigan (10.2%)	North Carolina (6.6%)	79.4%	1,095,700,000 lbs
Lettuce, head	California (72.0%)	Arizona (25.1%)	Colorado (1.0%)	New Mexico (0.9%)	New Jersey (0.5%)	99.5%	6,854,200,000 lbs
Lettuce, leaf	California (81.1%)	Arizona (17.0%)	Ohio (1.3%)	Florida (0.6%)		100%	924,100,000 lbs
Lettuce, romaine	California (68.5%)	Arizona (28.3%)	Florida (2.2%)	Ohio (1.0%)		100%	927,900,000 lbs
Onions	California (25.0%)	Oregon (16.9%)	Washington (11.7%)	Idaho (8.9%)	Colorado (8.4%)	70.1%	6,388,300,000 lbs
Potatoes	Idaho (29.5%)	Washington (19.2%)	Colorado (6.1%)	Wisconsin (6.1%)	Oregon (5.9%)	66.7%	45,991,200,000 lbs
Sweet potatoes	North Carolina (35.7%)	Louisiana (24.6%)	California (15.3%)	Mississippi (8.4%)	Texas (7.1%)	91.1%	1,302,500,000 lbs
Tomatoes, fresh	Florida (41.8%)	California (30.8%)	Georgia (6.0%)	Virginia (3.5%)	Tennessee (2.0%)	84.1%	3,780,900,000 lbs
Tomatoes, processing	California	Ohio (93.8%)	Indiana (2.5%)	Michigan (1.6%)	Pennsylvania (1.2%)	99.4% (0.3%)	9,973 tons
Wheat, all	Kansas (20.0%)	North Dakota (10.6%)	Montana (7.4%)	Oklahoma (7.1%)	Washington (6.7%)	51.7%	2,526,552,000 bushels

SOURCE: USDA Economics and Statistics System.

and responsibly to heal the land. Community-supported agricultural projects are one sort of example.

In 1990 Congress passed the Organic Foods Production Act.[71] This act standardized a confusing hodgepodge of local and regional organic certification plans that were developed in the 1970s and 1980s to promote and protect organic farming. The act set in motion the National Organic Program (NOP) and established the National Organic Standards Board (NOSB) to develop and implement a national standard for organic agriculture. In April 1995, the NOSB defined "organic" in the following way:

> Organic agriculture is an ecological production management system that promotes and enhances biodiversity, biological cycles and soil biological activity. It is based on minimal use of off-farm inputs and on management practices that restore, maintain and enhance ecological harmony.

> 'Organic' is a labeling term that denotes products produced under the authority of the Organic Foods Production Act. The principal guidelines for organic production are to use materials and practices that enhance the ecological balance of natural systems and that integrate the parts of the farming system into an ecological whole.

> Organic agricultural practices cannot ensure that products are completely free of residues; however, methods are used to minimize pollution from air, soil and water.

> Organic food handlers, processors and retailers adhere to standards that maintain the integrity of organic agricultural products. The primary goal of organic agriculture is to optimize the health and productivity of interdependent communities of soil life, plants, animals and people.[72]

The USDA developed final rules for implementing the act in 2000 and is currently putting these rules into effect. Organic standards have been developed for all aspects of organic food production. They place certain restrictions on the methods, practices, and substances that may be used in the production and handling of crops, livestock, and processed farm products. For instance, no organic food can be produced using genetic engineering, sewage sludge, or ionizing radiation. The organic standards contain a list

Fig. 19. *Organic produce on sale at the Knoxville Community Food Co-op. Copyright by Mignon Naegeli.*

that details what kinds of natural and manufactured products can and cannot be used in the production of organic foods. The standards require farmers and other handlers who sell over $5,000 worth of organic products a year to be certified by state or accredited private companies.

Organic farming harnesses natural components and processes of ecosystems in order to produce foods and other agricultural products in ways that are more ecologically sound than the practices typically associated with chemical agriculture. In essence, organic farmers use natural processes as farm tools. For instance, many organic farmers practice biological pest management. Instead of using dangerous and expensive chemical herbicides and pesticides, they provide food and shelter for predators and parasites of crop and animal pests. Crop rotation is another method organic farmers use to control pests. Composting uses natural soil organism activity to recycle animal and crop wastes. Compost not only fertilizes organic crops, it can also replenish the soil used to grow those crops. It is a very efficient, natural practice. Organic dairy products come only from cows that are raised as naturally as possible. This means that organic cows have access to fresh air, pasture, shade, exercise, shelter, and sunlight.

Organic farmers exclude the use of most synthetic substances—like chemicals, antibiotics, and hormones—in crop production, and prohibit the use of antibiotics and hormones in animal production. Organic crops—fruits, vegetables, grains, oilseeds, and legumes—can only be produced on land that has not had prohibited substances applied to it for at least three years before harvest. Organic dairy cows have to be fed rations composed of 80 percent organic feed for nine months and then 100 percent organic feed for three more months, or grazed on certified pasture, before their milk can be labeled organic. Organic beef, pork, and poultry is produced under similar restrictions and conditions.[73]

By nearly all measures of ecological health, organic agriculture is far superior to conventional techniques. A Washington State University study, for example, compared two adjacent farms in Washington, one of which had been operated organically for over eighty years, the other of which had used fertilizers and pesticides since 1948. (We would prefer to report regional research, but we have been unable to find studies of organic agriculture in Southern Appalachia. Funding for agricultural research is dominated by chemical companies and other corporate interests that do not support scientific investigation of organic techniques.) In the Washington study, soil samples were taken from adjacent areas with identical slopes on each farm. Soil from the organic farm contained much more moisture and was richer in nutrients (especially nitrogen and potassium) than the soil of the nonorganic farm. This was due in part to greater microbial activity. (Microbes, which generate nutrients that enrich the soil, are often killed—along with beneficial insects and worms—by conventional agricultural chemicals.) The organic farm had 60 percent more organic matter (which improves soil structure and increases the soil's ability to store moisture) at the soil's surface. It also had a lower "modulus of rupture"—a measure of how easily seedlings can break through the surface of the soil—and was superior in tilth. The topsoil on the organic farm was over six inches thicker than on the nonorganic farm. This was due in part to differences in the erosion rates; erosion was nearly four times greater on the nonorganic farm. But it was also due to the practice on the organic farm of plowing in cover crops to build topsoil. The researchers concluded that the nonorganic farm was gradually becoming less productive as a result of the erosion, though the impoverishment of the soil was masked by the use of higher-yielding plant varieties and more effective chemical fertilizers, but that the organic farm could maintain its productivity in the long term.[74]

In a similar, more recent, study, researchers compared the sustainability of organic, conventional, and integrated (organic with some conventional practices) apple production systems. Researchers found that both the organic and integrated systems are better for the soil and the environment than the conventional system. The organic system produced sweeter and less tart apples than either of the other systems, higher profitability, and greater energy efficiency. The researchers conclude that the organic system is more environmentally and economically sustainable than either of the other apple production systems.[75]

Most organic farms strive for diversity, growing many varieties of fruits and vegetables and rotating crops to keep pests from becoming established or from infecting the whole crop. Many use biointensive methods, which concentrate the plants into highly enriched beds of soil. Unlike traditional row cropping, this method of concentrating plants simulates natural growing conditions by creating a cool, shady microclimate near the surface of the soil that maintains constant moisture and discourages weeds. Though the biointensive method requires more labor than "conventional" methods, its yields are in many cases much higher.[76] Moreover, through the use of terracing, biointensive growing is adaptable to Southern Appalachian hillsides, where conventional row cropping creates unacceptable erosion, thus permitting the production of fruit and vegetable crops on lands that might otherwise be suitable only for pasture. Connie Whitehead, using biointensive methods at Planted Earth Farm in Strawberry Plains, Tennessee, has been easily able to grow on two acres nearly all the vegetables needed over a period of five months by twenty families, who participated with her in a community-supported agricultural project.[77] Her methods were strictly organic.

Organic agriculture also has a positive overall effect on biodiversity. A recent report produced in the United Kingdom, summarizing the results of nine studies, showed important differences between the biodiversity found on organic and conventional farms. Most of the studies showed both greater abundance and diversity of species in and around organic farms than conventional farms. Five times as many and 57 percent more species of wild plants were found in arable fields on the organic farms. Several rare wild arable species were found only on the organic farms. On the organic farms, 25 percent more birds were seen at the field edge, three times as many non-pest butterflies were found in crop areas, and one to five times as many numbers and one to two times as many species of spiders were found. There was a significant decrease in the number of aphids found on organic farms as compared to conventional farms. Although the highest levels of wildlife were found at the field boundaries, the cropped areas of the organic fields showed the highest increases in wildlife levels. These cropped areas account for 95 percent of the farmland in the United Kingdom, so the increases in biodiversity in these areas is extremely significant.[78]

There is no question that organic farms are superior in almost every environmental measure to "conventional" farms. The only serious objection to them is economic. Primarily because they require more labor than conventional farms, their produce usually costs more. This may have a positive overall effect, however, since, as several recent studies show, organic farming is as profitable, if not more profitable, than conventional farming.[79] This is good news for us, as it may attract farmers to this mode of production. In fact, the amount of U.S. certified organic cropland doubled to 1.3 million acres between 1992 and 1997, and estimates made in 2001 show another significant increase since 1997.[80] If not biointensive, organic farms may also have lower yields. But a number of recent studies show that organic production systems can actually produce higher yields of many fruits and vegetables.[81] In any event, any additional labor needed on organic farms is an advantage, as this creates jobs in this important but declining part of our culture. Cost is still a great deterrent, but Americans spent $7.8 billion on organic foods in 2000.[82] The organic food industry in the United States has been growing about 20 percent annually since 1990.[83]

The True Cost of Food

The true cost of food is hard to measure, but clearly only a part of it is reflected in the prices we pay in our grocery stores and restaurants. Our way of getting food in Southern Appalachia—in fact, in the whole United States—is loaded with hidden costs that never show up on price stickers or menu boards. Some of these hidden costs are tangible, and some are not, but they are all real, and we are paying them. We have to acknowledge this before we can even begin to make an accurate assessment of the health of our present way of life. Market price does reflect part of the true cost of food. It reflects some of the cost of the ingredients, machinery, and labor that are needed to produce the food. It includes some of the costs of packaging, transporting, and marketing food. But it leaves out a lot as well. Our food actually costs a lot more than most of us realize.

There are many tangible costs associated with our way of getting food that never show up on price stickers. We no longer have a healthy relationship with our food. We are getting fat, we are getting sick, and we are paying for it in higher medical bills. Clearly our food costs us more, in real dollars, than we pay in the stores.

There are also moral costs to the way we get food in the country. Our present food practices do cause a great deal of unnecessary harm, to both animals and human beings. Millions of animals a year are kept in giant CAFOs, (concentrated animal farm operations), that are dirty, crowded, and stressful. CAFOs and slaughterhouses are terrible places for people to work. The work environments are dangerous and the pay and incentives are poor. The giant CAFOs and slaughterhouses also harm the communities that surround them. The smells and wastes from these facilities seep unwanted into the lives of those who must live near them. Whenever we buy meat or other animal products that are produced in these ways, we support an unnecessarily harmful industry. Not only this, when we buy products from these industries, we implicitly condone the behaviors and practices associated with its production.

Then there are environmental costs. One important hidden cost is siltation. Since "conventional" agriculture increases erosion, it can substantially add to the sedimentation of streams and rivers. A 1981 study found that conversion of previously unfarmed land to conventional cropland in a Southern Appalachian Ridge and Valley landscape increases the influx of sediment into streams by over twelve tons per acre annually (about an inch of topsoil every fourteen years)—an unacceptably high figure by any standard.[84]

Silt makes rivers and streams less suitable for fishing and recreation, which can hurt the businesses that support these activities. Furthermore, as it settles into reservoirs, it hastens the day when the reservoirs become so clogged that they are no longer effective for flood control.

But silt is not the only problem caused by erosion. The loss of the soil is itself a cost. Early farming and logging in the Southern Appalachians were so reckless that in effect they mined the soil. As the steep hillsides were clearcut and then plowed and plowed again, more and more topsoil washed away, until sometimes only the subsoil remained. The losses were monumental. In 1977, long after most farmers had become aware of the need for soil conservation, a national resource inventory indicated that the soil of Tennessee's croplands was still eroding at an average rate of one inch every eleven years.[85] Earlier in the century, the losses were no doubt much greater. By now, much of the soil of Southern Appalachia lies beneath the waters of the Gulf of Mexico.

Though agricultural reforms and soil conservation programs have substantially slowed the rates, erosion still continues to be a problem.[86] (Recent advances in "no-till" agriculture can reduce soil losses still further, but no-till systems may also require larger applications of herbicides or pesticides.)[87] In 1992 Tennessee had the highest average annual sheet and rill erosion on nonfederal cultivated land of any state in the Union.[88] At that time Tennesseans were losing 9.3 tons of soil per acre. That same year, North Carolinians were losing 5.6 tons per acre. By 1997 Tennesseans had reduced soil erosion of these lands to 7.7 tons per acre, and North Carolinians to 5.0,[89] but this is still much too high. Soil erosion in Tennessee bioregion counties is not too bad, relatively speaking, as the worst erosion problems are in the more intensively farmed western part of the state.[90] This is not so in North Carolina. There erosion is most serious in the western part of the state.

Soil loss is partially counteracted by the creation of new soil due to the erosion of underlying rock and the deposit and decay of organic material, such as wood and leaves, but this takes a long time under the best of circumstances. Under warm, humid conditions it takes at least one hundred years to produce a thin, young, immature layer of soil.[91] Under less favorable conditions, it may take several hundreds of years to get this most basic layer. As generations of plants live and die the soil gets deeper and more flush with organic matter. In areas where there is plenty of rainfall, water moving through the soil carries some materials deeper into the ground. In this way different layers, called horizons, develop in soil. Mature soil, the most productive soil, contains a variety of horizons, and takes from one thousand to several thousands of years to develop.

Soil ultimately forms as rock, the "parent" material, breaks down.[92] Plants can speed up this process, by helping to break down the rock and by adding organic material to the soil.[93] Organic soils, those composed of more than 20 percent organic matter, typically form under water. Mineral soils, those with less than 20 percent organic matter, are typically formed on land. The mineral soils with the highest concentration of organic matter form under grasslands. The organic matter content of grassland soils is kept high by the constant growth and death of a dense mat of fibrous roots. Forest soils tend to have less organic matter than grassland soils. In forests the leaves and trees that die tend to create a surface layer of organic matter that does not mix with deeper layers. Nevertheless, both grassland and forest soil contain more organic matter, and thus form more quickly, than cropland soil. Cropland soil contains relatively little organic matter. Here the breakdown of rock is almost entirely unaided by plants, as most extraneous plants are removed from cropland to make room for the crops. This problem is compounded by the fact that cropland plants are harvested each year, further retarding the soil formation process.

Most forms of human land use (organic agriculture being a notable exception) erode soil more rapidly than it forms, creating net losses. Because we use virtually all the land and allow our remaining pastures to be overgrazed and eroded, the forests (when they are not being logged) are the main source of net gains of topsoil in Southern Appalachia. Since topsoil accumulates at a rate somewhere between an inch a century and an inch every three or four centuries, and since much of Southern Appalachia has already lost many inches of soil, merely to restore the losses we have already inflicted would take many centuries of forest growth. And because topsoil is essential for nearly all land-based life, this is another reason (if more were needed) for defending Southern Appalachian forests.

Loss of topsoil is, in effect, loss of capital; the land grows poorer each year, so that over time it becomes less suitable for growing crops and requires more chemical inputs. Food production thus becomes increasingly expensive, though we may not notice the loss or pay the price for many years.

And conventional agriculture imposes other hidden costs as well. It adds substantially to the nutrients (especially nitrogen and phosphorus from chemical fertilizers) and pesticides that contaminate streams and groundwater, increasing the risk of cancer for those who drink the groundwater and harming fish and wildlife that use the streams. The results are higher health-care costs (which must ultimately be borne by everyone) and a further generalized degradation of water quality, which is likely to reduce income from tourism, hunting, and fishing.

So how much does our food really cost us? When we add the tangible and moral costs to the market price of our food, we get a much better sense of what we are paying; we are paying a lot.

Genetically Engineered Organisms

The past decade has seen the massive and rapid introduction of genetically engineered organisms into American agriculture. Genetic engineering is an artificial means of adding new genes to an organism, altering its genetic blueprint, in order to produce new traits. In this way genetic engineering produces organisms with gene combinations and traits that do not occur in nature. Such novel organisms are variously referred to as "transgenic," "genetically engineered" (GE), or "genetically modified organisms" (GMOs).

One important way that genetic engineering differs from traditional agricultural practice is that genetic engineers can insert genes from one organism into an organism of a radically different type. On the traditional farm, as in nature, new gene combinations are produced through sexual reproduction, which limits combinations to organisms of the same, or very closely related, species. Genetic engineers are not so restricted. A genetic engineer who set out to make, for example, a yellow tomato, might splice a gene for yellow trait from a species of butterfly or daffodil into the genetic code of a tomato plant. One company, DNA Plant Technology, wanted to produce tomatoes more resistant to cold temperatures, so they inserted antifreeze protein genes from winter flounder into the tomatoes. Another company, Amoco Technology Company, increased production of sterol—an unsaturated solid alcohol product—in tobacco by introducing genes taken from the Chinese hamster.[94]

Most genetic engineering in agriculture has been applied to crops.[95] Genetic engineering research is being done on many different kinds of fruits, vegetables, and grains.[96] Table 8 lists some of the plants being manipulated by scientists today. Herbicide and pesticide tolerance and tolerance to pests (insects, fungi, viruses) are the traits most often engineered into crops. Certain desirable processing traits, like color, shape, and texture, and longer shelf life are also engineered into crops.[97] Some crops are now being engineered as biological factories.[98] These crops—called pharm or industrial crops—may be used to create such products as medicinal drugs or industrial chemicals.

Transgenic crops present environmental risks. One worry is that their new traits might allow them to become weeds that interfere with both agricultural and natural ecosystems.[99] It is also possible that transgenic crops could transfer pollen to wild relatives that would become weeds. Invasive weedy species can, as was noted in chapter 4, out-compete native plants and radically alter ecosystems. Table 8 shows that many of the transgenic plants under development have wild/weedy relatives in the United States.

It is not yet clear whether transgenic food crops present serious risks to people, but the experiment is underway—and we are the guinea pigs. Unless you grow the food you eat, you are probably eating food that has been genetically engineered. Soybeans are used, in one form or another, in about two-thirds of the processed foods available in our grocery stores.[100] Soy products are used, for instance, in bread, mayonnaise, canned tuna, ice cream, cookies, frozen yogurt, and baby formula. They are used in nutritional supplements and soy burgers. And in the United States, over 60 percent of the soybean crop is a genetically engineered variety produced by the chemical company Monsanto. Monsanto's Roundup Ready soybeans have been manipulated to be able to withstand large doses of Monsanto's weed killer Roundup. Much of our nation's corn crop is also genetically

Table 8

TRANSGENIC PLANTS UNDER DEVELOPMENT

Fruits: apple, avocado, banana, cranberry*, grape, kiwi, mango, melon*, papaya, pear, plum*, strawberry*, raspberry*

Vegetables: asparagus*, beans (Phaseolus), beets, cabbage*, carrot*, cassava, cauliflower*, celery*, chili pepper, chickpea, cowpea, cucumber*, eggplant*, horseradish, lentil, lettuce, onion, pea, potato*, soybean, squash*, sweet potato*

Grains and Seeds: barley*, coffee, linseed flax*, macadamia nuts, maize, oilseed brassicas, rice*, rye*, sorghum*, sunflower seed*, walnut*, wheat*

Others: alfalfa, carnations, chicory, Chrysanthemum, cotton*, foxglove, Morning Glory, orchard grass, Petunia, poplar trees, rose, spruce trees

NOTE: *Indicates existence of wild/weedy relatives in the United States.
SOURCE: From G. G. Khachatouris, A. McHughen, R. Scorza, W. Nip, and Y. H. Hui, *Transgenic Plants and Crops* (New York: Marcel Dekker, Inc., 2002), 94 and J. Rissler and M. Mellon, *The Ecological Risks of Engineered Crops* (Cambridge, Mass.: MIT Press, 1996), 92

engineered. In 2000 and 2001, about one-fourth of our corn was engineered either to be resistant to herbicides or to produce a toxic protein that kills pests.[101] This protein is produced by the bacterium *Bacillus thuringiensis* (Bt). Scientists create "Bt corn" by taking the gene that produces the toxic protein from the Bt bacterium and putting it into the corn. The corn plants then produce the same toxic protein.[102] In 1999 Consumers Union bought several common foods in grocery stores and had them tested for genetically modified ingredients. They found that several baby formulas, several soy burgers, and many popular processed foods, including Ovaltine Malt mix, Bacos bacon flavor bits, and Jiffy Corn Muffin Mix, tested positive for genetically engineered ingredients.[103]

Eating with the Seasons

Throughout all of human history up until the last few decades, diet had a seasonal rhythm. In Southern Appalachia, greens came in the early spring, then peas and other spring vegetables. Strawberries appeared in May, blackberries in June. July, August, and September were the months of the corn and bean harvest and of fresh tomatoes and watermelons, and later in the fall there were nuts, squashes, and persimmons. Meat, dried beans, and stored grains made up the bulk of the winter diet, though the careful gardener knew how to extend a fall crop of greens long into the winter.

But these rhythms, which once gave texture and tempo to life and provided cause for anticipation and celebration, have long been broken. Processed food, which makes up the bulk of our diet, has no seasonal rhythm and is available on demand, constantly. Even "fresh" produce of virtually any variety, is now available year-round, trucked in from Mexico or California, or shipped up from Chile, New Zealand, or Brazil—the only seasonal variation being a fluctuation in price.

What is true of the food supply generally is also true in particular for the bioregion's restaurants. The fast-food restaurants, of course, get their stocks from regional or national suppliers that standardize it so rigidly that it varies not at all from franchise to franchise or season to season. But even the more up-scale and unique local restaurants that offer changing menus usually buy from a few corporate suppliers, such as Robert Orr/

Sysco or IJ, which may truck in just about any food from just about anywhere in just about any season.

Perhaps this constant availability of everything has contributed in some measure to human happiness, but nearly everyone now takes it for granted, and many find it blasé. Correlatively, celebrations of the seasonal rhythms and harvests have for many lost their meanings.

However we assess its effect on the quality of our lives, this much, at least, is clear: this vast supply of exotic and luxurious foods is procured at great environmental cost. The energy required to transport and refrigerate all this food is tremendous, and most of it is generated by the burning of fossil fuels, with all the attendant effects described in chapters 2, 3, 7, and 9.

Progress toward a sustainable food supply would require, in part, a return to eating with the seasons. Much more of our produce would be locally grown, farm-ripened, transported only short distances, and consumed while still fresh. This would certainly enhance the health of the body and of the land. It is not unreasonable to hope that it might also improve the health of the spirit.

Growing Your Own

The ultimate local food source is the home garden. Its ecological advantages are manifold. Fruits and vegetables grown at home are eaten at home, with no consumption of fossil fuels for transportation. In fact, the home garden is usually small enough to be worked entirely by hand—especially by biointensive methods—so that fossil fuels need not be used at all; and most other inputs can be reduced or eliminated by organic techniques.

Yard waste, kitchen waste, and leaves, which might otherwise be landfilled or incinerated at considerable ecological cost, can, with very little effort, be made into compost that continually enriches the soil and eliminates the need for industrial fertilizers. A family can grow most or even all of its fruits and vegetables on an acre or two, saving hundreds or even thousands of dollars annually in food bills. And almost anyone can gain some advantage and make some contribution by gardening. Even apartment dwellers usually have access to a nearby patch of ground or, lacking that, a set of planters and pots, in which tomatoes, peppers, fresh herbs, or greens may be grown. Rainfall in Southern Appalachia is abundant in most years, and the climate is so mild that it is possible, using such simple methods as row covers or cold frames, to harvest something every month of the year.[104]

The home gardener enjoys exercise in the open air and the freshest and most healthful of foods. She eats with the seasons, rejoices in harvests, develops a weather eye, becomes acquainted with a whole world of small creatures (both helpful and frustrating) that once lived beneath her notice, learns discipline of keeping and improving the soil, and—in taking personal responsibility for the source of her nourishment—taps a wellspring of meaning that is inaccessible to those who consume only the cargo of eighteen-wheelers.

Chapter 7

ENERGY

John Nolt and Keith Bustos

Energy Consumption

We consume energy for heat, light, the manipulation of information, and it is the motive force in a myriad of residential, commercial, and industrial applications. This chapter considers the consequences of that consumption. But we also use energy to move people and things across the land; transportation, however, raises special issues, which are reserved for chapter 9. Here the topic is the non-transportation uses of energy, and, especially, of electricity.

It is a topic already broached, for consumption of electricity affects the water, the air, and whole system of life. Chapters 2 and 3, for example, traced the paths of acid, smog, and ozone pollution from the tall stacks of the coal-fired power plants to the imperiled spruce-fir forests of the Great Smoky Mountains National Park. This chapter traces other paths: those that run backwards from the ends of electric power lines of homes and businesses to the power stations—hydroelectric, coal-fired, and nuclear—and beyond, to the depths of the oxygen-deprived reservoirs; to the ravaged coal country; to the uranium mines, the contaminated enrichment plants, and the places where nuclear waste is to be buried deep in the earth.

We pride ourselves on having learned to control energy, yet our control is only local and brief. We can generate enormous power, send it across wires, and make it do almost anything. But in the process we set loose social, ecological, and even geological forces that we only dimly understand and have neither the knowledge nor the will to control.

The most immediate effect of our energy consumption, the one we notice, is the work that the power does for us. But even this effect sometimes has unwanted consequences. Take, for example, the cumulative effects of electric lighting. A few electric lights here or there may have little effect, but a landscape aglow with millions of streetlights, house lights, and advertising signs on poles fifty feet in the air is utterly transfigured. In such a landscape there is no darkness, for the light streaming upwards is reflected back from clouds or particulate pollutants in the air to illuminate everything—sometimes in unearthly tones of yellow or orange. Moths that instinctively navigate by moonlight become disoriented, spiraling madly and fatally around sources of brilliance for which nothing in their evolutionary past has prepared them. People lose touch with the stars, which can be seen, if at all, only faintly, as a few lonely points of light amid the brighter, flashier moving lights of airplanes—never as the scintillating celestial vault that inspired the ancients with thoughts of a transcendent heaven. Minor matters, one might suppose, in comparison with the lofty purposes served by the lights themselves—but perhaps worthy of mention.

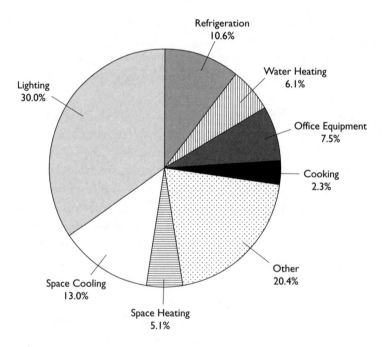

Fig. 20. *Total commercial sales of TVA electricity. Adapted from Tennessee Valley Authority (TVA), Energy Vision 2020: Integrated Resource Plan Environmental Impact Statement, vol. 2, (1995), p. T5.16.*

Electric lighting is, as figure 20 indicates, by far the single largest commercial use of electricity. Much of it is devoted to the all-night illumination of gigantic billboards and advertising signs that fill the skies above seemingly endless strip developments. Much is also consumed by interior lighting during the day—which on most days could be accomplished at no expense and with negligible environmental impact by the sun. Many businesses also keep their interior lights burning through the night, though the building is empty. This may provide a measure of advertising or security—but at what environmental and ultimately human cost? Equal security, moreover, can be provided at less cost in the long run by occupancy sensors that turn the lights on automatically when someone enters the building.

One of the most extravagant commercial uses of lighting in our region is the holiday-season display that runs for twenty-three miles from the Smokies to Interstate 40 through the towns of Gatlinburg, Pigeon Forge, and Sevierville, violating the night well up into the mountains.

Electricity is also used to power a vast array of residential appliances, many of which are inefficient and outdated. The least efficient of these is the electric water heater, still used in approximately 2,452,000 households (70 percent) in the TVA region.[1] To maintain a constant supply, electric water heaters heat a large volume of water even during long periods when no water is being used. The heat, moreover, is usually provided by electrical resistance, which is itself inefficient and expensive since the conversion from heat to electricity (in the power station) and back to heat is only about 25 to 30 percent efficient.[2]

Better alternatives are readily available. Merely switching to natural gas increases energy efficiency, though water is still being heated when it is not being used. Moreover, the combustion of natural gas releases considerably less sulfur dioxide and carbon dioxide into the air than the combustion of the coal that is usually used to generate electricity. Thus gas appliances are not only more efficient, they also contribute less to pollution, acid rain, and global warming. More efficient still are "on-demand" gas heaters, which have no storage tank at all, and heat the water only when the tap is turned on. Widely used in Europe, they provide piping hot water efficiently and almost instantly. Any water heater can in addition be made more efficient still by fitting it with a passive solar pre-heater. This simple device does no harm to anything and, once installed, heats

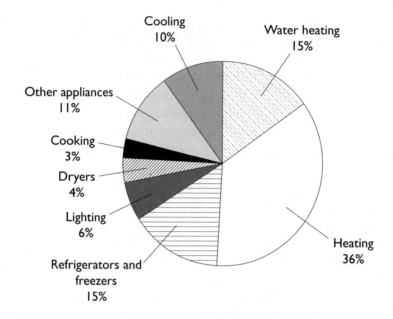

Fig. 21. *Total Residential Uses of TVA Electricity TVA, "Indoor Lighting,"*
<http://tva.apogee.net/res/relinfo.asp> (19 Mar. 2003).

the water for free, so long as the sun is shining.[3] In the early 1980s TVA actively promoted the residential use of solar energy, especially solar water heaters. But as the political winds shifted and the agency fell deeply into debt (matters soon to be discussed in more detail), these progressive initiatives were terminated.

Heating and cooling comprise the largest categories of home electrical use. About 14 percent of the homes (490,000 customers) in the TVA distribution area still use electrical resistance for space heating. Many of the 105,000 TVA customers that still use wood to heat their homes also rely on electrical resistance as a backup.[4] Electrical resistance is even less efficient for space heating than for water heating, and should be regarded as obsolete. Electric heat pumps are more efficient, but gas is more efficient still and generates significantly less pollution. Good insulation and passive solar architecture can reduce both the financial and environmental costs of heating to a minimum. Wood heat has the advantage of using a more-or-less renewable energy source that a person can procure independently, but unless the stack is equipped with a working catalytic converter, wood smoke adds significantly to air pollution.

Air conditioning has become so common that many people now regard it as a necessity—an opinion that would have been unintelligible to earlier inhabitants of this land. Roughly 99 percent of all homes in the South are air-conditioned.[5] But whatever common opinion may decree, the fact is that we keep ourselves cool primarily by burning coal and nuclear fuel, which entails certain risks and harms. The benefits of air conditioning should consciously be weighed against these risks and harms.

The "other appliances" category in the residential chart (figure 21) represents such uses as microwave ovens, dishwashers, clothes washers, and other electronic equipment. The relative amounts of electricity used by some of these common appliances are shown in table 9. Appliances that use resistance heating (electric water heaters, electric ranges, fully electric clothes dryers, blow dryers, most dishwashers, irons, toasters, hot plates, electric frying pans, and so on) are especially high consumers of energy. However, the actual energy used also depends on the amount of time an appliance is drawing power. A thousand-watt air-conditioner running twenty-four hours a day, for example, consumes forty-eight times as much electricity as a thousand-watt iron used only half an hour each day.

Especially significant in figure 21 are the entries for "Light bulb." Standard incandescent bulbs, which shine by electrical resistance, also produce much waste heat and so are highly inefficient. Equivalent compact fluorescent bulbs (available at most large hardware or department stores) shine coolly, but just as brightly, and so use one quarter as much electricity. Their purchase price is higher, but the investment is returned in savings on your electric bill—not to mention environmental savings.

All this power usage adds up to astronomical numbers. Table 10 gives the total electrical consumption for the TVA portion of the Upper Tennessee Valley region, excluding federal facilities. A kilowatt-hour is the amount of electricity used by a thousand-watt appliance (such as a typical room air conditioner) in an hour. Municipal systems include such distributors as the Knoxville Utilities Board. Cooperative systems, such as the Powell Valley Electrical Cooperative, operate mostly in rural areas.

But Table 10 does not specifically represent the energy consumed in our bioregion, since the TVA service area extends beyond the bioregion and also since it leaves out a part of western North Carolina, including Asheville, that lies in the upper Tennessee watershed but is serviced by Nantahala Power and Light, Carolina Power and Light, or Duke Power, rather than TVA. Still, it is evident that the quantity of energy consumed in our bioregion is enormous. And we are by any standards using more than our share. The average person in the United States consumes about seven times as much energy as the average person in Central and South America, and almost twenty-four times more than the average person living in Africa.[6]

Table 9

APPROXIMATE ELECTRICITY CONSUMPTION OF COMMON APPLIANCES

Appliance	Watts
Air conditioner, room	1000
Air conditioner, central	2000–5000
Hair dryer	1,200–1,875
Clock radio	10
Clothes dryer, electric heated	1,800–5,000
Clothes dryer, gas heated	300–400
Coffee maker	900–1,200
Coffee pot	200
Computer, laptop	50
Computer, pc	80–150
Dishwasher	1,200–2,400
Drill 1/4"	250
Drill 1/2"	750
Electric blanket	200
Electric heater, portable	750–1,500
Electric range (per burner)	1,500
Fan, ceiling	65–175
Fan, window	55–250
Frying pan	1,200
Furnace blower	300–1,000
Garage door opener	350
Hot plate	1200
Iron	1000–1800
Light bulb, 100w incandescent	100

Appliance	Watts
Light bulb, 100w equivalent compact fluorescent	25
Microwave	750–1,100
Stereo	100
Television, 19" color	100–150
Television, 27" color	150–200
Printer for PC	100
Refrigerator/freezer, 16 cu. ft.	725
Sander (3" belt)	1,000
Satellite dish	30
Saw, 7 1/4" circular	900
Saw, 8 1/4" circular	1,400
Shaver	15
Toaster	800–1,400
Vacuum Cleaner	1,000–1,400
VCR/DVD	17–21/20–25
Washing machine	350–500
Water heater (40 gallon)	4,500–5,500
Water Bed (with electric heater)	120–380

Wattage x Hours Used Per Day = Daily Kilowatt-hour (kWh) consumption.
(1 kilowatt (kW) = 1,000 Watts). To determine annual consumption, multiply the daily kWh consumption by the number of days the appliance is used in a year. Multiply annual kWh consumption by the cost for one kWh to obtain the annual cost.
SOURCE: Some information was adapted from the ABS Alaskan, "Power Consumption Table," accessible at http://www.absak.com/design/powercon.html (19 Mar. 2003).

Table 10

POWER DEMAND FOR TVA'S ENTIRE DISTRIBUTION AREA, FISCAL YEAR 2002

Power Purchase	Kilowatt-hours
Municipalities & Cooperatives	128,600,000,000
Industrial Direct Purchasers	26,478,000,000
Federal Agencies & Other	5,013,000,000
Total	160,091,000,000

SOURCE: TVA, 2002 *Annual Report*, 3.

Power Transmission

Some of the uses we make of electricity, such as commercial lighting, affect the environment directly, but all affect it indirectly. To trace the indirect effects, we must follow the incoming current backwards through the plug, into the junction box, back along the power line to the transformer, the substation, the generating station, and beyond.

Transformers, the dark cylinders high up on telephone poles or the larger devices protected by the chain-link and barbed-wire fences of the substations, have contaminated much of the soil and water of Southern Appalachia. Until the late 1970s, the cooling and insulating oil in nearly all transformers contained as much as a 40 percent mixture of fire-retardant chemicals known as polychlorinated biphenyls (PCBs), a highly persistent and hazardous group of environmental pollutants. Whenever the transformers were moved or accidentally damaged, this liquid leaked out onto the ground. Heavy industrial users of electricity, such as Alcoa Aluminum or the nuclear weapons plants at Oak Ridge, employed many large transformers and were especially troublesome sources of PCB leaks. When transformers wore out, they were scrapped at places like David Witherspoon, Inc., in the Vestal community of south Knoxville—where they were broken open, allowing the oil to spill out on the ground, in order to salvage the copper wire inside. PCBs, as was noted in chapter 3, are now widespread in the silt of regional waterways, and in some places they still contaminate the soil. In the 1970s it was discovered that ingestion of PCBs or PCB-contaminated food can cause cancer, reproductive problems, and a host of other maladies. Their manufacture was banned in 1976, and since then many older transformers have been replaced or provided with a substitute insulating oil. In 2002 TVA and Nashville Electric Services even commenced trials of biodegradable soybean-based transformer oil. Still, PCB contamination remains one of the most important residual problems of an incautious industrial era.

Beyond the transformers run the power lines. Power lines form a grid that distributes electricity everywhere the roads go—and some places where they do not. They, too, are not without effect. Some studies have suggested that exposure to the electromagnetic fields that surround these lines may cause childhood leukemia or other forms of cancer. But other studies have shown no effect, and the evidence remains weak and inconclusive.[7]

Power lines do, however, have obvious and visible effects on the environment. Where long-distance transmission lines cut across the land, vegetation must be cut or suppressed with herbicides in corridors a hundred or more feet wide and many miles long. These swaths are prominent from the air, running often in straight lines, even over rugged mountainous or forested terrain.

Power-line cuts fragment forests, facilitating entry for disease organisms, exotic invasive species, and the predators of songbirds. Many songbirds, as was noted in chapter 4, require deep forest habitat, well away from open edges, in order to have a fair chance at reproduction. Power-line cuts deny them that chance throughout much of our bioregion. Though recently TVA has begun to work with property owners to naturalize power line cuts by planting low-growing native plants and trees in portions of their 17,000 miles of cuts, maintenance of most cuts still causes significant environmental damage.[8] Vegetation must be kept constantly short to prevent interference with the wires. This used to be done almost entirely by tractor or by hand, but nowadays it is often accomplished by the spraying of herbicides—Arsenal, Accord, Escort, or Big Sur, for example—from helicopters. This can be a dangerous practice. In the fall of 1994, one of these herbicide-spraying helicopters crashed in the driveway of an upper East Tennessee residence.[9] Moreover, some of the herbicides used can have acute toxic effects on human beings and may harm fish or other wildlife. American Cyanamid, the manufacturer of Arsenal, warns against spraying the chemical on water, food, crops, wetlands, or gardens, and in winds greater than five miles per hour. A helicopter rotor generates a wind much stiffer than five miles an hour, and springs, crops, wetlands, and gardens are often located in or very near to power-line cuts. People in upper East Tennessee and western Virginia have complained of the spraying of streams and springs, sometimes despite repeated pleas to stop it.[10]

TVA

Tracing the power lines back far enough takes us back to the power's source: a generating station. Most of the generating stations that supply power for Southern Appalachia are owned and operated by TVA, and there can be no adequate understanding of power generation in the Tennessee

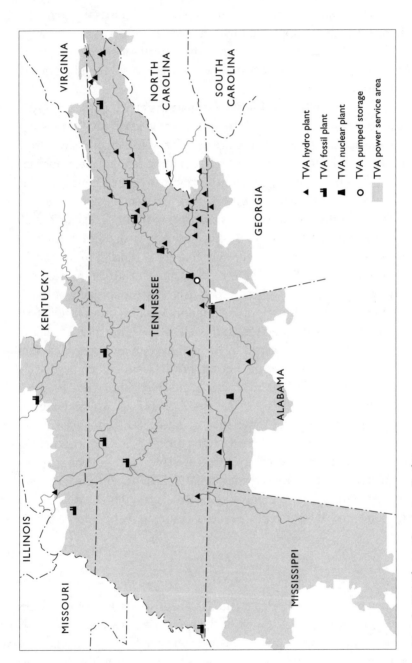

Map 2. *TVA Electric Generation Facilities.*

TVA hydro plant
TVA fossil plant
TVA nuclear plant
TVA pumped storage
TVA power service area

ILLINOIS

MISSOURI

KENTUCKY

VIRGINIA

NORTH CAROLINA

SOUTH CAROLINA

TENNESSEE

GEORGIA

MISSISSIPPI

ALABAMA

Valley without a sense of the history, politics, and economics of that unique agency.

TVA is the largest public power producer in the country, generating between 4 and 5 percent of all the electricity used in the United States. It serves more than eight million people in seven states and has an electricity-generating capacity of more than 31 GW (gigawatts or million kilowatts). In 2002, TVA produced over 151 billion kilowatt-hours.[11]

TVA was established in 1933 to help the Tennessee valley recover from the Great Depression and decades of environmental abuse. At that time, most of the Tennessee River watershed had been clear-cut, which, together with destructive mining and farming practices, was producing calamitous erosion and flooding. TVA's solution was to build dams on the Tennessee River and its major tributaries, not only to help control the flooding, but also to promote river traffic and generate electricity. By the beginning of World War II, the agency had completed twelve hydroelectric dams. The dams, together with exemplary programs to reforest the hillsides and promote responsible agriculture, succeeded in controlling the flooding—though, ironically, at the expense of permanently drowning some of the best farmland in the valley under deep permanent flood.

During World War II, regional demand for electricity soared as a result of wartime aircraft component production by Alcoa Aluminum and the enrichment of uranium for the first atomic bombs at Oak Ridge. In 1945 TVA completed its first coal-fired power plant at Watts Bar. Over the next twenty years, the agency built eleven coal-fired plants, which provided power at rates among the lowest in the nation.[12] Electrification proceeded rapidly. By 1970 about 30 percent of the homes in the valley were heated with electricity, and TVA customers used nearly twice as much electricity as the national average.[13]

In the 1960s it became apparent that the electricity demand in the Tennessee Valley could not be met by coal and hydroelectric power alone. TVA had already used most of the feasible hydroelectric sites and was under increasing pressure to reduce the emissions from the coal facilities. The agency saw nuclear power as the clean, cost-effective answer. By the late 1960s, TVA had announced plans to build seventeen nuclear units at seven sites. In 1967 it broke ground for Browns Ferry, the nation's largest nuclear power facility.

The energy crisis of the 1970s brought a new wave of conservationism to TVA, revitalizing the agency's vision. TVA's fifty-year retrospective, published in 1983, proudly proclaims that "Energy consumers throughout the Valley are now encouraged to use electricity efficiently and to substitute solar and renewable resources where possible."[14] During the early 1980s, TVA sponsored a million residential energy audits and provided many low-interest or no-interest loans to finance energy-saving improvements, including thousands of residential solar water heaters. Over six hundred thousand residences were weatherized, saving over 1.4 million kWh of power per year—the equivalent of two medium-sized nuclear reactors. These initiatives were internationally recognized, catapulting TVA to the forefront of the push for sustainable energy. Yet, despite their well-documented success, most were abruptly canceled in 1988 in cutbacks initiated by Chairman Marvin Runyon, a Reagan appointee. They have never been restored.

In the meantime, instead of introducing a new age of abundance and energy "too cheap to meter," TVA's nuclear program, tainted by mismanagement, became increasingly mired in debt and shaken by rumblings of financial disaster. In 1975 an electrical insulation fire at the Browns Ferry plant, caused by a worker using a candle to check for air leaks, destroyed several core-cooling systems. A catastrophic meltdown was narrowly averted. This harrowing accident caused the Nuclear Regulatory Commission to issue a host of new regulations, which significantly increased TVA's costs. The still more serious partial meltdown at Three Mile Island in 1979, though not involving TVA directly, produced still more regulations and increased costs further. In 1985 whistleblower complaints prompted the Nuclear Regulatory Commission to shut down TVA's entire nuclear system in order to correct repeated errors in operations and maintenance and flagrant safety violations. The reactors were eventually restarted, but delays and cost overruns plagued the entire program—especially the Watts Bar plant. Construction on the two Watts Bar reactors began in 1973, but Unit 1 did not come on line until 1996. Unit 2 was canceled in 1995. TVA originally projected the cost for both reactors at about $370 million.[15] The actual cost for completing Unit 1 alone was about $7 billion.[16]

By 1997 this nuclear debacle had deepened TVA's debt to $27.7 billion. Interest payments were running about $2 billion annually (about 35 percent of the agency's total revenues).[17] In July of that year TVA announced a ten-year business plan, aiming at halving its debt to $13.8 billion by

2007.[18] But by 2001 the debt had been reduced only to $25.4 billion, roughly $1.5 billion higher than the 1997 projection. This enormous debt has made it difficult for TVA to charge competitive rates for its power—a difficulty exacerbated by a tendency toward deregulation and intensified competition within the utility industry generally.

Deregulation and the debt have increasingly forced the agency to push consumption rather than conservation. In 1995 TVA launched a crass series of television advertisements, promoting the wasteful use of electricity. The ads, targeted toward residential customers—who already use more electricity per capita than the national average—and costing at least $4.6 million, showed an unattended television running as the voice-over proclaimed the value of cheap TVA electricity. This campaign was selected by a Washington-based coalition of consumer, health, and environmental groups to receive the tongue-in-cheek Hubbard Award as one of the "most misleading, unfair, and irresponsible ad campaigns" of 1995.[19] TVA terminated it soon afterward.

To its credit, TVA has in recent years adopted some so-called demand-side management (DSM) programs, ostensibly to reduce consumption, especially at times of high demand. But these have been largely ineffective. They reduced peak-time demand very slightly—by an annual average of 41,000 kW (0.001 percent of TVA's overall capacity) between 1996 and 2000—but total demand has continued to grow. The two DSM programs responsible for the modest energy savings between 1996 and 2000 were the Cycle and Save Program, which offered residential customers a bill credit for allowing TVA to switch off their air conditioners and water heaters during periods of peak demand, and the Energy Right Program, which encouraged residential customers to install energy-efficient heat pumps and other electric appliances in their homes.[20] Congress's General Accounting Office (GAO) has criticized TVA's Energy Right program by saying that it "actually increased consumption" in past years by encouraging consumers to use electricity rather than natural gas.[21]

In response to the GAO's disapproval, TVA plans to upgrade its DSM programs so that a savings of 396,000 kWh will be experienced between 2001 and 2005.[22] But even taking these improvements into account, the GAO concludes, "TVA's demand-side management programs are generally limited in scope, and they contribute little to moderate future demand. As a result, to meet its customers' growing demand for power, TVA will need to generate more power itself, or purchase more power from others. . . ."[23]

TVA itself predicts that electricity demand in its distribution region will increase about 1.7 percent each year through 2010. This means that the demand for TVA's electricity will jump from 161 billion kWh in 2001 to 188 billion kWh in 2010. To meet this demand, the agency says it will need to boost its generating capacity by 4.5 GW, roughly the equivalent adding two more generating facilities the size of the Sequoyah nuclear power plant.[24]

Power Generation

To generate electricity TVA uses twenty-nine hydroelectric dams, a pumped-storage hydro facility at Raccoon Mountain, eleven coal-fired power plants, forty-eight natural-gas or oil-powered combustion turbines, and five nuclear reactors. Each has specific advantages and disadvantages. TVA uses some as much as possible and others (such as combustion turbines and pumped storage) mainly during peak hours. Table 11 shows the contribution of these sources to total electricity generation.

Energy from Running Water

In 2002 TVA was operating a total of 109 conventional hydroelectric generators at twenty-nine different dams with a total capacity of 7.67 GW (5,673,000 kilowatts) and four pumped-storage generators at Raccoon Mountain with a capacity of approximately 1.5 GW. Not all the hydroelectric generators are located within our bioregion. Those that are in our bioregion are listed in table 12.

Some of the generators, particularly those on the Ocoee River, are threatened by siltation (see chapter 3). The damage occurs when sediment from upstream settles into the lake and takes the place of the water, reducing the holding capacity of the reservoir, eroding generating equipment, and ultimately clogging water intakes. Siltation forced the closing of the Nolichucky power station in 1972.[25]

Hydroelectric generation is TVA's cleanest and cheapest source of electricity. As a result, it might be expected that the hydroelectric units would be used as much as possible. However, hydroelectric generators are more easily adjusted to variable demand than are nuclear or coal-fired genera-

Table 11

TVA'S POWER PRODUCTION SYSTEM, FISCAL YEAR 2002

Generator Type	Capacity (megawatts)	% of total capacity	Energy produced (millions of kilowatt-hours)	% of actual generation
Hydroelectric dams/pumped storage	5,673	18	10,205	6.74
Fossil Fuel	15,443	49	94,930	62.66
Combustion Turbines	4,728	15	1,190	0.78
Nuclear	5,673	18	45,179	29.82
Green power*	8	N/A	18	0.012
Total	31,517	100	151,504	100

NOTE: *This has not yet been figured into the generation mix since it is so new.

tors. Moreover, TVA must regulate the flow of the rivers to control flooding and maintain high water levels for recreation in the summer, and these activities may prevent the release of water for generating electricity. Furthermore, some of the generators on the smaller tributaries do not have enough water flow to provide constant electricity. For these reasons, hydropower is used only intermittently.

While the impacts of hydroelectric generation have been small compared to the impacts of other TVA sources, they are not negligible. The most tragic social effect was the forced removal of thousands of families from their valley homes and farms as the dams were constructed. The largest ecological impacts have been the permanent submersion of large areas of rich agricultural bottomland and sensitive riparian habitat, and the transformation of swift-flowing rivers and streams into large, relatively stagnant reservoirs.

The damming of rivers, as was explained in chapter 3, deprives the deep waters behind the dams and the tailwaters that flow from the bottoms of the dams of oxygen, especially at the deep tributary reservoirs. As a result most of the dammed tributaries of the Tennessee River are less productive fisheries than they once were, and many species of native fish have declined as more adaptable exotic species have moved in. Damming and siltation have,

Table 12

TVA HYDRO FACILITIES WITHIN THE BIOREGION

Facility	River	No. of Generators	Capacity (kilowatts)
Apalachia	Hiwassee	2	93,600
Blue Ridge	Ocoee	1	22,000
Boone	Holston	3	81,000
Chatuge	Hiwassee	1	10,000
Cherokee	Holston	4	135,200
Chickamauga	Tennessee	4	160,000
Douglas	French Broad	4	145,800
Fontana	Little Tennessee	3	238,500
Fort Loudoun	Tennessee	4	145,000
Fort Patrick Henry	Holston	2	59,400
Hiwassee	Hiwassee	2	165,600
Melton Hill	Clinch	2	60,000
Nickajack	Tennessee	4	104,000
Nolichucky*	Nolichucky	0	0
Norris	Clinch	2	131,400
Nottely	Nottely	1	15,000
Ocoee #1	Ocoee	5	19,200
Ocoee #2	Ocoee	2	23,100
Ocoee #3	Ocoee	1	28,800
South Holston	Holston	1	38,500
Watauga	Watauga	2	57,600
Watts Bar	Tennessee	5	175,000
Wilbur	Watauga	4	10,700
Subtotal		59	1,919,400
Raccoon Mountain Pumped-Storage	Tennessee	4	1,530,000
Total		63	3,449,400

NOTE: *Nolichucky was closed in 1972 due to siltation.

SOURCE: TVA, *2002 Annual Report*, 3.

as was noted in chapter 4, also decimated the once-flourishing mussel beds of the Tennessee valley.

Another biological effect of the reservoirs is the fragmentation of habitats and populations. During dry periods the rivers of the Tennessee valley were once nearly all shallow enough at some places for animals to swim or wade across. But the big, deep reservoirs are, for many terrestrial species, impassable barriers, preventing dispersal, migration, and genetic exchange. This may hasten the extirpation or extinction of some species.

In addition to the dams, TVA operates a pumped-storage generating facility at Raccoon Mountain on the Tennessee River west of Chattanooga. This installation uses low-cost power generated during periods of low electrical usage to pump water uphill to a reservoir. The water in the reservoir stores potential energy, functioning in effect as a huge battery. During periods of peak demand when additional power is needed, the water in the reservoir is released back down the mountain to a bank of four generators, which convert its momentum back into electricity. The result is a net loss of power but an increase in TVA's peak generating capacity.

To create a pumped-storage facility, trees must be cut down and a mountaintop reshaped into a reservoir. The reservoir cannot be used for recreation or wildlife habitat because it is frequently emptied and refilled. Since pumped storage uses more energy than it produces, its environmental impact also includes the effects of whatever generating source (coal, hydro, or nuclear) is used to pump the water up the mountain.

TVA is also experimenting with other small-scale forms of energy storage, which could improve the efficiency of electric power distribution in the future. A superconducting magnetic energy storage device has undergone tests in Mississippi, and a 120,000 kWh bank of reversible fuel cells is under construction in Alabama. Another bank is proposed to accompany further wind farm development in East Tennessee.

Energy from Fossil Fuels

Most of TVA's electricity is generated by burning coal. Unfortunately, of all the common methods of power generation considered under normal operating conditions (that is, barring a catastrophic nuclear accident), burning

Fig. 22. *TVA's Kingston Steam Plant, a coal-fired power plant. Copyright by Mignon Naegeli.*

coal is the most destructive. In the upper Tennessee valley there are four-teen active coal-fired units at three sites and four idle units at the Watts Bar Reservation. These plants were completed between 1952 and 1967, and are therefore between thirty-five and fifty years old. Retrofits (e.g., installation of flue-gas desulfurization, selective catalytic reduction, or low-NO_x burn-ers) have been carried out to reduce pollutant emissions from some plants. But emission reductions have also been achieved without such technical improvements, for example, by burning low-sulfur coal. Table 13 lists the coal-fired plants in our bioregion. Still, the environmental problems asso-ciated with these plants are monumental. Some of their effects on the air and the water were discussed in chapters 2 and 3. We merely recount them here: acid rain; haze; carbon monoxide pollution; low-level ozone; thermal

Table 13

COAL-FIRED GENERATING PLANTS IN THE UPPER TENNESSEE VALLEY

Plant	Location	Number of Generators	Capacity (megawatts)	Coal Burned/ Day (tons)
Bull Run	Oak Ridge, TN	1	870	6,300
John Sevier	Rogersville, TN	4	712	5,700
Kingston	Kingston, TN	9	1,456	14,000
Total		14	3,038	26,000

SOURCE: TVA, "TVA Reservoirs and Power Plants," accessible at http://www.tva.gov/sites/sites_ie.htm (15 July 2003).

pollution of rivers; depletion of atmospheric oxygen; airborne emissions of radionuclides, mercury, and other toxic metals; global climate change.

There are other effects as well. One is the production of ash and slag, which pile up in huge mounds at the coal-fired plants. There are two kinds of ash: fly ash and bottom ash. Fly ash is material precipitated out of the coal smoke before it is released into the atmosphere. Bottom ash is the ash left in the furnaces, along with the slag. The primary components of ash and slag—silica (SiO_2), alumina (Al_2O_3), and iron oxide (Fe_2O_3)—are relatively harmless minerals, though smaller quantities of toxic metals and radioactive materials are also present. The use of scrubbers to absorb the sulfur from exhaust gases also creates many thousands of tons of flue gas desulfurization sludge. Coal-washing residue creates yet another form of waste. Some of these wastes are used to create cement products, asphalt products, roofing granules, road-bed and road-resurfacing materials, grit for snow control, or blasting grit. Some are also used for fill. But much of the waste must be landfilled or stored on the plant site in ash ponds. This landfilled or stored waste may contaminate groundwater.[26]

Coal-washing operations can produce even bigger problems. On 11 October 2000, in Martin County, Kentucky, a 72-acre coal waste impoundment owned by Martin County Coal Corporation (a subsidiary of A. T. Massey) burst, sending approximately 250 million gallons of coal sludge down the Big Sandy River. The release contaminated more than 75 miles of streams with measurable amounts of arsenic, mercury, lead, copper, and chromium; and tainted water supplies for riverside communities in Kentucky and West

Virginia—ultimately affecting more than 27,600 people. The lava-like spill killed countless fish and other animals in the downstream waterways. Some of the chemicals released, especially mercury, may linger for decades and may bioaccumulate up the food chain.[27]

Since coal contains relatively small amounts of radioactive materials, the burning of huge amounts of coal can also release significant quantities of radionuclides. An Oak Ridge National Laboratory report reckons that a typical coal-fired power plant releases 5.2 tons of radioactive uranium and 12.8 tons of radioactive thorium per year. Nearly all of this is concentrated in the ash. The report concludes that "although trace quantities of radioactive metals are not nearly as likely to produce adverse health effects as the vast array of chemical by-products from coal combustion, the accumulated quantities of these isotopes over 150 or 250 years could pose a significant future ecological burden and potentially produce adverse health effects, especially if they are locally accumulated."[28]

These releases do not exhaust the environmental impacts of coal-generated electricity, for the coal plants themselves are merely the end-points of yet another series of vast, destructive processes. The coal that fuels them is carried from the mines to the plants by truck or rail. As new mines are opened, new roads or railroad links must be built. These transportation corridors fragment forests, destroy wildlife habitat, and add to the general degradation of the land. The railroad engines and trucks burn diesel fuel, which is supplied by the usual array of energy-consuming and polluting processes (see chapter 9).

Tracing backwards along paths by which the coal arrives, we find an expanding lattice of highways and railroads leading back into the coal country of Tennessee, Illinois, or Kentucky or westward to the vast open mines of the Rocky Mountains (see table 14; note that the second and third largest sources of supply to TVA are both remote western states). At the coal yard, these roads and railroads way to dusty or muddy mining roads traversed by huge dump trucks with wheels taller than a man. The mining roads open out at last onto scenes of utter desolation. Especially when the coal is taken by strip mining or by mountaintop removal, the living hills, farms, and forests are reduced to bare earth, caked mud, protruding rock, subsiding land, and flowing silt on a geological scale.

It is not as bad as it once was. The federal Surface Mining Control and Reclamation Act of 1977 outlawed some of the worst mining abuses and

Table 14

TVA COAL RECEIPTS, FISCAL YEAR 2001

State	Tons Received	Percent
Alabama	14,764	0.0
Colorado	5,508,955	12.3
Illinois	3,514,396	8.0
Indiana	53,298	0.1
Kentucky	21,532,494	48.1
Mississippi	48	0.0
Pennsylvania	946,859	2.1
Tennessee	476,725	1.1
Utah	2,063,264	4.6
Virginia	3,008,154	6.7
West Virginia	1,432,091	3.2
Wyoming	6,185,211	13.8
Total	**44,736,259**	**100**

SOURCE: Myra Ireland, TVA, personal communication, 8 July 2002.

required coal companies to restore the land to its original contours before closing a mine. (Here again, federal regulation has ameliorated catastrophic damage with some success—contrary to the fashionable myth that governmental restraint of industry is counterproductive.) But the new law did nothing to restore the older mining areas, particularly in West Virginia, eastern Kentucky, and parts of East Tennessee, where operators simply pushed the overburden downslope from mountain mines, causing landslides, acid drainage, massive erosion, sedimentation, and flooding. There, the mining companies pillaged, profited, and proceeded on to new ground, leaving behind thousands of square miles of tortured moonscape—land that will never be reclaimed and may take centuries to heal.[29] But the law at least ensured that future strip mining would be monitored more carefully and that destroyers of the land could be penalized.

Yet strip mining and mountaintop removal are inherently destructive, even if the law is followed to the letter; the best reclamation does not wholly heal the wound. A strip mine begins with a clear-cut, with all the attendant environmental effects described in chapter 4—more than a clear-cut, in fact, since *all* the vegetation is bulldozed down to the bare earth. Then the topsoil is scraped away and (if the mine is well-operated) stored or used separately. This exposes the overburden (subsoil and rocks overlying the coal seam), which is drilled and blasted. The pulverized overburden is removed to expose the coal, which then is fractured by blasting and hauled away. After the coal is gone, dump trucks and bulldozers replace the overburden, then the topsoil. (The original overburden and topsoil are usually used elsewhere, so that the replacing overburden and topsoil come from a newly mined area.) Since the blasting decompresses the overburden, its volume increases, so that the excess must usually be deposited in a fill somewhere else. Finally, the topsoil is sowed and replanted.[30]

This process greatly increases erosion and siltation, even under the best of conditions. Watercourses and aquifers are disturbed or destroyed, and water both on the surface and beneath the ground may be contaminated with acid drainage from sulfur-bearing rocks, or with toxic metals or minerals. Aquatic life may disappear as springs and streams turn red with iron oxide or yellow with iron hydroxide (called "yellow boy"). Nearby wells may become cloudy, dry up, or be poisoned. Residents are sometimes forced to choose between leaving their homes and living with water they cannot use—water that looks like apple cider, kills house plants, and burns eyes, nose, and mouth when they take a shower.[31] Coal-hauling trucks create potholes and other road hazards. The blasting and moving of earth may buckle or crack the foundations of buildings. The noise produces stress for both miners and residents and can lead to loss of hearing. Sometimes the blasting is deadly—and not only to miners. In 1994 a sixteen-year-old boy was killed when rocks and debris from a blast at the Flatwoods Mine in Campbell County hit the car in which he was riding with his family on Interstate 75.

Mountaintop removal (or cross ridge mining, which is the currently popular variant of it) has become increasingly common throughout Appalachia. This practice involves blasting entire hilltops in order to expose the low-sulfur coal that lies underneath. The spoil that is blasted away is often bulldozed into nearby valleys and streams. According to a recent federal study, this practice resulted in 724 miles of streams in Kentucky, West Virginia, Virginia, and Tennessee being covered by valley fill between 1985 and 2001.[32] While our bioregion had until recently been spared these depredations, in the summer of 2003 an enormous cross ridge mine (2,100 acres) was permitted at Zeb Mountain in Tennessee's Campbell and Scott counties, over the vocal objections of nearby residents.

When a mine opens, the land and the people suffer stress, damage, and disease for many years; but, in the end, the mine falls silent, and the land, at least, begins to heal. A few years after replanting, a reclaimed strip mine may look healthy and green, but the land is never the same. The new topsoil may be mixed with less fertile subsoil, decreasing fertility. The activity of beneficial insects and microorganisms, which were killed during the mining, may not be entirely restored.[33] Acid drainage is likely to continue. A 1992 survey of twelve reclaimed sites in the Sewanee coal seam of southeastern Tennessee found indications of continued acid mine drainage at ten of them.[34]

Revegetation of reclaimed mines usually involves some tree planting, but to obtain a quick, complete cover, many operators use a seeding mixture dominated by Kentucky-31 fescue, a tall grass. The fescue shades the tree seedlings and competes with them for moisture. After a few years, it forms a thick sod cover that discourages both forest growth and the return of wildlife such as quail and wild turkeys. Moreover, the soil compaction that occurs as the topsoil is hauled in and bulldozed into place can stunt tree growth by more than thirty years.[35] As with clear-cuts, if the land is allowed to revert to forest, the understory vegetation probably takes well over a century to recover—if it ever does.[36]

These damages occur at the best of mines, under conditions of compliance with the law. Unfortunately, some coal companies have enough money and political influence to stave off law enforcement, while others move in quickly, violate the law, and then vanish into a cloud of legal shenanigans. From the time of its passage in 1977 until 1984, the Surface Mining Control and Reclamation Act was supposed to have been enforced in Tennessee by the state's Division of Surface Mining. The enforcement, however, has been so lax that in 1984 the federal Office of Surface Mining Reclamation and Enforcement stepped in and took over the administration of the state's permitting program.

Underground mining is less damaging than surface mining, since it disturbs less surface area, though the harm from mining roads and initial cuts may still be considerable. Moreover, like surface mining, underground mining can disrupt aquifers. Old underground mines may fill with water, which builds up pressure and may eventually create a "blow-out," sending toxic or acidified water into aquifers or streams. Further, underground mines may collapse unpredictably decades after their abandonment, disturbing water flow, opening sinkholes, or damaging buildings on the surface.

More sinister problems have occurred in the transfer of the coal to TVA power plants. Much of the coal of Southern Appalachia is so high in sulfur that it creates unacceptably high levels of sulfur dioxide (which contributes to acid rain and smog) when burned. To meet federal regulations, TVA must burn relatively low-sulfur coal. In 1996 investigators from the Tennessee Valley Energy Reform Coalition and the Foundation for Global Sustainability videotaped a large and politically well-connected coal operator loading trucks bound for TVA power plants with high-sulfur coal and then placing a layer of low-sulfur coal on top to conceal the substandard shipment.

Coal is not the only fossil fuel used to produce electricity in the Tennessee Valley. TVA also operates forty-eight combustion turbine generators. Twenty-eight of these can burn either natural gas or fuel oil. The other twenty burn oil only. Due to their relatively high fuel costs, these generators are used only during peak demand. They are located at four of TVA's coal-fired plants, none in our bioregion.

Fossil fuels are also used to generate energy in a variety of other ways. In 2001, for example, the University of Tennessee consumed almost 32,000 tons of coal at its Physical Plant on Neyland Drive.[37] (This plant, incidentally, occasionally emits heavy clouds of thick, black smoke, reportedly the result of technology failures.)[38] Many industries use boilers or furnaces that burn coal, oil, or natural gas. Some use high-sulfur coal. Natural gas, propane, and coal are also used to heat homes. Diesel, gasoline, and propane-powered generators are legion. These all, to various degrees, add to air pollution, ground-level ozone, haze, oxygen depletion, and global warming.

There is, moreover, some question about what we really gain in burning fossil fuels—coal, in particular—for energy. Extracting and transporting the coal, reclaiming the strip-mined land, and maintaining the infrastructure by which all this is accomplished are themselves energy-consuming processes. Since the current cost of coal is about one dollar per gigajoule, about half the cost of the liquid fuels used to mine and transport the coal, economic production of coal probably requires an energy output/input ratio greater than, say, 3:1. However, such an output/input ratio ignores many of the externally borne energy and economic costs of coal mining. Including such "externalities" would result in an even lower energy ratio. By comparison, many renewable energy sources display an output/input ratio of at least 10:1.[39]

Energy analysis is a tricky business, and much depends on how you cut the pie. Yet it is clear that to generate energy from coal, we must also consume vast quantities of energy. And, whatever the output/input ratio, when we consider the totality of effects on the health of people, the air, the water, and the land, and the frivolous and wasteful purposes for which much of the energy is used, the propriety of generating electricity by burning coal is profoundly questionable.

Energy from Nuclear Fission

The final source of TVA electricity is atomic energy. TVA currently operates five nuclear reactors at three sites on the Tennessee River. Two of the sites are in our bioregion: Sequoyah, on Chickamauga Reservoir above Chattanooga, and Watts Bar, near the Watts Bar Dam. Sequoyah has two working reactors with a net capacity of 2.28 GW.[40] Watts Bar has one reactor with a capacity of 1.16 GW.[41] The remaining two reactors are at Browns Ferry near Decatur in northern Alabama. Though TVA began work on seventeen reactors, all the others—with the exception of Browns Ferry Unit 1, discussed below—have been mothballed or abandoned.

Since the normal operating and maintenance costs of a nuclear power plant are relatively inexpensive and since nuclear reactors cannot readily be switched off and on, TVA runs the nuclear power plants, like the big coal plants, as much as possible.

All nuclear power plants routinely release small amounts of radioactive material into the air and water. The evidence is inconclusive as to whether these routine releases can harm people living nearby; some studies show no effect, while others find elevated rates of cancer or leukemia. (It is worth noting, however, that until recently nearly all significant research on the health effects of radiation was funded by institutions ideologically committed to the proliferation of nuclear energy.)

The greatest concern with nuclear plants, however, is the possibility of a major release of radionuclides. The worst example of this sort was the

Fig. 23. *TVA's Watts Bar nuclear plant. Copyright by Mignon Naegeli.*

explosion and meltdown that occurred at Chernobyl present-day Ukraine in 1986. Thousands of people died from this accident, and over three hundred thousand had to be relocated. About 3,900 square miles of land were heavily contaminated, and much of Europe received dangerous levels of fallout.[42] Though American reactors are better designed than the Chernobyl reactor, they are not foolproof. TVA has had one near-meltdown (at Browns Ferry in 1975), and an American-built reactor at Three Mile Island in Pennsylvania underwent a partial core meltdown in 1979, exposing thousands of people to dangerous radiation. These incidents indicate that the risk of a serious accident in the Tennessee valley is not negligible. The possibility of terrorist attack compounds the peril. American nuclear reactors are not designed, for example, to safely withstand the impact of a crashing airliner, and their temporary stores for highly radioactive materials (e.g., cooling ponds) are even more vulnerable. Efforts have been made to quantify these risks, but these depend heavily on guesswork. Single accidents and terrorist attacks are by nature unpredictable.

The damage to Southern Appalachia from a large radiation release at Watts Bar or Sequoyah could be enormous. Signs along the roads near both plants mark the evacuation routes, grim reminders of the scramble that must ensue if the sirens sound. If an air release occurred at Sequoyah when the wind was blowing down the valley from the northeast, the cloud of radiation would move directly into Chattanooga twenty miles away. A southwest wind blowing up the valley could drop radiation from a release from Watts Bar onto Knoxville and Oak Ridge, sixty miles downwind. If the wind were blowing from the west, a major release from either plant could fall out largely in the Great Smoky Mountains National Park and the surrounding national forests. The extent of the damage would depend on many factors besides the wind—most crucially on the amount of radiation released. In the worst case, thousands of lives, billions of dollars, and the health of much of the land of Southern Appalachia could be lost.

Not all the harms from nuclear power are mere possibilities. TVA's nuclear power plants use partially enriched uranium for fuel. Together all five reactors consume about 1,875 tons of uranium per year (each of TVA's nuclear units requires about 750,000 pounds of natural uranium per year).[43] The mining of uranium, like the mining of coal, does damage both to people and to the land. In the 1970s the nuclear industry's greed for uranium, like the forty-niner's greed for gold in the gold rush of the previous century, prompted thefts of Native American land. Efforts by TVA, former Oak Ridge DOE Contractor Union Carbide, and other large corporations to secure uranium rights on or near tribal lands in South Dakota were strands in a complex pattern of intrigue, oppression, and violence that led to the bloody siege at Wounded Knee in 1973 and the shoot-out at the Pine Ridge Reservation in 1975. The details of this sad, tangled, and little-known story are recounted in the book *In the Spirit of Crazy Horse* by Peter Matthiessen. One of the FBI agents in charge of the suppression of Native

American resistance to these land grabs was Norman Zigrossi,[44] who subsequently served as the chief administrative officer for TVA from 1994 until his retirement in December 2000—when, incidentally, he was awarded one of the largest pensions in TVA history, totaling almost $1.5 million.

Today, the uranium that fuels TVA's reactors comes from the United States, Australia, Russia, and Canada.[45] The uranium is mined in open pit, deep pit, or leaching operations. The mining and milling processes produce heaps of radioactive tailings and release carcinogenic radon, which is a danger primarily to miners. In the 1950s and 1960s, before the potential hazard of these materials was widely understood, uranium mill tailings were used as fill material under and around new buildings, for example, in Grand Junction, Colorado. Later surveys identified hundreds of buildings, including schools, with excessive radiation levels due to radon production. Uranium tailings were also used by Native Americans for domestic construction. Publicly funded remediation work to clean up these sites was expensive and took many years. Some tailings are unaccounted for, and over twenty-five million tons remain in unstablized piles in several western states.

The commercial-grade uranium that is used in the United States is transported to one of two conversion plants in Illinois and Ontario for processing, then transported again for enrichment to one of two federal facilities at Paducah, Kentucky, or Portsmouth, Ohio. Commercial uranium may in the future also be derived from nuclear warheads decommissioned as a result of post–cold war treaties.[46]

TVA claims that the reuse of weapons material in civilian reactors is "in the true sense of swords to plowshares."[47] Given the state of Southern Appalachian agriculture (see chapter 6), the reference to plowshares is disingenuous; microwave ovens, air conditioners, and advertising signs would be nearer the truth. Moreover, the agency is just as eager to turn "plowshares" back into swords. In December 1999 TVA signed an agreement with the Department of Energy to assist in the production of tritium at Watts Bar Nuclear Plant, with Sequoyah Units 1 and 2 serving as backup reactors.[48] This is the first use of civilian reactors for nuclear weapons production, a move that is likely to weaken the American argument for nuclear weapons nonproliferation. The operation also entails the transportation of more nuclear weapons materials on the region's railroads and highways.

Nuclear energy is touted by its proponents as "clean," yet it inevitably creates large volumes of radioactive waste. Radioactive waste is classified as either low level or high level in the United States (many other countries distinguish a third type, known as intermediate-level waste). Low-level radioactive waste (LLRW) consists of such objects as filters, cloth wipes, paper wipes, plastic shoe covers, tools, water purification devices, and various residues—all contaminated with varying quantities of radioactivity. Much of the radiation decays away within a decade after disposal, though some persists for much longer. TVA used to send its low-level radioactive waste to a land burial facility in Barnwell, South Carolina, but this became so expensive that in 1999 the agency began storing the waste at two TVA plant sites.[49] In the wake of the September 11 terroist attacks, the exact location of the two LLRW storage sites is not available to the general public.

High-level waste, which currently consists mostly of spent fuel rods, is another matter. Each of TVA's five reactors produces about 27 metric tons of used fuel per year—an annual total of 135 metric tons.[50] A large additional volume of high-level waste in the form of irradiated bulky metal components and chemical sludges (classified as intermediate-level waste in Europe, Australia, and other countries) is likely to be created within several decades when the plants wear out and are dismantled; there will also be significant additional quantities of low-level waste. High-level waste, when first removed from the reactor, is extraordinarily "hot" and must be handled with utmost caution. Some shorter-lived isotopes decay relatively rapidly (months to years), but much of the waste remains deadly to humans for tens of thousands of years—longer than civilization has existed on earth.

There is no practical way to make the waste non-radioactive. Other forms of disposal, such as shooting it into space or injecting it deep into the earth, are too expensive or too dangerous. Storage is, for the present at least, the only acceptable option. Storage facilities must be safe from earthquake, climate change, rising water tables, and so on, and must also be constantly guarded against terrorists, since the waste could be used to make nuclear weapons or (more likely) "dirty bombs" that spread radioactivity. Whether this can be done successfully over a period of tens of thousands of years is doubtful.

There is no permanent disposal facility for high-level nuclear waste anywhere in the world. The only place in the United States currently being considered for such a facility is Yucca Mountain, which is located on Shoshone tribal lands in Nevada about one hundred miles northwest of Las Vegas. At Yucca Mountain, the U.S. Department of Energy plans to store up to 77,000 tons of radioactive waste produced over the past fifty years from defense activities, nuclear power plants, and the reduction of the United States' nuclear arsenal.[51] The waste is to be transported there in 4,300 highway and rail deliveries over a twenty-four-year period.[52] Thousands of casks of high-level waste are to pass through our bioregion, exposing its inhabitants to the attendant risks of traffic or rail accidents and terrorist attacks.[53]

The geological suitability of the Yucca mountain site is a matter of controversy. There have been 621 seismic events, with magnitudes greater than 2.5 on the Richter scale, within a 50-mile radius of Yucca Mountain since 1976. In July 1992 an earthquake measuring 5.2 on the Richter scale occurred about 8 miles from the proposed repository.[54] A decade later another quake measuring 4.4 occurred 12 miles east of the site.

While TVA waits for the DOE to build a national nuclear repository, it continues to store all of its high-level nuclear waste on site—in storage pools, which were originally meant for temporary storage and are now quickly approaching capacity. Browns Ferry and Sequoyah storage pools are expected to reach capacity in 2006 and 2004, respectively; Watts Bar will not meet capacity until 2018.[55] In order to manage the excess high-level nuclear waste after the storage pools crest, TVA plans to store the waste above ground in steel casks placed in concrete containers.[56] Back in 1995 TVA stored 760 metric tons of spent fuel at its Browns Ferry nuclear plant and 417 metric tons at Sequoyah; but, again, due to heightened national security concerns, TVA does not disclose information about current quantities of spent fuel being stored at its facilities, though it produces about 135 metric tons of high-level radioactive waste per year.

Eventually, nuclear plants wear out. Over time, the high temperature and extreme radiation inside a reactor cause metal parts to weaken and become brittle. If operation continues beyond this point, the reactor becomes increasingly dangerous. Many researchers think that the appropriate life span for a nuclear reactor is thirty to forty years. The two Browns Ferry reactors (TVA's oldest) came on line in 1975 and 1977. Their operating licenses expire in 2014 and 2016, and TVA is aggressively lobbying for a twenty-year license extension for the two reactors, which will suspend their decommissioning at least until 2034 and 2036 respectively. In any event, these reactors will have to be closed and decommissioned within a few decades. Although there is some limited experience with decommissioning experimental reactors, and the oldest British reactors are now being dismantled, nobody knows exactly how expensive it will be to dismantle American commercial reactors and dispose of the resulting waste. TVA says that the decommissioning cost for one nuclear plant is approximately $804 million,[57] but TVA cost estimates in nuclear matters have a history of being optimistic. Independent estimates have run as high as $5.86 billion.[58] The money TVA has set aside for decommissioning all five reactors amounted to $600 million by the end of fiscal year 2001.[59] If TVA's deep debt history continues, there may well be financial pressure to scrimp on this fund. Who, then, will pay for decommissioning?

In March 2002 TVA's board approved the restart of Browns Ferry Unit 1, which had been unused since 1985. The repairs and upgrades needed to bring this nuclear reactor back on line will add about $1.8 billion to TVA's current $25.4-billion debt during the five-year project.[60] But TVA expects the plant eventually to contribute enough generating capacity to lower the average cost of power production, ultimately reducing its debt. Whether the investment will succeed remains to be seen. There is substantial risk. A disaster on the scale of Chernobyl in the United States could wreck the nuclear power industry.

It should be noted, by contrast, that TVA is not at all disposed to deepening its debt to develop sustainable energy sources. The explanation for this comparative reluctance lies largely in the history of federal subsidies to the nuclear industry. Between 1947 and 1999, the federal government spent approximately $151 billion to support the development of wind, solar, and nuclear power. Of this $151 billion, the nuclear industry received more than $145 billion—about 96 percent.[61]

Sustainable Energy Sources

None of TVA's three main modes of generating electricity (hydroelectric, coal, or nuclear) offer much hope for the future. Though hydropower generation could be modestly increased by modernizing existing hydropower projects,[62] there is little room for new dams, except perhaps for some small generators on tributary streams. But these tributaries harbor almost all that is left of Tennessee valley's originally rich aquatic biodiversity. Damming them would threaten even these last remnants. All the reservoirs, moreover, will eventually silt up and lose their utility for power generation.

Fossil fuels offer still less hope. For one thing, fossil fuel reserves are nonrenewable and finite. We have already used up much, if not most, of this nation's once vast oil supply. Imports are approaching 60 percent of U.S. demand. Considering all the world's known reserves and estimating the unknown reserves, at current rates of consumption, all the world's oil will be gone by the middle of this century. The "Hubbard Peak" of world-wide oil production is expected to occur some time around 2005–10, after which it is downhill all the way. Natural gas reserves in the United States may last another sixty years at current rates, but much less if consumption increases. Coal is more abundant; it may last three hundred years—if consumption rates are steady (which is doubtful).[63] But the consequence of burning all of our remaining fossil fuel would be immense: mining operations of increasing complexity, unprecedented pollution, and (perhaps most dangerously) climate change so rapid and extreme that it would threaten the stability of civilization itself. This is simply not an option.

Nuclear energy has proven to be unexpectedly expensive and less than reassuringly safe. Yet, on the whole, even considering Chernobyl, nuclear power plants have probably done less harm than coal plants. Still, any technology that passes radioactive trash on to generations tens thousands of years into the future is fraught with moral problems (see introduction). If we do not know how to clean up our mess, but only how to shovel it under the rug (or under Yucca Mountain, as the case may be), perhaps we have no business making it.

The need for new sources of energy that are safe, reliable, and sustainable is therefore obvious. In response largely to pressure generated by the Southern Alliance for Clean Energy, a coalition of nonprofit organizations, TVA began in a small way in the late 1990s to generate power using sustainable sources under a program that is now called the Green Power Switch. But, in stark contrast to its nuclear power investments, TVA is paying for its new "green" energy facilities largely though voluntary contributions made by business and residential customers on their monthly electric bills. The burden for developing sustainable energy in the Tennessee valley thus has fallen disproportionately upon the conscientious.

Among the most promising of sustainable or "renewable" energy sources are bioenergy fuels, solar panels, and wind turbines. Under the Green Power Switch, TVA is developing all three.

One of many types of biological fuel is the methane gas that is released from landfills, sewage plants, and composting operations. Methane burns relatively cleanly, releasing only water and carbon dioxide. And, since it is a very powerful greenhouse gas (up to thirty times more effective at trapping heat than carbon dioxide),[64] collecting and burning it in a small gas-fired generator, rather than releasing it into the atmosphere, helps to mitigate global climate change. What is otherwise an obnoxious waste is thus converted into a relatively clean and beneficial source of energy.[65] The only major drawback to landfill gas is that not all landfills emit enough methane to produce electricity and those that do may only produce at a level that is economical for ten to fifteen years.[66]

Landfill methane was already in use in the bioregion before TVA became interested in it; methane collected at the Chestnut Ridge Landfill in Anderson County, Tennessee, runs four electrical generators, whose output is sold to the Knoxville Utilities Board. But TVA has also begun using methane gas to generate electricity at two sites—one in Murfreesboro, Tennessee, and another in Memphis. The Middle Point landfill in Murfreesboro, which came online in February 2001, has a generating capacity of roughly 2,600 kilowatts and is expected to generate more than 16 million kilowatt-hours annually.[67]

Among other forms of bioenergy, the most practical for large-scale power production consist of various forms of solid biomass fuels. Three general types are readily available: 1) waste or residue consisting of wood or the byproducts of agriculture and food processing, 2) energy crops, and 3) whole logs.

Wood residue from pulp operations is already being used by the regional paper industry to generate thermal energy. In addition TVA has successfully experimented (independently of the Green Power Switch) with burning a mix of wood waste and coal at its coal plants. This practice, called "co-firing," decreases coal consumption, reduces emissions of sulfur and toxic metals, and may divert some wood waste from landfills. But the source of the "waste" is important. If it includes material from forest sites, then energy is being generated at the expense of the forests. Materials which used to be sent directly to the landfill, such as clean (untreated) wood from demolition sites, are beginning to be used as biomass fuels.[68] With these, there is the additional advantage that, unlike coal ash, the ash from the burning of biomass materials is a good fertilizer, which can be returned to the land to enrich the soil.

A future source of biomass energy will be energy crops. Typically these are made up of fast-growing trees and grasses, particularly switch-grass, poplar, and willow—native species that have been selected by the Bioenergy Feedstock Development Program at Oak Ridge National Laboratory. Since these energy crops actively remove carbon from the air while growing, the net carbon emission from such a fuel system is effectively nil; as much carbon is absorbed by the growing plants as is released in combustion. Thus, energy crops add very little to global warming. (There is, however, an increase in carbon emissions relative to other possible uses of the land; for example, if the land were left as undisturbed forest, the trees would continue to remove carbon from the atmosphere, until the forest matured. Some carbon is also released in harvesting the crops and transporting them to the power plant—especially if these operations are powered by fossil fuels.)

Other environmental effects of energy crops depend on where and how the crops are grown. Transportation impacts (and costs) are higher if the crops are not grown close to the power plant. If industrial fertilizers, pesticides, and herbicides are used at levels higher than alternative agricultural land use, then the air, water, and soil suffer pollution. Heavy machinery may compact soils and enhance erosion. Moreover, though the burning of biomass fuels releases very little sulfur or toxic metals, it does, like coal combustion, produce NO_x and volatile organic compounds and therefore may contribute to smog. Finally, given the precipitous loss of regional farmland (see chapter 6), there is a trade-off between using the remaining farmland for energy crops, for other non-food uses, or for enhancing the local food supply.

The least acceptable source of biomass fuel consists of whole logs from natural forests. Southern Appalachian forests are already under intense pressure from pulp and paper and lumbering operations and from a host of other insults (chapter 4). Any increase in the cutting of existing Southern Appalachian forests is therefore unacceptable. One proposed type of biomass operation can use whole trees—wood, limbs, roots, leaves, and all—which would be even more destructive to forests than chip mills, which use only logs.

Wind energy is the fastest-growing renewable energy source worldwide, and already cost-competitive with fossil fuels and nuclear in many situations. Although some of the early wind turbines were noisy, and they can be hazardous to birds, their overall environmental impact is much smaller than that of biomass, hydroelectric, coal, or nuclear power generation. Wind power is, however, by nature an intermittent source of energy, generating only when the wind is blowing. Its primary use is therefore for "fuel saving," i.e., offsetting electricity from existing power systems (though for home or farm use, a small windmill plus batteries may be adequate). However, up to 20 percent or more of regional electricity demand can be met in this way (as is the case in parts of Denmark, Germany, and Spain) without significant changes to the power distribution system. Under the Green Power Switch program, TVA has erected three windmills on a reclaimed strip mine in Rosedale, Tennessee, which is now called Buffalo Mountain Wind Farm. Together, the three windmills generated approximately 4.8 million kilowatt-hours of electricity during their first year of operation and are expected to produce 4.6 million kilowatt-hours each year.[69] TVA plans to increase its wind-generating capacity from 2,000 kilowatts to approximately 22,000 kilowatts by 2003; but this depends on a significant increase in demand for green power.[70]

The ultimate source of biomass, fossil fuel, hydropower, and wind energy is the sun. Fossil and biomass fuels store energy in a chemical form that originated as sunlight captured by plants. The sun's heat drives the rain cycle that fills the rivers that generate hydropower. And it is the sun's heat, again, that stirs the atmosphere and moves the wind. Why not, then, use the sun to generate power directly?

Photovoltaic cells do just that. Recent advances in photovoltaics have made them economically competitive with conventional power sources for remote, rural applications. Increasing use will bring the price down still further. Current commercial photovoltaic panels can generate 120 watts per square meter in the noonday sun. They are less effective, but still operable, on cloudy days and when the sun is low, and (of course) they do not provide power at night. Photovoltaic panels must therefore be combined either with power-storage devices, such as batteries, or some other form of power generation to provide a continuous supply of electricity.

Unlike most other power sources, photovoltaic panels are poorly suited for large, centralized power plants. A plant equaling the hundred-million-watt output of Norris Dam, for example, would require a square mile of panels.[71] But one of the beauties of solar panels is that they need not be centralized into immense power plants to operate efficiently. Indeed, they work best if not thus centralized. The roofs of buildings already provide a vast, though scattered, area, much of which is well-suited for solar collection.

Utilizing this roof area, each household can generate much or (if energy is carefully conserved) all of its own electricity.

Regional demand for electricity is highest on hot, sunny summer days—precisely the times at which photovoltaic panels reach maximum output. Moreover, in hybrid systems that integrate photovoltaic panels with conventional power lines from a central utility, there is no need for storage batteries, because when the solar panels are not generating, the system automatically reverts to the conventional power source. During periods when the solar panels are generating more power than is being consumed, the excess can be fed back into the commercial grid. Federal regulations require utilities to buy back power generated in this way, effectively running the electric meter in reverse and subtracting dollars from electric bills.

Fig. 24. *TVA's solar power array at Ijams Nature Center. Copyright by Mignon Naegeli.*

Photovoltaic systems work even when there is a general power failure. In the blizzard of 1993, power was out for nearly a week in Hogskin Valley in Grainger County. The only electric lights in the valley on during all that time were at Narrow Ridge Earth Literacy Center, which generates its electricity from solar energy.

Solar power has become increasingly cost-effective and will continue to be so as rates for conventionally generated power rise. The main obstacle to increased use of solar energy is the initial cost of installing a home system, which can run into the tens of thousands of dollars. If low-interest loans were made available for buying such systems, increasing usage would bring down the price still further and make them yet more widely affordable. As part of its Green Power Switch, TVA has installed twelve solar panel arrays at various sites in Tennessee, Virginia, Kentucky, Mississippi, and Alabama. Combined, these arrays have a generating capacity of approximately 300 kilowatts, and during fiscal year 2002 they generated a little more than 335,000 kilowatt-hours.[72]

These are steps in the right direction, but still the generating capacity of the entire Green Power Switch was by the summer of 2002 a mere 8,000 kW—approximately 0.01 percent of TVA's total generating capacity. TVA expects the residential demand for green power to be about 6,460,200 kWh per year by 2007. But even if this optimistic prediction is realized, meeting this future demand will still keep the green power production well below one-half of 1 percent of TVA's overall generating.[73] The Green Power Switch is thus no panacea; substantial reductions in the destructiveness of our power system will be attained, if at all, only through energy conservation. There are several solar power sites within our bioregion:

- American Museum of Science & Energy—Oak Ridge, TN.

- Cocke County High School—Newport, TN.

- Dollywood—Pigeon Forge, TN (Dollywood has 2 solar generating systems).

- Duffield Primary School—Duffield, VA.

- Finely Stadium—Chattanooga, TN.

- Ijams Nature Center—Knoxville, TN.

- Oak Ridge National Laboratory—Oak Ridge, TN.[74]

Energy Conservation

In the first section of this chapter, we surveyed some of the purposes for which energy is used and found many of them wasteful, frivolous, or unnecessary. In the succeeding sections, we traced that energy back to its sources and surveyed the damage that is done and the risks undertaken to produce and deliver energy. It is evident that the damage is immense and the risks considerable. There is, therefore, a moral imperative to reduce energy consumption.

But there are economic imperatives as well. Demand for electricity is increasing. New generating facilities are expensive and financially risky, as the history of TVA's nuclear power program clearly illustrates. Some utilities have therefore found it less risky and yet still profitable to increase the overall efficiency of the entire power system by simultaneously raising rates and offering conservation programs aimed at decreasing demand. These conservation programs provide ways for customers to become more efficient energy users, which would, of course, decrease the utility's profits if the rates were not raised. The purpose of raising the rates, then, is to ensure that the utility can still make money. But because customers use less electricity, their bills remain about the same, despite the rate increase. Thus the utility retains its profits and customers on average pay no more, but the whole power system becomes more efficient, less electricity is generated and used, and less damage is done to the environment. Such programs may provide discounts on or even giveaways of compact fluorescent light bulbs, offer low-interest loans for weatherization, install solar energy systems or solar water heaters, promote passive solar architecture, provide incentives for industries to replace old electric motors with newer, more efficient ones, and so on. TVA pioneered some of these strategies in the early eighties, but abandoned them a few years later as its nuclear debt skyrocketed.

TVA's leadership rightly fears that if it raises rates on a deregulated electricity market, it will lose customers to more competitive (and less debt-burdened) utilities. This concern is valid, and little can be done on a regional level to address it. TVA can return to its conservation mission only if the federal government develops a strong, far-sighted energy policy. That may take years.

In the meantime, the responsibility for using energy sanely and wisely falls by default to individuals and to a few responsible businesses. There are three things anyone can do: 1) support and vote for public officials who understand the energy dilemma and are willing to take strong steps to move TVA back toward conservation, 2) complain to businesses that misuse electricity, especially for ostentatious lighting displays, and refuse to patronize those that continue to do so, and 3) reduce home energy use.

One of the simplest and most effective ways to reduce energy use at home is to obtain a solar clothes dryer—otherwise known as a clothesline. A conventional clothes dryer, particularly one that uses electrical resistance heating, is one of the most wasteful home appliances, and its solar replacement costs a few dollars at most. The solar clothes dryer pays for itself within a month or two in reduced electricity bills, and thereafter saves both money and energy.

Slightly more expensive initially, but equally cost-saving in the long run, is the replacement of incandescent lights with compact fluorescents. Compact fluorescents last longer than incandescent bulbs, and they provide the same amount of light for less than one-third of the electricity. Of course, forming the habit of turning off all lights (and televisions) that are not being used is an easy way to conserve that requires no investment and begins to save both energy and money immediately.

The largest home savings can be realized in water heating, but these may require a larger investment. The simplest step is to wrap the water heater and hot water pipes with additional insulation. Replacing electrical resistance heaters with electrical heat pumps or gas heaters saves more energy, installing a timer that turns the heater off when hot water is not needed saves still more, and an on-demand heater more yet. Supplementing any of these with a passive solar pre-heater reduces energy consumption dramatically.

Opening the windows and using fans instead of air conditioning not only saves energy but reduces indoor air pollution. Other effective cooling measures include installing a white roof, planting shade trees near the house, and using awnings or other forms of window shading.

These are some of the easiest and most effective ways to reduce home energy use. There are hundreds of others. Most homes and businesses could probably cut their electric bills in half with no serious loss in quality of life and palpable improvements to the health of the air and land.

Chapter 8

consumption and waste

Athena Lee Bradley and John Nolt

Everything that has a front has a back
The bigger the front, the bigger the back

–Taoist saying

Hyper-Consumption

Consumption is our most conspicuous cultural activity. We crave not just high consumption, but ever-rising consumption. Last year's luxury becomes a necessity, while we pursue new luxuries in an epidemic of stress, overwork, waste, and indebtedness that has been aptly called "affluenza."[1] But, beyond our usually momentary delight with our new "stuff" (which all too often ends up collecting dust in an attic, basement, or commercial storage facility), lies pollution, waste, and the degradation of environmental and spiritual health.

Consumption so pervades our values that the term "consumer" has become synonymous with "person." Buying things has displaced many of the functions of neighborhoods, churches, and families. Today, family time often means shopping to buy children new video games and toys. Consumerism dominates our religious holidays. Though we are told that "the borrower is servant to the lender,"[2] every Christmas families go deeper into debt just to buy gifts. While our vehicles, tools, entertainment centers, and computers have provided some degree of independence, we have lost many of the benefits of working together and socializing with our families, neighbors, and churches. Instead of turning to families and friends for social comfort, often we turn on the computer or television. Many of us cannot sit in the quiet of nature or even our own homes without the distraction of electronic entertainment.

To pay for all this consumption, we work longer and longer hours. Each year, the average American works 137 hours more than the average Japanese, 260 hours more than the average Britain, and 499 hours more than the average German.[3] The more we work, the more we neglect our families and ourselves. We increasingly turn to convenience in our foods and home life, producing less at home but consuming more. Even in rural Southern Appalachia, once known for home-cooked, country meals, many families have all but abandoned cooking from scratch, turning instead to packaged goods, microwave meals, and "take-out." Today's home no longer has a pantry, but merely a compact kitchen, complete with a built-in shelf for the microwave. Once the producers of homespun crafts and handmade clothes, increasingly we consume the offerings of the chain discount store.

Locally owned shops in neighborhoods and downtown areas are steadily replaced by shopping malls, "strips," and superhighways. The mall has become our favorite gathering place. The places we shop are increasingly impersonal: Wal-Mart, Target, Big Lots, and Kroger. The "Kroger Garden" is hardly a garden, unless one considers vegetables grown with pesticides and petroleum-based fertilizers hundreds or thousands of miles away a garden.

With respect to groceries, Wal-Mart now controls almost 30 percent of Tennessee's retailer market share, while the share of regionally owned retail stores, such as White Stores and Food City, is declining. In Knoxville Kroger and Wal-Mart have almost 53 percent of the market share, followed by Food City with 20 percent and the little remaining market share by the smaller, independent retailers serving the area.[4]

Other big "superstores" out-compete with the local hardware store, home and garden centers, and other locally owned businesses. Wal-Mart is now branching into the "Neighborhood Market" with its "new, smaller" concept.[5] But such stores' profits, unlike those of authentic neighborhood markets, will leave the region and there will be few if any locally produced goods for sale.

Moreover, while the big discount stores facilitate consumption by those who previously could not afford common luxuries, they promote products that may be cheap, but often do not last. The VCR we buy one year often breaks the next, and buying a new one is cheaper than repairing it. The result is an ever-rising consumption of resources to make new replaceable products that quickly become nonrecyclable and often hazardous waste.

Unsustainable Growth

To support this hyper-consumerism, natural resources are extracted at unsustainable rates. The United States uses more than 3.6 billion metric tons of new mineral materials each year, or about 40,000 pounds per person. About half of this is in the form of mineral fuels (petroleum) and the other half is metals and nonmetals. Over a lifetime the average American uses:

- 4,550 pounds of aluminum.

- 1,050 pounds of zinc.

- 1,050 pounds of lead.

- 1,750 pounds of copper.

- 27,000 pounds of salt.

- 91,000 pounds of iron and steel.

- 360,500 pounds of coal.

- More than 1 million pounds of stone, sand, gravel and cement and clay.[6]

Resource use has moved from renewable and replenishable to nonrenewable and depletable. In the early part of the twentieth century, the United States acquired 75 percent of its material from renewable sources, with the remaining 25 percent met by nonrenewable minerals. By the mid-1980s, renewables met only 32 percent of this nation's nonfood and nonfuel uses, with 68 percent being obtained from nonrenewable sources.[7]

Growth in materials consumption has outpaced population growth. The population of the United States grew slightly more than threefold between 1900 and 1991, while the total use of virgin raw materials in 1991 was fourteen times greater than in 1900.[8]

From 1970 to 1991, total world consumption of raw resources increased 38 percent. The use of nonrenewable organic chemicals—the feedstocks for synthetics materials production, including plastics—rose

69 percent, agricultural materials grew 44 percent, nonmetallic minerals 39 percent, and metals 26 percent during that same time period.[9] A recent report by the World Resources Institute (WRI) found that the total output of wastes and pollutants in Austria, Germany, Japan, the Netherlands, and the USA has risen by as much as 28 percent since 1975, despite efforts by these nations to use natural resources more efficiently. According to WRI, from one-half to three-quarters of the annual resource inputs used in these five countries are returned to the environment as wastes within one year.[10] In 1999, some 9.6 billion tons of marketable minerals were extracted from the planet, nearly twice as much as in 1970.[11]

Resources to Trash

Our increasing consumption has generated increasing waste. Municipal waste generated per person in the industrialized nations has increased threefold in the past twenty years, with two-thirds of this being dumped in landfills. If current trends continue, the Organization for Economic Cooperation and Development (OECD) predicts an additional increase in waste generation of 70 to 100 percent in industrialized nations by 2020 and a 200 percent increase in developing nations.[12]

Collectively, our trash or garbage, known as "Municipal Solid Waste" or MSW, includes wastes generated typically by households and businesses. MSW ranges from food wastes and furniture, to clothing, appliances, and packaging. Industrial wastes, sewage sludge, and construction and demolition wastes are not typically included as MSW. Table 15 lists the EPA's estimate of the typical composition of municipal solid waste.

In 1960 Americans generated an average of 2.7 pounds of MSW per person per day. This increased to 3.7 pounds per person per day by 1980, and 4.5 pounds per person per day in 1990. Through the 1990s and into the new millennium, trash generation has fluctuated slightly, but has remained close to the 1990 figure.[13]

Nationally, we generated some 231.9 million tons of MSW in 2000, an increase of 0.9 million tons over 1999. In the United States 30.1 percent of MSW is recycled or composted, 14.5 percent is burned at combustion facilities, and the remaining 55.3 percent is disposed of in landfills.[14]

Table 15

COMPOSITION OF MUNICIPAL SOLID WASTE BY WEIGHT,
2002 NATIONAL AVERAGES (BEFORE RECYCLING)

Material	%
Paper and paperboard	37.4
Yard trimmings	12.0
Food Scraps	11.2
Plastics	10.7
Metals	7.8
Rubber, Leather, Textiles	6.7
Glass	5.5
Wood	5.5
Other	3.2

SOURCE: EPA, "Municipal Solid Waste—Basic Facts," accessible at
http://www.epa.gov/epaoswer/non-hw/muncpl/facts.htm (5 Sept. 2002).

In Tennessee, the Solid Waste Management Act of 1991 (SWM Act) made sweeping changes in the way counties in Tennessee handled trash. Prior to the act, many counties did not have any disposal options for their residents, other than the local "holler." The SWM Act required that local governments assure, through direct services or contract with a private hauler, that the solid waste collection and disposal capacity meet the needs of its citizens. It required that public agencies adopt a "materials management" approach to quantify waste generation and ultimately to reduce waste and the toxicity of wastes being disposed.

Each solid waste region (comprised of one or more counties) had to develop a ten-year plan describing the ways in which the region would meet the goals of solid waste management and waste reduction. For the first time, each region had to conduct planning, keep records of its solid waste activities, and establish a citizen planning board to have input into solid waste activities. In addition, each county was required to provide at least one collection site for waste tires, lead acid batteries, and used oil. Each county also had to provide at least one collection site for recyclables. To assist in raising public awareness of solid waste issues, a public information campaign and a K-12 solid waste education program were also developed.

All this was intended to reduce the amount of waste going into land-fills, but success has been mixed. Originally, the act mandated a 25 percent per capita reduction in waste going to Class I landfills or incinerators for all counties by 1995, using 1989 waste-disposal figures as a baseline. But this goal was not met. In some counties, per-capita waste disposal actually increased over this period.[15] A new version of the act was passed in 1999, resetting the baseline year to 1995 and mandating a 25 percent reduction by 2000. This goal, too, was missed, though not by much; by 2000 the state as a whole had achieved—at least on paper—a per-capita reduction of 22.6 percent.[16]

These figures can, however, easily be misinterpreted. It is important, first, to recognize that they refer to all categories of solid waste, not just MSW. Though some of the reduction is due to improvements in recycling of MSW (the largest category being yard waste), some is simply the diver-sion of construction waste from regular Class I landfills to construction and demolition landfills. In other words, some of the waste involved in the supposed "reduction" is still landfilled; it is just landfilled in a different place. Moreover, these are *per-capita* reductions. But population is growing.

The Three R's

The EPA developed a national strategy for handling municipal solid waste over a decade ago. At the heart of this strategy is what has come to be known as the "three R's": reduce, reuse, recycle. The order of the terms is significant. "Reduce" is listed first to indicate that the highest priority is source reduction: generating less waste to begin with. Second in the hier-archy is reuse—an idea aimed at reversing the momentum of the throw-away culture. Recycling is third. These three primary strategies are followed by composting, incineration, and landfilling—in that order. Landfilling, be it noted, is the last resort.

Unfortunately, the two top priorities—source reduction and reuse—have largely been ignored. Most communities in our bioregion continue to generate large quantities of waste and send most of it to landfills. There have, indeed, been notable improvements in recycling and composting, but we have yet to adequately deploy the most effective strategies: reduction and reuse.

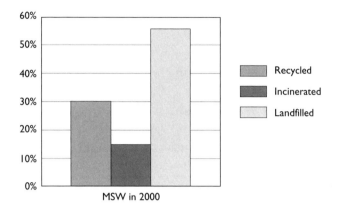

Fig. 25. *Disposal of Municipal Solid Waste in the United States in 2000.*

Reduction and reuse have been difficult to achieve because our current waste-disposal system provides few incentives to undertake them. Most localities still pay for residential (and in some places even commercial) disposal through general tax revenues. The cost of disposal is the same no matter how much garbage you throw away. This "trash welfare" allows for virtually unlimited trash generation and disposal, with no incentive to reduce, reuse, or recycle. Those who do recycle or otherwise minimize their waste (many of whom tend to be senior citizens) thus subsidize the wasteful habits of others. Most private residential services also charge a flat rate, regardless of the amount of garbage customers generate, once again providing no monetary incentive to reduce.

Incentives can be introduced by various "pay-as-you-throw" schemes, which charge for waste disposal either by weight or by volume. One simple way to charge by volume is to require that the garbage be placed in special bags, the purchase price for which is the disposal fee for the volume they can hold. Paying by weight is more complicated, since it requires garbage trucks to be fitted with scales, but is still a workable and cost-effective strategy, especially as more communities turn to automated garbage collection systems.

Pay-as-you-throw policies may cause a temporary increase in illegal dumping, especially in rural communities, as freeloaders seek to evade the fee. But ultimately, as the fees gain acceptance, they promote a community ethic of thrift.[17] Moreover, with the proper enforcement in place, illegal dumpers can be found and fined.

Pay-as-you-throw plans have beneficial ripple effects throughout the whole economy. Aiming to reduce their garbage bills, people seek products that are reusable, recyclable, or less wastefully packaged. This in turn impels producers to increase product quality, reduce packaging, and eliminate throw-aways. Consumers also have greater incentives to reject planned obsolescence, motivating manufacturers to build sturdier, longer-lasting products. These adjustments increase efficiency in the use of materials, promote conservation, and help protect the land from which raw materials are extracted.

Pay-as-you-throw pricing also promotes reuse, since money is saved by repairing, selling, or donating items rather than throwing them away. Reuse was, until fairly recently, a tradition imposed by hard necessity on Southern Appalachia, but that tradition has faded with burgeoning wealth. Yet age-old patterns of reuse still survive in garage sales, flea markets, and donations to Goodwill, the Salvation Army, and other charities. A new twist to these patterns is the notion of a waste exchange, in which businesses or individuals can deposit items they no longer want so that others can use them. Tennessee operates a statewide waste exchange, which has proven useful as an inexpensive source of raw materials for businesses; and smaller waste exchanges (such as the Teacher's Depot, operated by the Knox County Parent Teacher Association) have promoted the same ideal on a more local level.

Pay as you throw programs offered in other communities around the country, including some in Georgia and South Carolina, effectively reduced trash disposal and increased recycling rates. Pay-as-you-throw pricing also encourages the third process in the EPA hierarchy: recycling. People are motivated to recycle when they can lower their garbage bills by doing so.

Recycling

The benefits of recycling are many. Communities that recycle can sell recycled materials to offset waste collection and program costs, reduce their budgets for collection and disposal of non-recyclable waste, postpone investment in siting and construction of new landfills, and create jobs as businesses develop to collect and process recyclables and remanufacture them into new products. Manufacturing products from recycled materials can also reduce environmental and energy costs associated with logging, mining, and other extractive industries.

By using recycled materials in manufacturing, industries can reduce materials and energy costs, as well as the cost of complying with environmental regulations—and reinvest the savings in regional economies. The recycling of aluminum, for example, has dramatically lowered the consumption of energy (and associated costs) at the Alcoa Aluminum Plant. Recycling also benefits regional economies by contributing valuable resources for use in manufacturing of new products with recycled content. Loudon County, for example, boasts a paper-manufacturing facility that uses only recycled paper to make toilet tissues, paper towels, and other paper products. Several paper mills in the region use recycled paper, including Sonoco in Newport and RockTenn in Chattanooga.

Small recycling businesses provide opportunities for sustainable growth throughout the region. In Hamblen County a materials-recovery facility (MRF) operated by Goodwill Industries, Inc., of Knoxville processes some three hundred tons of materials each month. The facility accepts both residential and commercial materials. Companies such as John Deere, Shelby Williams, and Mahle have realized thousands of dollars in savings from sending truckloads of unsorted wastes to the facility. Employees at the MRF then sort through the loads and sell a variety of materials, including plastics and paper.

Unfortunately, recycling is often hampered by unfair competition from suppliers of virgin materials. Federal subsidies or tax incentives for lumbering, mining, and oil and gas exploration, for example, keep the prices for virgin materials artificially low, discouraging the recycling of paper, metals, and plastics. These subsidies include depletion allowances for the extraction of oil, natural gas, and various minerals; below-cost timber sales, road building, and other federal subsidies for timber operations—especially in the national forests;[18] tax provisions for energy development, including the funding of exploration and development, and tax-exempt bonds; and various other tax provisions.[19] Inheritance laws, for example, often treat timber as an investment, which encourages logging as a way of avoiding inheritance taxes.[20]

Changing the tax structure could, therefore, reduce the extraction of virgin materials and increase recycling. Much higher rates of recycling with much greater economic efficiency have been achieved in the more advanced waste-management systems of Western Europe and Japan. We could do as well or better if the disincentives were replaced with positive incentives.

Even so, as a nation we are recycling more. Recycling rates have increased from 10 percent of MSW generated in 1980 to 16 percent 1990, to its current rate of 30.1 percent in 2000. The national recycling rate in 2000 was 1.4 pounds per person per day. Since the average person today generates about 4.6 pounds of MSW daily, discards after recycling were 3.2 pounds. The highest rates of recovery were achieved with yard trimmings, paper products, and metal products. About 57 percent (15.8 million tons) of yard trimmings were recovered for composting in 2000. This represents nearly a fourfold increase since 1990. About 45 percent (39.4 million tons) of paper and paperboard, much of it packaging, were recovered for recycling in 2000. Recycling these organic materials alone diverted nearly 24 percent of municipal solid waste from landfills and combustion facilities. In addition, about 6.4 million tons, or about 35 percent, of metals were recovered for recycling; 23 percent of glass was recovered; 27 percent of aluminum was recovered; and 5.4 percent of plastics was recovered.[21]

We are also doing better regionally. The number of recycling collection and processing facilities around Tennessee has climbed markedly from just 160 in 1992 to more than 700. Almost thirteen million scrap tires have been shredded as a volume-reduction technique for landfilling; an additional fifteen million tires have been diverted from disposal for reuse and recycling opportunities. More than 800 used oil collection centers have been established across the state.[22]

Recycling programs around the bioregion vary considerably. Knox County and city of Knoxville residents can bring their recyclables to attended drop-off centers located throughout the community. The city of Knoxville and Knox County operate a joint permanent, state-of-the-art household hazardous waste facility—as does the city of Chattanooga. Both Knoxville and Knox County have also ventured into the composting business by sending their residential yard wastes (and pallets from businesses) to compost operations. Knoxville has also become a leader in the South by offering computer recycling events to their residents twice a year. The computers are reused by Goodwill Industries and recycled by a local company in Oak Ridge, the Oak Ridge National Recycle Center.

Johnson City, Morristown, Hamblen County, and Kingsport offer curbside recycling services. Kingsport provides weekly curbside collection of both garbage and recyclables to more than 16,000 households, serving a total of 41,200 people. Since the beginning of Kingsport's recycling program, more than 5,000,000 pounds of recyclable material have been collected, averaging more than 100 tons per month. Residents can recycle paper, metals, plastic, and glass. Used oil can be taken to convenience centers for recycling.[23]

Many large regional manufacturers and office complexes have realized substantial savings by recycling cardboard and office paper, but this typically involves paying a hauling fee, which discourages smaller businesses and restaurants. Moreover, many of the more rural parts of the region offer minimal recycling opportunities, mostly for newspaper and plastics, which are accepted at some county convenience centers.

Nonetheless, much more can be done to reduce, reuse, and recycle. Americans still throw away enough aluminum to replace our entire commercial aircraft fleet every three months.[24] And there are some unsettling trends. Aluminum can recycling actually declined in 2001, falling below 60 percent—down 6.7 percent from 2000.[25] As we rush toward becoming a "fast food nation"—spending more time eating, working, and playing away from home, and tossing our trash into the nearest receptacle—recycling suffers neglect.

Yard "Waste"

The greatest progress in diverting waste from landfills has been achieved with yard waste (grass clippings, leaves, tree trimmings, etc.). In Knoxville, for example, yard waste, which used to be sent to the Chestnut Ridge landfill, is hauled to the Knox Ag facility just off Central Avenue, where it is converted to compost and sold inexpensively for lawn and garden use. This facility diverts about a hundred dumptruck loads of yard waste from the landfill each day. Knox County has started a similar yard waste mulch facility, which receives yard waste from residents at three of its convenience centers.

This is a commendable improvement, convenient for urban and suburban residents, but not an optimal solution. Its main disadvantages are air pollution and the consumption of nonrenewable fossil fuels. The hundred diesel truck trips a day needed to haul the material burn a lot of petroleum and add to regional smog—though, of course, the same number of trucks would have had to travel at least as far to take the yard waste to the land-fill. Large, polluting diesel motors are also used to operate the chippers that grind the yard waste into mulch. The composting operation itself releases methane, a greenhouse gas (though this effect is balanced in part by the fact that the compost is used to grow plants that remove greenhouse gases from the air). And, finally, the front-end loaders used to load the finished compost and the pickup trucks that customers use to haul it away both consume petroleum and emit noxious effluents.

Arguing that the total air pollution from this method is excessive, neighboring Blount County officials have adopted a different procedure: incineration. Yard waste from Blount County is hauled to Alcoa and dumped into a high-tech Pactherm air-curtain incinerator. The incinerator burns the organic matter thoroughly and emits (according to the incinerator's manufacturer) mainly carbon dioxide and water.[26] But this process, too, requires diesel trucks to bring the waste to the incinerator, and the carbon dioxide it generates is also a greenhouse gas. With respect to carbon, at least, incineration is directly counterproductive. There is too much carbon in the air already (see chapter 2). We should be removing it from the air and feeding it into the biotic system, not removing it from the biotic system and injecting it into the air. In the long run, moreover, burning yard waste worsens the problem of soil-nutrient depletion (discussed in chapters 4 and 6).

Yard "waste" is in fact not waste at all but nutritious organic matter—an essential commodity in the economy of life. Without such "waste" returning to the soil year after year, the land would eventually grow barren. When people remove leaves, twigs, and grass cuttings from their land, they interrupt the natural cycle of growth and decay. Growth continues for awhile, but the decay that replenishes the soil ceases and fertility is gradually lost. Eventually, to maintain healthy greenery, landholders must import nutrients—which all too often take the form of petrochemical fertilizers, the products of another long chain of polluting industrial processes.

This is folly upon folly. The only ecologically sound way to deal with yard "waste" is to compost it on site—that is, simply to pile it up and let it rot, thereby returning it to the soil. Everything necessary for home composting (bacteria, rainwater, and a little heat) is available everywhere free of charge, and the process is neither time-consuming nor difficult. Even apartment dwellers can compost indoors, using worm bins. The worms eat the food scraps and produce clean, sweet-smelling, nutrient-rich soil, which can be added to beds for shrubbery, vegetables, or flowers. To cycle organic matter through polluting transportation and industrial processes is, by comparison, enormously wasteful and inefficient. To bury it in a landfill or burn it is even worse. As things stand today, however, still an unacceptably large volume of our trash is simply buried in the ground.

Landfills

In sheer volume modern solid-waste landfills are among the largest structures ever created by human beings. The Chestnut Ridge Landfill, owned by Waste Management, for example, located in Anderson County, Tennessee, near the intersection of Interstate 75 and Raccoon Valley Road, has grown to the size of many of the surrounding mountains. Its eastern flanks—treeless, despite its name (chestnut trees having long ago died off anyway)—now loom over the interstate, seeded with close-mown grass, studded with groundwater-monitoring wells, and frequently crowned by a circlet of wheeling buzzards.

Because landfills are so big, because unused land has become increasingly scarce, because many places are geologically unsuitable for landfills, and (not least) because people will fight to keep landfills out of their neighborhoods, it is more and more difficult to find new landfill space. Yet, so far at least, it has been found. We are not in danger of running out of places to put our garbage any time soon. New landfills, however, are often less conveniently located than the older ones, forcing garbage trucks to travel longer distances to the dumping ground. The increased transportation distance entails an array of negative environmental consequences (see chapter 9).

Increased transportation also increases costs, both for the transportation itself and for "host fees" which some communities that have landfills charge other communities to dump there. Anderson County, for example, charges host fees to Knox County for depositing its trash at Chestnut Ridge. Some of Knox County's trash goes to private landfills in Loudon County and even to Athens. The landfill in Hawkins County, owned by BFI, accepts trash from all over East Tennessee. Some landfills are more local. The Solid

Waste Authority for Morristown runs a landfill just for Morristown and Hamblen County. Waste Management operates one for Johnson City. As in other parts of the country, however, most of our bioregion's landfills have become more regionalized, thus serving multiple counties.

We haul trash all around the region and even to other states. This creates additional air pollution and degrades roads. Once landfills close there are high closure costs and maintenance. Opening new landfills or expanding existing ones has significant social effects. Neighbors usually oppose them, and sometimes violence ensues. When Chambers Development Corporation announced plans to build the Shoat Lick Hollow landfill in the early nineties, residents quickly organized and took their concerns to the Anderson County Commission. After one commission meeting, an elderly man who had opposed the landfill was beaten by thugs, some wearing Chambers t-shirts.[27]

Opposition to landfills may exemplify the NIMBY ("not in my back yard") syndrome, but it is not irrational. Besides the obvious noise, odor, and aesthetic problems, landfills bring with them a high volume of dangerous truck traffic (and garbage trucks are not the most pleasant of trucks). Moreover, many landfills leak, contaminating groundwater with whatever toxic stuff oozes out of the garbage. Such contamination threatens the health of nearby residents who get their water from wells, and who may not want to move. More than 20 percent of listed Superfund sites, the country's most toxic and contaminated locations, are former municipal landfills.[28] At Roane County's landfill, which is located in the Midtown area, rain has carried trash from the landfill into the Tennessee River, and leachate has contaminated residential water wells. Monitoring wells have revealed mercury contamination of the groundwater.[29]

Some cities are doing much better. The Bristol, Virginia, Integrated Solid Waste Management Facility opened in 1998. This state-of-the-art facility includes a landfill, as well as yard waste composting operation. The city diverts its brush to the wastewater treatment plant, jointly owned with Bristol, Tennessee, where it is processed into wood mulch or combined with treated, dried sewerage sludge to produce a dry compost product. (However, the use of sewage sludge—which may contain a variety of toxic materials—as fertilizer or compost is controversial.) Stumps, bulk wood, and trees too large to compost are burned in a permitted incinerator on-site. The facility also features a recycling process to grind waste tires with

Fig. 26. *A junkyard pond in East Tennessee. Copyright by Mignon Naegeli.*

a multiple shredder into usable end products, such as materials for landfill construction.[30]

Yet, despite some real progress and despite Southern Appalachia's relatively cheap trash disposal, the entire bioregion is plagued by illegal dumps. Much of this surreptitious dumping is the work of individuals or small businesses seeking a way to discard construction debris, landscape waste, tires, and gasoline tanks—and leave the cleanup bill for the rest of us. Many of these dumps contain dangerous substances; all are ugly.

Then there is the still more pervasive problem of litter.

Litter

Along the berm of any highway, in the dust or mud of any urban alley, in gutters and parking lots, in fields, along streambanks, riverbanks, and the shores of reservoirs, on the most remote trails of the Great Smoky Mountains—anywhere, in fact, where people do not regularly clean up the land—lies the debris of a throw-away culture. Bottles, cans, soft drink cups, wrappers of endless composition and variety, bags (both paper and plastic), small plastic articles, bits of metal, broken glass, chewing gum, food remnants, old tires, cigarette butts innumerable—these are now as prominent or more prominent in the landscape than flowers or mushrooms. Future archaeologists will be able easily to identify soil layers that begin about the middle of the twentieth century by their large component of nondegradable trash.

A mile of an average highway contains about sixteen thousand pieces of litter—mostly cigarette butts. These are surprisingly durable. It takes about fifteen years for a cigarette butt to decompose. Styrofoam cups last ten to twenty years, plastic milk jugs fifty to sixty years, aluminum cans several centuries. Glass, though it may be broken into smaller pieces, leaves sharp shards that last practically forever.[31]

If natural environments lift the spirit, trashed environments depress it. The effect may not be measurable, but it can hardly be doubted.

Toxic Waste

In addition to municipal solid waste, the region generates large volumes of industrial, mining, processing, military and construction wastes. According to the EPA, "large quantity generators" reported generating 40 million tons of regulated hazardous waste in the United States in 1999. The five states that contributed the most to our nation's hazardous waste (65 percent of the national total quantity generated) include Tennessee, ranked fourth at more than 2.2 million tons. These figures do not include hazardous waste generated by small businesses, including dry cleaners and auto service centers. Nor does the figure include all the toxic waste generated by households, most of which ends up in regular landfills.[32]

In East Tennessee, Eastman Chemical in Kingsport, Yale Security in Lenoir City, Unisys Earhart Site in Bristol, and Cannon Equipment Southeast in Chattanooga are among the ten largest hazardous waste generators in the state.[33]

Sites contaminated with toxic waste are common throughout the region. Two Knox county sites, one in Heiskell, the other in south Knox County contain hazardous waste hauled there from Knoxville's Coster Shop redevelopment by Burnett Demolition and Salvage Company, a contractor working for the city. Studies conducted on soils and rainwater around the Heiskell site show elevated levels of diesel fuel, lead, thallium, polychlorinated biphenyls (PCBs), and other contaminants. Flooring blocks found at the site and traced to Coster Shop contain dioxins and other pollutants. The contaminated waste lies next to a tributary of Bull Run Creek and near several residences.[34] Burnett also dumped more than 800 truck loads of toxic debris from the Coster Shop into a sinkhole off Sevierville Pike in south Knox County, contaminating at least seventeen residential wells.[35]

In July 2002 the Tennessee Department of Environment and Conservation notified the city of Knoxville, Burnett Demolition and Salvage Company, and the project's environmental compliance contractor that they had violated state law. All three parties denied responsibility for the dumping, though they have agreed to help remove the toxic material dumped in the sinkhole. A regulatory investigation, a criminal investigation, and several lawsuits are in progress.[36]

Mines are another major source of hazardous waste. Waste may be generated from the mining process itself as the overburden (the soil and rock covering the ore) is removed. The slurry and wastewater from mining and mineral washing may also be toxic. Concentrating the ore creates additional waste, called tailings. Finally, smelting and refining may produce large amounts of slag. When toxic, these wastes can poison the soil and water. Even when not toxic, they can be so voluminous as to clog rivers and streams. The Nolichucky Reservoir south of Greeneville, Tennessee, has been almost completely filled by mine tailings.

Aside from coal, which is mined in large quantities on the western slopes of the upper Tennessee valley, there are mines in our bioregion for mica, olivine, crushed granite or limestone, marble, shale, clay (for bricks), gemstones, and feldspar. Sand is mined by instream dipping. Barite mining occurs in the Sweetwater area, zinc in the Cherokee Dam area, silica sand mining to the northeast. Shale is mined at the far eastern tip of Tennessee.

Mine tailings are normally stored near the mine behind dams (called tailing or slurry impoundments), which are usually constructed in hollows, depressions, or valleys and left as an after mining "new contour" of the land. Where mining is well-regulated, the surfaces of these dams are covered with topsoil and vegetated when the mine site is closed. Still, wastewater, tailings, and slimes sometimes find their way into surface water or groundwater if the impoundments are inadequately lined or if cracks develop in the embankments or underlying rock.[37]

Nuclear Waste

The most egregious dumping in our bioregion, and the most expensive to correct, occurred on the Oak Ridge Reservation during World War II and the cold war years that followed. At Oak Ridge nuclear weapons—and, later, commercial nuclear fuels—were produced by novel and sophisticated industrial processes that brought some of the brightest minds in the country to the region. This genius was mustered, however, in the service of urgent and immediate goals; less thought was given to the more mundane-seeming matter of waste disposal. Because Oak Ridge was cloaked in the secrecy of "national security," there were few to raise questions when orders were issued to take the stuff out back and bury it. So take it out and bury it they did—often in unlined trenches dug with a backhoe, sometimes penetrating the water table.

On the grounds of the Oak Ridge National Laboratory (ORNL), for example, there are at least six solid-waste disposal areas, all located in the drainage basin of White Oak Creek. An almost unbelievable variety of toxic and radioactive trash has been dumped there, contaminating the soil, the groundwater, the creek, White Oak Reservoir, the Clinch River, and Watts Bar Reservoir. The contaminants include biological wastes, mercury, toluene, naphthalene, and both high-level and low-level radioactive wastes contaminated with such radioisotopes as tritium, strontium 90, cobalt 60, cesium 137, and plutonium.

In the late 1950s ORNL mixed radioactive waste with concrete and injected it into fractured rock deep underground. This practice, called hydrofracture, continued until 1984, when it was discovered that groundwater near the hydrofracture sites had become contaminated. The current disposal practice at ORNL is to store waste above ground in concrete casks within concrete tombs.

Bear Creek Valley, just west of the Y-12 plant, was used for nuclear and chemical waste disposal beginning in 1951. From then until 1984, four unlined seepage pits, known as the S-3 ponds, were used for the disposal of over 2,700,000 gallons of liquid wastes. These wastes contained concentrated acids, caustic solutions, mop waters, and byproducts of uranium recovery processes. The pits were designed to allow the liquid either to evaporate or percolate into the ground. From there it flowed into the headwaters of Bear Creek and East Fork Poplar Creek. The S-3 ponds were closed in 1984 and paved over to create a parking lot.

Also located in Bear Creek Valley was the Oil Landfarm Site. A "farm" only in the twisted nomenclature of the bureaucrats' imagination, this thirteen-acre site was, in fact, a dumping ground for PCB-contaminated oily wastes, which were spread out and plowed into the ground. The oil also contained toxic beryllium compounds and radioactive uranium. This "farm" has been closed and covered with a waterproof cap.

Like ORNL, Bear Creek Valley also has burial grounds. These are located along the southern flank of Pine Ridge. They received PCB-contaminated oils, radioactive mop waters, radioactive asbestos wastes, and items contaminated with beryllium, thorium, and uranium. In 1961 an open-topped tank was placed in one of the burial grounds and filled with 180,000 gallons of various toxic liquids, which were then set on fire. After the burning, the residues were dumped into nearby trenches. In 1990, monitoring wells south of Burial Ground A revealed PCB contamination 274 feet below the surface at a concentration of 29,000 parts per million—almost three thousand times the acceptable limit. All the Bear Creek burial grounds are now closed and covered.

In 1951 the Atomic Energy Commission began to use the abandoned Kerr Hollow Quarry, off Bethel road, as a dumping ground for explosive materials. Containers of these materials were dropped into the quarry from the ridge above and then shot to release their contents, which sometimes exploded. Disposal continued at Kerr Hollow until 1988.

The K-25 complex southwest of Oak Ridge is dotted with more unlined open-trench burial grounds. For many years radioactive scrap metal was stored there in a twenty-two-acre site along the banks of the Clinch River. The river flooded in 1984, inundating this scrap yard and carrying radioactive materials downstream. The scrap has since been moved to higher ground.[38]

The most abusive practices on the Oak Ridge Reservation have ceased, and remediation is proceeding, but much of the reservation is likely to remain contaminated for the foreseeable future, a testimony to our monumental lack of foresight.

Moreover, some of the nuclear waste found its way well beyond the reservation. A particularly striking example is the case of David Witherspoon, Inc. For many years beginning in 1948, Witherspoon used three parcels of land along Maryville Pike in the economically depressed south Knoxville community of Vestal to process scrap-metal parts, many of which were contaminated with radioisotopes, asbestos, and various toxic chemicals. Much of the metal was sold to Witherspoon by the Atomic Energy Commission (AEC)—later, Department of Energy (DOE)—nuclear weapons facilities at Oak Ridge.[39]

Internal DOE memos obtained by the Foundation for Global Sustainability through the Freedom of Information Act also reveal that Witherspoon may have handled scrap contaminated with plutonium—one of the deadliest of radioactive metals. A memo dated 21 April 1969, states that "the purchaser [Witherspoon] should emphatically be made aware that the material he is contracting to handle does contain a plutonium potential and we cannot guarantee a specific level below which all the material will read."[40] A second memo, dated 4 June 1969, said that although most of the potentially plutonium-contaminated material sold to Witherspoon remained in Oak Ridge, "four or five pickup truckloads have been taken to Knoxville."[41]

Radioactively or chemically contaminated metal in the form of large pieces of equipment, pipes, parts, chips, or tailings was hauled to the Witherspoon scrap yards from Oak Ridge and elsewhere by truck and by rail. Using Geiger counters, neighborhood women working for slightly more than minimum wage—often with no protection other than gloves—sorted the radioactive metal by hand, placed it into barrels and carried it into a warehouse. This operation was performed in an area called the "hot field." Metals were identified by using a grinder to remove the outer layer of rust and dirt and then applying acid to see what color the metal would turn. The grinding operation undoubtedly released radioactive particles into the air. Dorothy Hunley, who worked at this job for twelve years, died in 1985 of osteogenic sarcoma, a form of bone cancer associated with ingestion or inhalation of radioisotopes. Her doctor thought it likely that exposure to radiation was the cause.[42]

Former workers also report that transformers shipped from various locations were smashed open with a wrecking ball in order to salvage the copper wire inside, and the liquid they contained (presumably consisting largely of PCBs) was spread on the ground to control dust. Neighbors and former workers tell stories of large pieces of heavy equipment being off-loaded from trains and buried. Tires—and just about anything else combustible—were piled up and burned in the open air. Tire fires continued sporadically, at least until the fall of 1992, when Witherspoon finally received some heavy fines for air pollution.

Through the years, Witherspoon resold a great deal of the contaminated scrap to the Knoxville Iron Company. A 1981 investigation by the Nuclear Regulatory Commission (NRC) found that Witherspoon had bought over two hundred thousand pounds of steel contaminated with uranium 235, never reported the shipment, and then resold the steel to the Knoxville Iron Company, all in violation of NRC regulations. NRC was unable to determine what happened to the steel after that.[43]

Sampling conducted by the state and by the Department of Energy has revealed widespread contamination, including radioactive contamination as high as 129,700 picocuries per gram.[44] The applicable state guideline demands cleanup at thirty-five picocuries per gram. In places where PCB-contaminated items were burned, the state has found dioxin, which suggests that residents were exposed to airborne dioxin (a severe carcinogen) while the burning was occurring. There are also excessive mercury and PCB levels both in the soil and in the sediments of Goose Creek, which flows through the site. The groundwater at the Candora site is contaminated with lead, mercury, chromium, beryllium, antimony, and several organic compounds in excess of state standards. Small amounts of plutonium have also been detected in the soil at the Candora site.[45] Concerning radiological and chemical risks to human health at that site, a 1996 Department of Energy study concluded, "In almost every current and future scenario, the EPA's target risk range is exceeded, indicating unacceptable current and future risk."[46] After many years of negotiations, the DOE finally agreed to removing surface contamination for the sites and expects the job to be completed by 2006.[47]

In addition to the waste produced by the manufacture of nuclear weapons, there is the high- and low-level nuclear waste that TVA creates in generating electrical power (see chapter 7). The nuclear waste problem is global. The world's 440 commercial nuclear reactors generate more than 11,000 tons of radioactive spent fuel—all at risk from terrorist attack and accidental leakage. The United States is host to almost one-quarter of these reactors. Some 161 million Americans live within 75 miles of an above-ground nuclear waste storage site.[48] Above-ground nuclear storage facilities are located at three places in our bioregion: the Department of Energy's nuclear weapons complex at Oak Ridge, the Watts Bar nuclear plant near Watts Bar Dam, and the Sequoyah nuclear plant on Chicka-mauga Reservoir.

A Healthy Economy

We live, it would seem, in order to consume; yet, too often the end result of that consumption is waste. The Gross Domestic Product (GDP)—a leading indicator of the "health" of our economy—is defined by how much our economy produces. In 2002 the United States produced $10.4 trillion of stuff and services. To economists, however, the absolute number is not so important; it's the rate of growth that really matters. If our purchases do not grow at least 3 percent each year then the economy is not considered "healthy."[49]

But economic health so conceived is often in conflict with spiritual, environmental, and bodily health. A growing GDP does not discern the difference between productive and destructive activities. It does not reflect the destruction of the environment or loss of resources. Indeed, destruction of the environment is reflected as a benefit through the growth of remediation businesses.

A genuinely healthy economy would exhibit sustainable integrity. It would be robust and hale, beautiful, and invigorating to the spirit. It would generate little or no waste. It would be self-healing. It would develop, but in a dynamic equilibrium. It would not impose its pattern on everything else. It would not grow without limit. In a healthy economy, the pursuit of material gain would be subordinated to deeper and more wholesome values: the love of family, friends, and neighbors; responsibility to community; work that is meaningful, healthful, and beneficial to the whole; and respect for a living world that is larger, more consequential, and more lasting than we are.

Chapter 9

Transportation

John Nolt with research assistance by Sarah Kenehan

Masters or Slaves?

It takes no science, no computers, and no satellite-borne instrumentation to detect the trouble wrought by our transportation system. Anyone can see it. The roads are widened and widened again. Still they fill with traffic, and still the traffic slows to a crawl. Asphalt proliferates across the land, baking in the summer sun. The farms are swallowed up, the mountains hemmed in. Noise and fumes are ubiquitous. The highway culture blurs local distinctions, making every place like every other. The pace of life constantly quickens. Cities, neighborhoods, homes, and individual lives shape themselves around the automobile.

It is not unreasonable to suppose that the average American spends about an hour a day driving or being driven somewhere. Over a lifetime of seventy-two years, that amounts to three years sitting in the confinement of a motor vehicle. Since the interior of a motor vehicle is scarcely more commodious or healthful than a prison cell, it is questionable whether these three years are well spent—even if the cell is equipped with a cellular phone.

To the years spent in cars and trucks, we might also add the time served working to pay for them—and for their financing, their upkeep, their fuel, their parking, their licensing, and their insurance. Between the automobile and its owner, then, we may wonder which is the servant and which is master.

The rewards of service to the automobile are, moreover, not always what we hope: hurling across the land in vehicles of steel, plastic, and glass, we routinely risk impacts that no human body is built to sustain, and many bear scars or sorrow as a result.

Yet many still uncritically regard the automobile as essential to freedom. Consider, for example, this excerpt from the 1994 *Tennessee Transportation Plan:* "Our society is based upon the principle that people must be free to choose among its opportunities. . . . Adequate personal transportation or mobility is one of the basic requirements for personal freedom. The provision of such mobility can be a significant means for enhancing the quality of life."[1]

There is no question that mobility enhances freedom. Yet our current transportation system reduces all forms of mobility (on land at least) to one: automotive transportation. It is now more difficult than in former times to get anywhere by walking, taking a train or streetcar, riding a horse, or biking—all healthier forms of transportation, in an ecological or in a human sense, than the automobile. Because cities are designed for cars, most people are compelled by the structure of the roads and the elimination of neighborhood schools and businesses to drive just about everywhere they go—and to pay all the costs associated with all this driving. Children, especially, must constantly be shuttled here and there, because so few places can safely be reached on foot. All things considered, it is doubtful that the automobile has made us more free.

Driving and Consumption

The statistics confirm what we know already: we drive too much and burn too much oil—and the quantities are constantly increasing. In 2002 Tennessee had 88,066 miles of roads,[2] 5.6 million registered automobiles,[3] and 4.1 million licensed drivers[4]—72 percent of the state's population.[5] Tennesseeans drove over 68 billion miles in 2002[6]—a distance equivalent to more than a 140,000 trips to the moon and back.

Knoxville, especially, has become a city of drivers. The average Knoxville household owns 1.84 cars. The national average is 1.65. Knoxvillians average 35.6 miles of automotive travel per day—more miles driven than the average resident of Atlanta or Los Angeles.[7] Metropolitan Knoxville is rated as the eighth-most sprawling metropolitan region in the nation;[8] largely as a consequence, Knoxville ranks among the ten worst cities in the nation for ozone pollution.[9]

Advances in engine design and the use of lightweight materials have made cars and trucks more fuel-efficient, but these advances have had to

Fig. 27. *A traffic jam on Interstate 75 south of Lenoir City, Tennessee. Copyright by Mignon Naegeli.*

compete against increases in population, driving distance, and vehicle size. The increasing popularity of large vehicles—SUVs, big pickup trucks, and vans—has, over the past decade, essentially negated simultaneous progress in engineering. While fuel efficiency for new passenger cars has risen slightly (from 28.4 to 28.6 miles per gallon from 1991 to 2001), mileage for new light trucks (including pickups and SUVs) has over the same period *decreased* from 21.3 miles per gallon in 1991 to 20.9 miles per gallon in 2001. Since population is steadily increasing and so are driving distances (largely because of escalating urban sprawl), total fuel consumption is rising relentlessly. Table 16 summarizes consumption for Tennessee over the past four decades.

This upward trend shows no signs of abating. In 2002 the U.S. General Accounting Office projected that highway travel in Tennessee would continue to increase at an average annual rate of about 2.29 percent over the next eighteen years.[10] Nationally, fuel consumption is expected to increase

Table 16

MOTOR GASOLINE CONSUMPTION ESTIMATES FOR TENNESSEE

Year	Consumption (barrels)
1960	26,468,000
1970	41,241,000
1980	54,446,000
1990	56,954,000
2000	68,252,000

SOURCE: U.S. Dept. of Energy, Energy Information Administration, "Transportation Energy Consumption Estimates, 1960–2000, Tennessee," accessible at http://www.eia.doe.gov/emeu/states/sep_use/tra/use_tra_tn.html (4 Aug. 2003).

at an average annual rate of 2 percent between 2001 and 2025[11]—a projection that assumes continued improvements in fuel efficiency.

Transportation accounts for about a quarter of our total energy usage,[12] but not all of the energy used for transportation directly powers cars and trucks. For every barrel of oil burned by motor vehicles, about a quarter of a barrel is used to produce, refine, and distribute the oil. Additional energy is expended in manufacturing and repairing vehicles and in building and maintaining roads.[13]

Output Effects

The transportation system exacts a heavy toll on health—environmental, personal, and spiritual. The most obvious environmental consequences are output effects—most notably those caused by the hot medley of pollutants and poisons that pulses from the exhaust pipe. Chief among these is the stifling gray smog that hangs over cities on still, hot summer days. Smog is a mixture of various noxious substances, many of which are generated by the exhausts of traffic. These include nitrogen oxides (NO_x) and volatile organic compounds, which contribute to ground-level ozone. Nitrogen oxides also react with moisture to form nitric acid, creating acid rain. In addition, automotive exhaust contains smoke particles, which contribute to respiratory disease (see chapter 2),[14] and carbon monoxide, which can damage the nervous system. At high concentrations, as in a closed garage, carbon monoxide is quickly lethal. Most of the carbon in the exhaust of a well-tuned motor vehicle, however, takes the form of carbon dioxide, a greenhouse gas.

Nationwide, mobile sources account for significant proportions of the various air pollutants: 81 percent of the carbon monoxide, 27 percent of the particulate matter, 36 percent of the volatile organic compounds, and 49 percent of the NO_x.[15]

Older, dirtier, and less-efficient cars contribute disproportionately to automotive pollution; 10 percent of the vehicles on the road produce 56 percent of the pollutants.[16] But a car's polluting level does not depend on its age alone. Poorly maintained newer cars pollute more than well-maintained older cars.[17] Many states mandate inspection and maintenance programs to control excess tailpipe emissions from year to year. Though such programs are clearly needed in Southern Appalachia—especially in Knoxville, which has some of the worst air in the nation[18]—no city in the bioregion requires emission testing.

Speed is an important factor in automotive air pollution. According to models constructed by the EPA, NO_x emissions increase gradually with speed above forty-four kilometers per hour (twenty-seven miles per hour) and much more quickly above seventy-six kilometers per hour (forty-seven miles per hour). Emissions of volatile organic compounds, by contrast, decrease with speed until about eighty-eight kilometers per hour (fifty-five miles per hour), but rapidly increase thereafter.[19] Fuel consumption also increases with speed. As speed is increased from fifty-five to seventy-five miles per hour, fuel consumption increases 30 percent.[20]

Driving with underinflated tires has a surprisingly large effect on fuel consumption. If the rolling resistance of tires is decreased by only 10 percent, fuel economy can increase by 3–4 percent. The Department of Energy estimates that underinflated tires waste a hundred thousand barrels of oil nationally each day.[21]

Exhaust is not the only problematic output of motor vehicles. The fluids in the gas tank, radiator, transmission, windshield washer reservoir, braking system, and oil sump are all, to varying degrees, toxic. As vehicles age, these fluids tend to leak onto pavement, from which they are washed by rain through storm sewers into rivers and streams. Older brake linings may contain asbestos, which sloughs off as brakes abrade and may accumulate in street dust. Asbestos exposure increases the probability of several forms of lung disease, including lung cancer.[22]

Cars and trucks also add significantly to the problem of solid waste disposal. Tires, used oil, and lead-acid batteries, and many other worn-out or broken automotive components, are difficult and costly to dispose of properly, and so are often improperly handled, at considerable environmental cost.

Input Effects

Some of the less-obvious effects of oil consumption come to light when we turn from output to input, starting at the gas station and tracing the input process backwards, toward the oil fields where it begins. As gas is pumped into the fuel tank, vapor escapes into the air. One significant component of this vapor is benzene, inhalation of which can cause cancer and leukemia.[23]

The fuel is pumped from underground storage tanks beneath the asphalt of the filling station. Sometimes these tanks leak, contaminating soil and groundwater. Dribbles from the gasoline hose or overflows from the fuel tank splash onto the asphalt and are eventually washed down the gutter and into the nearest river or stream.

The fuel comes to the filling station in tank trucks, which burn more oil and belch black diesel exhaust. The trucks themselves are loaded at petroleum storage depots, such as the one just off Middlebrook Pike in Knoxville. Fuel leaks have been common at these depots, and the soil beneath them is contaminated with petrochemicals.

The depots themselves are supplied by pipelines, trucks, trains, or barges from refineries where the fuel is distilled from crude oil. Much of the gas used in Southern Appalachia comes from heavily polluting refineries along the Gulf Coast of Texas and Louisiana—an area known as "cancer alley" because it is so polluted that it has the highest cancer rate in the nation. Because domestic supplies of oil are nearly gone, much of the crude refined in cancer alley arrives aboard supertankers from foreign oil-producing nations. These supertankers occasionally leak or break up—with familiar environmental consequences. And the extraction of the oil, whether it takes place in foreign lands, at home, or on offshore rigs, does still further environmental damage. All these environmental and health costs accrue on the input side, just to deliver the gasoline to the fuel tank. And there are similar input costs for each of the other fluids that we pump or pour into our cars—oil, brake fluid, antifreeze, etc.—as well as for all the other materials needed to make, repair, or maintain automobiles.

Other Costs of the Transportation System

There are also indirect health costs. People who drive or ride too often tend to grow fat and thus become more susceptible to a variety of diseases, including diabetes and heart disease—unless they make special efforts to exercise. But places suitable for exercise are typically distant or inaccessible to pedestrians, so to get there people drive.

Motor vehicles fill the air with noise. Levels of sound associated with traffic and road construction often reach nearly ninety decibels in cities— loud enough to cause progressive loss of hearing with prolonged exposure.[24] Traffic noise often drowns out the voices of birds or the sighing of the wind—an effect that is subtle, but perhaps not entirely irrelevant to spiritual health.

Other health effects are direct and unsubtle; motor vehicle accidents kill or maim hundreds of people and millions of animals in Southern Appalachia each year. Many of these accidents are due to speeding. But even when no accidents result, speeding takes its toll in frayed nerves. Rare is the driver who consistently stays within designated speed limits. Truck drivers, especially, are pressed to speed to deliver their cargo quickly. The newer rigs are lighter and longer than the older ones, so that they can carry more cargo at higher speeds. The sight of a monster truck suddenly filling the rearview mirror, then blasting by at seventy-five miles per hour, is now routine.[25]

Increases in fuel consumption are also costly to the economy and to national security. Domestic oil production peaked in 1970 and has been steadily declining ever since, as well after well runs dry.[26] Imports have long exceeded domestic production, and our dependence on foreign oil is only increasing. Economically, this addiction to foreign oil worsens the trade deficit. Strategically, it embroils us in the politics of oil-producing nations, making us targets for terrorism.

The largest remaining oil reserves lie in the Middle East. We have recently fought two wars there—each not unconnected with efforts to secure oil supplies—and our continuing military presence consumes both dollars and lives. A full accounting of the cost of our transportation system must therefore include the costs of war and occupation.

Offroad Engines

Cars and trucks are not the only culprits with regard to oil. All petroleum-fueled internal-combustion engines contribute to the problems described above. All, for example, pollute the air. Surprisingly, diesel emissions pose some of the most serious risks to human health. In 2002 the EPA announced that exposure to diesel exhaust can cause lung cancer.[27] This has sparked special concern for children who ride to and from school in diesel busses, many of them aging. Recent research suggests that these children may be exposed to dangerous quantities of diesel exhaust.[28]

Air pollution has been compounded by rapid increases in the use of construction equipment, forklifts, and other off-road vehicles, marine motors, and small machinery—leaf blowers, chainsaws, brush cutters, air compressors, snow blowers, lawnmowers, rototillers, weedeaters, and the like. The small motors of these heretofore largely unregulated devices have often produced more pollution (per motor) than the engines of cars and trucks. Recently the EPA has begun to phase in a series of regulations designed to reduce emissions from these engines.[29] But as with automobiles, advances in law and technology are vying for clean air against the countervailing pressures of increasing population and burgeoning engine use.

The Proliferation of Roads

Perhaps the most profound *environmental* effect of the transportation system is the proliferation of roads. We use more land for roads and for the parking, fueling, and repairing of automobiles than for housing. Nationwide, over sixty thousand square miles are paved, about 2 percent of the total surface area of the country.[30] Southern Appalachia has kept pace with

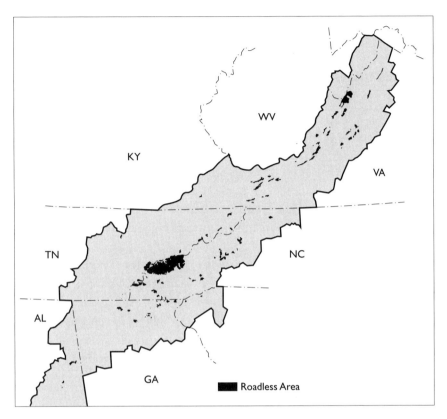

Map 3. *Roadless areas in the Southern Appalachian region. The big roadless area on the Tennessee/North Carolina border is the Great Smoky Mounatins National Park. (Dark outline designates the Southern Appalachian Assessment area, which is larger than the bioregion covered in this report.)*

this trend. The only large roadless areas in the bioregion occur within the Great Smoky Mountains National Park.

An area counts technically as "roadless" if it is relatively untouched and contains no more than half a mile of improved road for each thousand acres.[31] Hence an area that is technically roadless may contain dirt roads that allow motor vehicle access—and even some stretches of paved roadway. Yet even by this relatively lax criterion, there are no roadless areas at all in the upper Tennessee valley (see map 3).

The few places that remain unscathed by roads are confined to the roughest terrain, high in the mountains. These comprise about 3 percent of the land of Southern Appalachia—by far the greatest part of which lies within the bounds of the Great Smoky Mountains National Park. No other roadless area is even a twentieth the size of the park.[32]

The most important ecological effect of roads is habitat fragmentation. Large roads with heavy traffic flows are almost impossible for animals to cross. Thus, today migration and movement across the landscape are restricted to a degree unprecedented, so far as we know, in ecological history. And, just as the penetration of the skin opens the body to infection by microorganisms, so the penetration of the landscape by roads opens ecosystems to invasion by non-native organisms, furthering the homogenization of nature (see chapter 4).

Though roads are virtually everywhere already, the drawing boards of the planners are replete with many more. New areas of pavement accompany each new housing development or industrial park. Where traffic is most dense, new road projects are planned to relieve the congestion. But historically such projects have been only a temporary fix. New roads bring new development, which in the long run worsens the problem, and the cycle of road building and development rapidly gobbles up the land.

The Tennessee Department of Transportation is, for example, in the process of widening U.S. 321 from Townsend to Cosby along the northern border of the Great Smoky Mountains National Park. The predictable result will be intensification of development along the park borders, further isolating the park and destroying the buffer lands that help ensure its integrity. Between Gatlinburg and Cosby, there are places at which widened road will impinge on the park itself.

Even more threatening to the park is the proposed North Shore Road, a highway through the park along the northern shore of Fontana Reservoir. The controversial federal plan, which would destroy thousands of acres of park land, is currently (summer 2003) under review. If built, it would constitute a historically significant failure of our generation to defend besieged park lands.

Construction of the Foothills Parkway just north of the Great Smoky Mountains National Park is already disrupting the quiet and ecological integrity of the Smokies' northern foothills. As currently planned, this road would extend seventy-two miles around the northern and western borders

Fig. 28. *Landmovers widening Route 62 at the site of the University of Tennessee Arboretum near Oak Ridge. Copyright by Mignon Naegeli.*

of the park from Interstate 40 to Chilhowee Lake. Several sections have already been built; more are under construction. By further fragmenting the park's already severely fragmented buffer zones, the Foothills Parkway is constricting wildlife habitat and opening new routes for invasive species, impairing the park's function as a haven for native biodiversity. It is also degrading the quiet and historic quality of foothill communities.[33]

Roads imperil not only the Smokies but much of the rest of the biore-gion as well. A wide swath of south Knoxville, for example, is threatened by the big four-lane extension of the James White Parkway, formerly known as the South Knoxville Boulevard. Both names are deceptive; the road is de-signed not as a parkway or boulevard, but as an interstate-style freeway. The road was conceived a quarter century ago to shoot traffic from I-40 and I-75 more rapidly toward Sevier County and the Smokies—a function that since has been rendered largely superfluous by the construction of State Route 66 from Interstate 40 to Sevierville. Moreover, it is doubtful that either Sevier County or the Smokies needs more cars more quickly. The extension would conduct fifty-five-mile-per-hour traffic close by South Doyle Middle School (some of whose land would be taken), divide long-established neighborhoods, fragment one of Knoxville's few big stands of urban forest, and degrade one of the last remaining habitats of the endangered Berry Cave Salamander (see chapter 4).

Equally destructive are the proposed beltway loops around Knoxville, the extension of Pellissippi Parkway (another "parkway" in name only) into Blount County, and the portion of what is to become Interstate 26 that is being blasted through the mountains from Asheville, North Carolina, into upper East Tennessee. The latter will fragment and degrade (among many other places) the Pisgah National Forest, yet another buffer land for the Great Smoky Mountains National Park.

The National Environmental Policy Act (NEPA) was designed to prevent the kind of environmental degradation by highways and other federally funded projects that is now common across the bioregion, but it has largely failed to do so. One reason is that state departments of transportation have developed strategies to circumvent federal law. A standard ruse is to divide a single road project into a series of smaller projects in order to avoid environmental assessment of the project's cumulative effects. Often this strategy is combined with the claim that no federal funds are being spent on one or more of the project segments in order to evade NEPA requirements on that segment. The Tennessee Department of Transportation has used both strategies to circumvent federal regulations in the widening of U.S. 321 and the construction of the Pellissippi Parkway south and west of Knoxville.

Unfortunately, the roads mentioned in this section are merely examples; to describe all of the bioregion's environmentally detrimental road projects would take a chapter in itself. The asphalt network is expanding everywhere and contracting nowhere, and with each new expansion something of value is lost.

Asphalt

Whether it is homes, woods, mountains, wetlands, farms, or fields that are bulldozed to make a road, what is laid down in their place is usually asphalt. Asphalt surfaces, while well suited for cars and trucks, are poorly suited for just about everything else. Biologically, they are dead zones, fatal to almost all organisms that venture onto them.

From a hydrological perspective, their most important property is imperviousness to water. Nearly all soils are porous and capable of quickly absorbing water from rain or melting snow. Vegetated soils can absorb even more, discharging the water slowly into creeks or underground water systems. But asphalt absorbs no water. Rain or melting snow rushes off in sheets, quickly swelling streams and increasing the potential for floods.

Runoff from asphalt is contaminated with whatever drips, falls, drifts, or runs onto the pavement: dirt and dust, rubber and metal deposits from tire wear, asbestos from brakes, antifreeze, grease, gasoline, engine oil, road salt, heavy metals, and litter.[34] Toxic oils may also seep from the asphalt itself, which is composed largely of sticky residues from the oil-refining process.

When first laid down, hot asphalt releases a noxious vapor known as "blue smoke." This vapor contains two important precursors of tropospheric ozone formation, nitrogen dioxide (NO_2), and volatile organic compounds (see chapter 2). Prominent among the volatile organic compounds in blue smoke is benzene, inhalation of which may cause cancer, leukemia, and suppression of the immune system. Other carcinogens—including benzo-(a)pyrene, benzo(k)fluoranthene, ideno(1,2,3-cd)pyrene, and naphthalene—are present in smaller amounts. There is evidence that some components of blue smoke—e.g., fluoranthene and benzo(a)pyrene—interact synergistically, becoming more carcinogenic in combination than in separate exposures.[35] Given these facts, it is troubling that road-construction workers and others who are regularly exposed to asphalt fumes almost never wear respirators.

Barges

Roads are not the only means of transportation in our bioregion. There are also about 652 main-channel miles of commercially navigable waterways along the Tennessee River, beginning a mile above Knoxville and ending at the Ohio River in Paducah, Kentucky. These include secondary channels that extend 21 miles up the Hiwassee River, 61 miles up the Clinch River, and 29 miles up the Little Tennessee River.[36] The barges that use these waterways are the most efficient means we have of moving large quantities of raw materials, including logs, wood chips, asphalt, coal, and coke, at low cost.

River traffic provides access to international markets via the Tennessee–Tombigbee Waterway, a 470-mile-long canal extending from Pickwick Reservoir on the Tennessee–Mississippi border to Mobile, Alabama, a port on the Gulf Coast.[37] Before the opening of this waterway in 1985, barges could reach the Gulf only by traveling the length of the Tennessee River to its juncture with the Ohio and then down the Ohio and the Mississippi to New Orleans. The "Tenn-Tom" reduced the distance to the Gulf by 882 miles and greatly increased barge traffic.[38] Net tonnage of river freight traffic grew from 26,006,000 tons in 1981 to 45,374,400 tons in 2001.[39]

Barges use fuel efficiently and are a relatively minor source of pollution. Thus, given that a system of locks and canals is already in place, they have important advantages over other less-efficient and more-polluting forms of transportation—especially truck transportation. Barge traffic has, however, one grave drawback from an ecological point of view: it facilitates the exploitation of Southern Appalachian forests (see chapter 4). The advantages of any attempt to increase river traffic must therefore be weighed against the disadvantage of an increasing assault on regional forests.

Trains

Civilization is shaped in part by the possibilities of transportation, and transportation in turn is sculpted by regional geography. The early economic development of West Tennessee, for example, was based largely on Mississippi river traffic. But the swifter and shallower rivers of Southern Appalachia, though serving as conduits for settlement, were not easily navigable and did not play as large a role in economic development. The mountainous terrain of Southern Appalachia became economically accessible only after the introduction of railroads.

By the beginning of the twentieth century, the United States had the most extensive railroad system in the world. But now, a hundred years later, the railroads have declined and extensive sections of track have been abandoned or dismantled. No longer, for example, can tourists ride the Smoky Mountain Railway from Knoxville to the Smokies, as they could earlier in this century. The track is gone, many of the rights-of-way no longer exist, and even the old railbed is becoming difficult to discern as it winds its way through a bulldozed and rebulldozed countryside.

Yet, from an environmental standpoint, trains are superior in almost every respect to cars and trucks. Railroad tracks generally do not fragment habitats as severely as do big roads, or the canals and reservoirs that support barge traffic. Animals can cross railroad tracks more easily than they can cross big highways or waterways. And trains are more efficient and less polluting than either cars or trucks. As Gus Welty points out, "If railroads are used as a baseline of 1, studies have shown trucks coming in at 13.6 on diesel particulates, 12.5 on hydrocarbons, 3.7 on nitrogen oxides, and 3.6 on carbon monoxide. It is much the same when fuel efficiency is considered, with railroads producing about 2.5 times the ton-miles of trucks per gallon of fuel."[40] Yet trucks predominate in the transportation of goods, and (except for tourist-oriented sight-seeing trains in Knoxville and Chattanooga) there is, so far as we are aware, no passenger rail service anywhere in our bioregion.

Tennessee has, however, recently begun to study the feasibility of an intercity passenger rail system, using the three thousand miles of active rail lines that remain.[41] The plan would make Tennessee a member of a multistate rail coalition (Tennessee is surrounded almost completely by participating states), allowing for rail travel throughout the country. Most of the major cities in Tennessee are already linked by functional railroad track (primarily used for shipping freight), except for thirty-one miles of track that would be needed to allow a direct connection between Knoxville and Nashville.[42] Service from Bristol, extending east into Virginia as far up as Washington, D.C., may soon be the first step in initiating passenger rail in Tennessee.

In addition to passenger transport, the new rail system would also be capable of transporting enormous amounts of freight. This addition would take a huge burden off the interstate highways, as the rail system would lure companies to ship freight via rail rather than by trucks. Trucks compose 41 percent of the vehicles that travel on I-81.[43] With the improvements that the proposed rail plan would make, about a thousand truckloads of freight could be diverted from I-81 each day.[44]

Not only would an accessible, statewide rail service lessen stressful traffic congestion, but it would also help to improve air quality, as well as helping to reduce the need to build new highways and expand existing ones. These advantages are coupled by the large economic advantages that implementing an intercity rail system would bring. For example, the costs of building a new freeway lane usually average around $8 million per mile. However, the costs to install a mile of new railroad track are substantially lower, averaging only around $1 million per mile.[45]

On a smaller scale, light rail lines could help relieve traffic congestion in crowded urban areas. The average daily traffic count at Interstate 40 at Hollywood Road in Knoxville, for example, has skyrocketed, rising from 24,170 in 1967[46] to 140,167 in 2001.[47] The inevitable consequence is frequent, anxiety-provoking slowdowns and traffic jams. If a light rail line or some other form of public mass transportation were built in this area—say, between Knoxville and Oak Ridge—traffic, transportation times, and fuel consumption would all decrease, the air would become cleaner, and West Knoxville's way of life would become palpably more sane.

Toward a Healthy Transportation System

Any transportation system encumbered with the side effects discussed in this chapter is destructive to both human and ecological health. The challenge, then, is to construct one that is healthier.

The single most important variable in the whole thorny transportation problem is how much people drive. Even the most progressive technologies imaginable (solar-powered electric cars—or cars powered by natural gas, fuel cells, or hydrogen) can only reduce pollution and oil dependence; they cannot remedy the broader social and ecological damage caused by the proliferation of traffic and asphalt. We will not, therefore, achieve a healthy transportation system until we can reduce the number of motor vehicles on the road and the number of miles they travel. Such a reduction, however, runs contrary to the inertia of historic forces. It will not be easy.

The most important of these forces is the flight from cities that is creating suburban sprawl. Unlike the farm homes that they have replaced, suburban homes lack any semblance of self-sufficiency. Their existence depends on a constant inflow of products, services, and cash and a constant movement of people to and from other places: stores, schools, recreational facilities, and places of employment. And, unlike traditional urban homes, suburban homes are generally located far from these other places. People who cannot provide for their own needs at home and who live far from other sources have no other option: they are forced to drive—far and often.

The waning of non-automotive forms of transportation is due in part, of course, to the distances created by sprawl. If you live twenty miles from where you work, you cannot walk there and back every day. But such distances are no hindrance to trains or buses. What prevents *their* use is a general disorganization and lack of planning.

While twenty miles is near the outside limit for bicycle commuting, five or ten miles is a practical distance for a well-conditioned biker, and up to about two miles is practical for walking. Yet even people who commute short distances usually drive. According to 2000 census data, in Knox County, for example, 156,194 workers usually drove to work alone and 17,017 carpooled, but only 4,113 usually walked and only 1,164 usually road a bike.[48]

One of the main impediments to cycling or walking is the road system itself, which is for the most part ugly, dangerous, and full of obstructions for pedestrians or cyclists. That problem, however, has a remedy: comprehensive networks of sidewalks, bike lanes, and greenways in all urban areas.

Greenways, especially, can enhance not only possibilities for healthy transportation, but also the beauty, air quality, peace, and habitat potential of cities. A greenway is a linear park, often containing a path for walking, jogging, biking, or horseback-riding, that serves to link together more traditional parks, schools, residential areas, businesses, or other important destinations.

Greenways already exist in Chattanooga (along the riverfront and Chickamauga Creek), Knoxville (along the waterfront and up Third Creek), Knox County (between Knoxville and Oak Ridge along the Pellissippi Parkway), Maryville, Sevierville, Townsend, and elsewhere. The region's most visionary greenway scheme is the Great Smoky Mountains Regional Greenway System, a plan to link Knoxville to the Smokies and various other regional locations, including Gatlinburg and Newport, via an interconnecting greenway network. If the plan—which is supported by all the relevant local and federal agencies—can be funded and implemented, it will become possible to bicycle between these locations safely, enjoying the green countryside, and avoiding the nerve-wracking traffic snarls that plague Sevierville, Pigeon Forge, and Gatlinburg.

Still, it is unlikely that large numbers of people will soon forgo the comfort and convenience of the private automobile, either for healthy muscle-powered transportation or for somewhat less-convenient public transportation. The transition to a healthy transportation system will take time, for it will require changes in the way we think and, more important, in the way we live.

Suburbanization, as was noted earlier, has two consequences that practically necessitate the automobile: the decline of self-sufficiency and the increase in driving distances. Long-range solutions to the transportation problem must address one or both of these causes of automobile dependence.

To address the first is to make home a more self-sufficient place—a place which we need not always leave to obtain food, clothing, education, recreation, or work. The farmers who first settled this land—and the Native Americans who preceded them—did not need cars, because they could grow, find, or make virtually everything they needed close to home. Thus one approach to transportation troubles is simply to stay put.

Staying put, of course, cannot now take the forms it took in earlier days. But the principle is still applicable. Some people, to take a high-tech example, can now "commute" electronically, performing their work tasks through the internet from their homes. A simpler instance of the same principle, applicable to virtually everyone, is to plan automobile use carefully, reducing the number of shopping trips by bulk buying, for example, or by eliminating trips to purchase one or two small items.

A more potent way of staying put is to move toward independent living. People who grow some of their own food or make some of their own necessities are likely to spend more time at home and need less from outside. They are also, if part of a family, less likely to need two paychecks, and so to have still less need of automotive transportation. Such efforts toward homemaking and simple, independent living, though they may seem unrelated to the transportation problem, are at the heart of its most hopeful and healthful solutions. They will be further discussed in chapter 11.[49]

A second approach to the transportation problem is to reduce the distances created by sprawl. Ideally, homes should be within safe walking or biking distance of places of work, shopping, education, and recreation. Where this cannot be achieved, they should be reachable by inexpensive, swift, and reliable mass transportation. Since individuals have little control over the organization of cities, solutions on this scale can be achieved only by collective political effort. But there is no reason, in principle, why such an effort could not, by ecologically and socially sound planning, reshape the senseless and unsightly crazy quilt of the urban and suburban landscape, restoring a measure of efficiency, beauty, and reason.

Chapter 10

Future prospects

John D. Peine

So Who Cares

Looking to the future is not a popular sport. Most people are uncomfortable thinking about the future. Haven't we got troubles enough? Who knows what's going to happen anyway! To quote Alfred E. Neuman, "What, me worry?" Visionary leadership is all about looking to the future and taking action today to prepare for that future. Bruce Babbitt, former secretary of the U.S. Department of Interior, wrote in his foreword to the book *Ecosystem Management for Sustainability: Principles and Practices Illustrated by a Regional Biosphere Reserve Cooperative:*

> The President of the United States' Council on Sustainable
> Development recently stated that human survival may well
> depend on widespread acceptance of principles of sustainable
> living and ecosystem management. When we act locally, we
> must think globally. We must also think holistically, considering
> the relationships between Nature and economic, cultural, and
> spiritual life of our communities and societies. As the 21st
> century and the next millennium approach, the need to put
> these principles into practice has never been more obvious.[1]

Bruce Sterling in his new book, *Tomorrow Now: Envisioning the Next 50 Years,* paints a vision of the future world in which people are different, in some cases *really* different.[2] He indicates that biotechnology will "drive a crossover effect between genetics, cybernetics, and cognition studies. . . . Once we industrialize those processes, all bets are off."[3] This chapter offers a look to the future environmental conditions in the Southern Appalachians highlands in which some of these *super* humans, along with our children, will be living.

The Importance of Learning from the Past

It is instructive to look to the past for perspective when contemplating the nature and enormity of future change.[4] The Southern Appalachian region is internationally recognized as one of the most intact temperate forest biomes on the planet. The region has been designated by the United Nations as an International Biosphere Reserve and the Smoky Mountains National Park a World Heritage Site. This seems to imply that not much has changed in the regional ecosystem.

Nothing could be further from the truth. Since the end of the last ice age approximately eight thousand years ago, the evolution of forests has been dramatic. The Appalachian Mountains provided a refuge and movement corridor for northern biomes during that period of rapid climate change, thereby preserving biodiversity on the eastern North American continent. The spruce-fir forest retreated from the Piedmont to the peaks of the Appalachian Mountains during this period of warming.

The earliest human inhabitants were Paleo-Indians, who were predominantly hunters and gathers. By 2500 B.C., most tribes had incorporated some degree of agricultural activity.[5] Once Native Americans adopted agriculture into their subsistence lifestyles, contiguous forests began to give way to clearings for agriculture. Eventually, fire became a commonly utilized management tool to clear land for agricultural activity and promote grassland habitat to attract and increase populations of game animals. Much of the pine forests distributed throughout the Southern Appalachians today are remnants of land management by Native Americans.[6]

The complex cultural framework that had evolved during the Mississippian cultural period that featured mound builders was dramatically altered by the introduction of European diseases such as smallpox, diphtheria, whooping cough, bubonic plague, chicken pox, influenza, and others by early European explores and settlers.[7] One of the most notorious among them was the Spanish explorer Hernandez de Soto who traveled throughout the Southeast (during 1539–43), including Southern Appalachia, searching for gold. The cumulative massive loss of life from these

European diseases dramatically altered Native American culture. The Joyce Kilmer Memorial Forest, on the route that de Soto traversed, has an even-aged stand of five-hundred-year-old tulip popular trees growing on a Native American ceremonial ground. This early-succession tree species began growing on the abandoned ceremonial ground soon after de Soto's visit.[8] Is this a coincidence?

European settlers accelerated land conversion from forests until, by the end of the nineteenth century, the landscape was largely devoid of mature forests due to agriculture activity, clear-cutting to produce charcoal to fire iron smelters, and commercial timber harvest for lumber. By the late nineteenth and early twentieth centuries, coal mining, centered in eastern Kentucky and western West Virginia, became highly mechanized, resulting in an enormous ecological disaster covering thousands of square miles. As building materials, energy sources, and land-use priorities changed, reforestation throughout the eastern United States became the greatest environmental recovery of the twentieth century in North America. A significant contributor to this recovery was the federal government, which purchased over six million devastated acres of land during the Great Depression to begin reforestation and create national parks and forests throughout the Appalachian Mountains. The vision for creating national forests in the eastern United States came from President Theodore Roosevelt. Fire suppression had been practiced on these federally owned lands until very recently, resulting in the unnatural buildup of fuel material increasing the potential for intense fires during drought periods. As a result, fire-dependent and fire-enhanced species have been in decline.[9]

Reflecting back on this litany of environmental changes, there is a central theme of environmental resiliency, of cycles of devastating disturbance and recovery. As we look to the future, there are numerous new stressors that greatly reduce the potential for the environment to recover to a previous "natural" state but rather to evolve to a dramatically different and alien future. By looking to the past and future, stewards of environmental conservation, which includes all of us, will gain invaluable insight into the need to take *definitive action now* to sustain the *natural* environment into the future.

Drivers of Environmental Change
and Mitigation Strategies

To understand the future, we need a clear view of the past and present. The preceding chapters aim to provide that. But to *predict* the future—especially as it involves ecosystem dynamics—we must identify and describe the drivers of change. These may exert their influence at the biome, ecosystem, or species levels. This chapter concerns those drivers most likely to damage the natural environment in the Southern Appalachians and strategies to mitigate their impacts. The real challenge in this effort is to estimate their *collective* influence.

Global Warming Problems

Climatic diversity due to slope, aspect, and elevation gradients is a primary driver contributing to species richness and biodiversity occurring in the Southern Appalachian highlands. This region has been designated an International Biosphere Reserve and Great Smoky Mountains National Park a World Heritage Site, largely due to its world-renowned species richness and biodiversity. During the recurring ice ages of the Pleistocene epoch, the Appalachian Mountains provided a refuge for northern realms of biota. Eastern North America has more tree species than western Europe, largely because the Appalachian Mountains provided a corridor for movement of species during periods of climate change, whereas in western Europe the Mediterranean Sea blocked such movement.[10] During the last ice age, which ended about 8,000 years ago, spruce-fir forests covered the Piedmont region in the Southeast.[11] This forest has now retreated to the Southern Appalachian highlands above 5,000 feet,[12] and its co-dominant Fraser fir trees occur nowhere else. The degradation of this forest foreshadows the probable future of the Southern Appalachian highlands. Like a canary in a coal shaft, the spruce-fir forest is an early indicator of the potential for extreme decline in environmental health.

In 2002 the National Assessment Synthesis Team for Climate Change reviewed literature for regional climate variability and change. For the southeastern United States, the National Assessment Team utilized results of the two principal climate models, the Canadian Model and the Hadley Model, to predict impacts. For the Southeast, the predictions of these models differ: "The Hadley model projects a warmer and slightly wetter future, with serious but manageable impacts; the Canadian model projects a much warmer and drier climate, with almost calamitous effects."[13] But they also have much in common. Both predict warmer temperatures in the region over the next century. Both predict an increased risk of drought and flood, which will alter water management and usage. And both predict more frequent and extreme changes of weather (violent storms, variable periodicity of frost-free days, etc.).

A shift in hydrological cycles with periods of extended rainfall and drought would affect the distribution of brook trout, as well as other biota, by reducing suitable aquatic habitat. According to the National Oceanic and Atmosphere Administration, El Niño and La Niña are both responsible for global weather-related changes due to prolonged periods of above-average sea surface temperatures.[14] For example, the Southern Appalachian region experienced abnormally dry conditions during the winter of 2002 as a direct result of El Niño.

Climate-sensitive ecosystem dynamics. The predicted increase in climate variability will stress native species and increase the influence of ecological stressors such as pests, pathogens, and exotic species invasions. The ramifications of these changes may, in the long term (over the next fifty to two hundred years), be globally devastating to human society and to ecological biomes The most vulnerable species will be those that are climate-sensitive, endemic, rare and threatened, or imperiled by pests, pathogens, or exotic species.[15] Again, the spruce-fir ecosystem is a prime example of how the *collective effects* of stressors can devastate an ecosystem.

Climate-sensitive species. Most of the endemic and rare vascular plants that contribute to the biodiversity of the Southern Appalachians are found in the high-elevation forests and balds. An example of a less than obvious native species that is vulnerable to climate change is the endemic southern strain of brook trout. This species inhabits second- and third-order mountainside streams, primarily because the lower-elevation streams are continuously stocked with exotic game fish (such as rainbow trout) that out-compete with the native trout. During droughts, the flow rates of these upper-elevation streams are greatly reduced, imperiling the reproductive cycle of the fish.[16] Moreover, during violent storms, these fish risk being flushed by fast-flowing water to lower-elevation rivers, where they face the jeopardy of competition. The twenty-three species of salamanders in the Southern Appalachians are also sensitive to forest understory moisture regimens and/or stream dynamics. Even black bears are vulnerable. In 1998 a late frost killed oak blossoms, resulting in a fall acorn crop failure that drove hungry bears out of public lands and into the built environment, where they were often killed by hunters or traffic. Some became habituated to humans and human-source food.[17]

Global warming trends, drivers, and consequences. The Intergovernmental Panel on Climate Change has concluded that contemporary rapid global climate change is occurring, largely as a result of human activity.[18] Global temperature data from NASA's Goddard Institute for Space Studies shows that the first 11 months of 2002 averaged 58.37 degrees Fahrenheit, just under the record 58.44 degrees Fahrenheit set in 1998, the warmest year on record.[19] Global warming is clearly a key driver, and probably the *preeminent* driver of environmental change in the next fifty to one hundred years. Current global warming trends will continue and accelerate unless greenhouse gas emissions are dramatically reduced. Within fifty to one hundred years, they are likely to become irreversible.[20]

Global warming effects on humans. Climate change will threaten human health and safety and decrease the availability of natural resources for human use. More violent weather will increase the likelihood of flash flooding, which has always been a problem in the Southern Appalachian highlands.

Gatlinburg, Tennessee, the primary gateway community to Great Smoky Mountains National Park, is a good example of a climate-related disaster waiting to happen. The West Prong of the Little Pigeon River flows through the center of town, which is situated at the foot of the steepest watershed in the entire Appalachian Mountain range. During the height of the summer tourist season, the traffic is frequently gridlocked, both in town and inside the park along a nearby road following the edge of the Little Pigeon River. An intense and lingering storm in the mountains has the potential to create an intense flash flood. An intense storm occurred in the Great Smoky Mountains National Park in 1998, causing over ten million dollars in damage[21]—an amount equal to the park operations budget for that entire year—and closed the Little River Road, a primary access road to the park.

In addition, periods of extreme drought could result in intense forest fires, like those that became so common throughout the West during the very dry summer of 2002. There is precedent; in 1988 a prolonged drought resulted in forty-eight forest fires occurring in Great Smoky Mountains National Park.[22] The forested lands surrounding the park contain a plethora of rental cabins and tourist resorts in the forested landscape surrounding the national park. This potential for human disaster will intensify as development density grows exponentially and climate variability increases in the future.

Global Warming Solutions

Global commitment. Reducing global climate change requires a global commitment to reducing emission of greenhouse gasses. The Kyoto Treaty on Climate Change, for which the original negotiations were completed in 1997, is a remarkable example of worldwide recognition of the seriousness of the threat of global warming to human society and the natural environment.[23] This treaty is about multinational leadership, temporary economic sacrifice, and visionary thinking for a sustainable future. Though understandably less than all-inclusive, the treaty provides specific goals requiring real sacrifice to reduce greenhouse gas emissions to 1990 levels.

More is needed than the reductions called for in the Kyoto Treaty but it is a substantial first step in addressing the problem. Strategies to attain the goals of the Kyoto Treaty include energy efficiency and conservation, conversion to non-fossil fuels via new technologies, and sequestration of carbon dioxide by such means as sustainable forestry. As of this writing, 109 countries have approved the treaty.[24] The United States signed the Kyoto Protocol in 1998 under the Clinton administration, but the Bush administration has since withdrawn approval, and the United States is now the only large industrialized nation that is still withholding its assent.

Leadership. In the Southern Appalachians, many of these dimensions of responsible energy policy have been adopted in a limited way. Some examples include the following.

Southern Alliance for Clean Energy (SACE). This nonprofit organization conducts programs concerning the environmental, public health, and economic impacts of energy policies on citizens in the Southeast. In addition, an education fund provides a forum for public discussion of these issues and advocates for energy plans, policies, and systems from Southeast utilities that best serve the social, economic, and environmental interests of

Southeast communities. SACE helped TVA launch the Southeast's first renewable energy program, Green Power Switch, on Earth Day 2000. Green Power Switch gives customers a chance to support the development and generation of cleaner energy for a minimal monthly participation fee. Businesses and individuals in Gatlinburg, Tennessee, are the leaders in making the Green Power Switch in the Southern Appalachian highlands.

Alternative energy vehicles. The United States spends about two billion dollars a week on oil imports, mostly for transportation fuel. Vehicle emissions are the leading source of U.S. air pollution, which jeopardizes the environment. In response to growing concern over air pollution and reliance on imported oil, the U.S. Department of Energy has been working with automakers and industry partners to develop technologies that reduce emissions and that are powered by abundant and renewable resources. There are a wide variety of tax incentives offered at the federal, state, and local levels to encourage the expanded use of alternative fuel vehicles.[25] The Great Smoky Mountains National Park is considering purchasing alternative fuel vehicles as a demonstration/education project.

Climate sensitivity analysis. Federal land managers in the Southern Appalachian highlands should be more cognizant of how climate affects natural and cultural resources. All national parks and forests should conduct climate sensitivity analysis of natural and cultural resources in order to understand what is most vulnerable to climate change. Such an analysis would help to target limited funds for environmental monitoring to species and communities vulnerable to climate change and its effects. Most endemic, rare, and/or threatened species cannot easily adapt to climate variation. Such an analysis should include the following:

- Identify ecosystems, ecotones and species that are particularly sensitive to climate conditions.

- Evaluate national park and forest resource monitoring programs as to their capability to detect ecosystem dynamics associated with climate change.

- Utilize results of the analysis for education materials and services.

- Devise a research and monitoring agenda concerning climate change on federal lands and throughout the Southern Appalachian region.[26]

Climate models are being utilized to predict the probability of an increase in violent storms in the Southern Appalachian region. The National Climate Data Center, an affiliate of the National Oceanic and Atmosphere Administration, has a branch office located in Asheville, North Carolina. The insurance industry is a leading financial supporter of climate change research because it sharpens predictions of risk to humans and the built environment. Such research makes a compelling case for all stakeholders to pay more attention to climate change. The Federal Emergency Management Agency is utilizing this information to assess its preparedness and to formulate preventive measures to avoid natural disasters. Natural resource managers should do the same.

Bottom line. Time is fast running out to reverse the trend of global climate change driven by human activity. No one is sure when the trend becomes irreversible, but it is likely to occur within the next one to three human generations.[27] So far, climate change has been a nonissue among many politicians and citizens in the Southern Appalachian region. For many "leaders," instant gratification rules; sustainability is for the next generation to worry about. The real leadership on effectively dealing with this *critical issue* is coming from all around the earth. Progressive leaders understand the urgency of definitive action at all levels of government, corporations, small businesses, and nongovernmental and religious organizations. British Prime Minister Tony Blair, for example, has stressed the need for an "international consensus to protect our environment and combat the devastating impacts of climate change." "Kyoto is not radical enough," he said. "Ultimately this is about our world as a global community. . . . What we lack at present is a common agenda that is broad and just. . . . That is the real task of statesmanship today."[28] *Follow these leaders!*

Human Population Growth and Land-Use Problems

Rate of growth. Human population growth is accelerating in the Southern Appalachian region. In the 1980s the majority of visitors to Great Smoky Mountains National Park came from the midwestern states.[29] Now, the majority of visitors are from the Southeast, particularly Florida. This reflects the dramatic shift in population in the eastern United States from the Northeast and Midwest to the Southeast. Population growth in the Southern Appalachian highlands is being driven primarily by the tourism industry and by retirees seeking escape from urban/suburban sprawl to a highly scenic and culturally rich landscape. Sevier County, Tennessee, the epicenter of the tourism economy in the Southern Appalachian highlands, is a hotspot of this growth. The county population grew 46 percent from 1990 to 2000.[30]

Sense of place. The Southern Appalachian highlands are one of the most scenic landscapes in North America. The landscape exudes a distinctive sense of place grounded in rich endemic early European American and Native American culture. Rapid growth is disrupting the scenic integrity of this landscape and marginalizing its indigenous culture.

Cultural implications. Development dramatically increases real-estate values. In hotspots of growth, local families of modest income can find it difficult to maintain their lifestyle. Not only is their buying power reduced, but their culture is eroded as well. These people are more likely to be the ones working in the tourist industry on a seasonal basis, earning low wages with no health insurance. Affordable housing becomes a problem as well.

Land-use conversion. Rapid population growth leads to expansive land-use conversion from agriculture and forests to the built environment. Over the last twenty-three years, forestland conversion to agriculture or urbanization in the southern United States totaled twenty-five million acres.[31] The patterns of development are influenced by the proximity to metropolitan areas and roads, topography, and riparian areas.[32] Most of the land in the region remains rural with minimal land-use change, but hot spots of population growth such as the metropolitan areas of greater Knoxville, Tennessee, and Asheville and Hendersonville, North Carolina, have recently experienced rapid land-use conversion. The impacts of development on natural resources are dramatic. Development adjacent to Great Smoky National Park is steadily creating a biological island effect. Particularly in Gatlinburg and Wears Valley, the growth has created numerous problems. Examples follow.

Human-black bear interaction. Bears wander out of the park seeking food or movement to other natural areas. There are conflicts among residents, tourists, and hunters.[33]

Edge effect along the park boundary. Dense development effectively reduces the size of the protected areas in the park boundaries. For instance, wood thrush populations are reduced by parasitism of their nests by the non-native cowbird, which penetrate into the margins of contiguous forests.

Introduction of pests, pathogens, and invasive species. Sevier County is a hotspot for gypsy-moth invasion due primarily to campers transporting the insects from infected areas to the north.

Night sky degradation. The lights from development reduces the boundary in the national park and adjacent national forests. These federal lands are the last refugia of dark skies in the southeastern United States. Viewing the heavens has become a rare privilege.

Bio-continuum interruption. Development interrupts ecosystem continuums along gradients such as elevation. For instance, rivers flowing out of the national park are free of pollution but become very polluted almost immediately.

Highways and traffic. Building and improving highways in the Appalachian region have produced enormous social and economic benefits,[34] but there are also costs. Hotspots of growth often either follow or require road-construction projects. In Sevier County, Tennessee, for instance, there are currently eighty million dollars committed to road construction projects with several more planned. Most projects are targeted to relieving tourism traffic congestion. The overriding problem is a lack of traffic-circulation patterns.[35] Widening the roads through the twenty-four-mile tourism corridor from Interstate 40 terminating in Gatlinburg, which then narrows down to the two-lane road inside the national park is effectively a dead end for tourists.

Some of these road construction projects are particularly disturbing in that they will attract increased development next to the northern boundary of the Great Smoky Mountains National Park. In addition, local officials in Bryson City, North Carolina, have renewed their request to build the proposed north shore road in the national park along Fontana Lake.[36] That project would most certainly attract new tourism-related development on the south boundary of the national park.

Human Population Growth and Land-Use Solutions

Think regionally, act locally. There are numerous agents of change influencing the sustainability of cultural, social, economic, and environmental resources in the region. The following are a few examples to illustrate the range and depth of the paradigm shift underway. These examples illustrate dynamic leadership, the most precious resource of all. Website addresses are found at the end of the chapter.

Managing community growth. Pittman Center, Tennessee, has engaged in extensive public visioning and from that has created regulations and advisory boards to sustain their community vision—"A community dedicated to conserving our mountain heritage." There are regulations on development density, commercial signage, river protection, and slope protection. They are contemplating regulations on ridgetop development, control of invasive species, and protecting the night sky. Community leaders have been persistent in focusing on sustaining their vision.

Watershed associations. The Little River Watershed Association is actively protecting the river by stewardship education on keeping the river clean and conserving water use.[37] The Little Tennessee River Watershed Association is responsible for restoring and stabilizing streambanks in both Swain and Macon counties in North Carolina since 1998.[38] This organization works with the local soil and conservation district to provide funding and assistance to landowners who want to improve their own stream banks.

Community empowerment. The Gatlinburg Gateway Foundation provides a neutral forum for citizens to pursue initiatives in sustainability and helps facilitate their initiatives. "The Mission of the Gatlinburg Gateway Foundation is to advocate positive action and civic responsibility to achieve an environmentally sensitive and economically prosperous gateway community to the Great Smoky Mountains National Park." The foundation supports twenty-two community action groups focusing on various projects to improve community sustainability in harmony with the adjacent national park.

Regionalism. Citizens from nine counties adjacent to Knoxville County, Tennessee, formed a not-for-profit organization, Nine Counties One Vision, to identify regional issues and provide a vision and support for sustainable regional growth. Task forces focus on the environment, cultural heritage, tourism, education, transportation, technology, rural preservation, social services, downtown revitalization, and quality of senior life. A series of goals, action plans, and committees have been formed. The regional perspective has been enlightening for all.[39]

Countywide zoning. Blount County, Tennessee, has instituted countywide zoning to address a variety of issues related to development and growth. This initiative was begun by involving county commissioners in a series of listening-sessions to ascertain the opinions and concerns of local citizens. Two straightforward questions were asked: What do you like and what would you like to change? From a series of these sessions held in the districts of all county commissioners, a consensus was defined leading to the zoning strategy.[40]

Highway design. The community of Townsend, Tennessee, partnered with state engineers to design a highway through town that provided aesthetic refinements and pedestrian pathways and public access to the Little Pigeon River. The design was the culmination of a community visioning and planning effort.

Sold waste management. Sevier County, Tennessee, has adopted a solid-waste system that sorts waste materials and creates marketable compost. This technology is feasible due to the large quantities of food waste from the restaurants in that booming tourism area.

Historic preservation. The village of Flat Rock, North Carolina, home of the Carl Sandburg National Historic Site, has developed regulations and programs to protect and rehabilitate historic structures and to protect the viewshed from the historic site.

Nuisance black bears. Gatlinburg, Tennessee, has begun an aggressive program to control nuisance bear activities, the centerpiece of which is to require bear-proof dumpsters and residential garbage containers. Education for tourists, residences, and businesses is equally important.

Sustainable development. The Balsam Mountain Preserve near Sylva, North Carolina, includes 4,400 acres with 360 two-acre home-sites and a nature preserve of 3,300 acres. The sale of real-estate funds natural science research and ecological restoration throughout the property, as well as a nature center and environmental education programs.[41]

Cultural heritage. The Cherokee Nation has launched a comprehensive cultural-preservation initiative for members of the tribe as well as the general public, which includes native crafts, language, cultural traditions, and story telling.

Bottom line. The regional landscape is being nickel-and-dimed to death by incremental change that collectively marginalizes the cultural and environmental integrity of the region. Follow those in the region who are leading by example and create a vision, mandate, and mechanisms to facilitate balanced regional growth and prosperity while sustaining environmental, cultural, and historic heritage.

Pests, Pathogens, and Exotic Species Problems

Waves of pests, pathogens, and exotic species have invaded the Southern Appalachian region during the last century, with devastating ecological consequences. Many have been introduced via commerce and trade from other continents into the northeastern United States and migrated following the Appalachian Mountains to the Southeast. Organisms in a distressed state due to other environmental stressors, such as pollution, climatic abnormalities, isolation, and/or fire, may be more susceptible to pests, pathogens, and exotic species. This broadly defined topic illustrates the interconnectedness of stressors to environmental health. Unfortunately, it is most difficult to measure the interconnectedness of stressors to environmental health in field conditions. A few examples of key stressors follow. Some have been on this continent for 150 years or more, and some have just appeared. This is just a small sample of some of the most devastating perturbations. More are discussed in chapter 4. A casual observation is that the number of these perturbations seems to be escalating exponentially. This trend will likely continue to escalate for the above-mentioned reasons.

American Chestnut Blight. Arguably, the most ecologically devastating pathogen to eastern temperate forests is the America Chestnut Blight, which has eliminated the most prolific nut-producing tree in mid-elevation temperate forests. The disease was introduced to North America in shipments of Asian chestnut nursery stock in 1904.[42] Most of the Chestnut stands were dead by the 1930s, but individuals remain today. Some of the root systems of old stumps still produce shoots, preserving the genetic diversity of this native species. Loss of the abundant and consistent fall mast crop from this species has adversely affected mammal populations

Gypsy moth. Since the demise of the American Chestnut, upland forests in the Southern Appalachian highlands have been dominated by oak-hickory assemblages. This forest is particularly vulnerable to attack by the European gypsy moth, the most destructive insect pest in the eastern hardwood forests. The insect was imported in 1869 in hopes of establishing a domestic silk industry. By 1991 gypsy moths had infested an estimated 125 million acres nationwide, of which 4.1 million acres were defoliated.[43] Gypsy moth defoliation causes trees to use energy reserves in attempting to produce new leaves. A healthy tree can withstand several consecutive defoliations, but the tree resilience is diminished.[44] Recently, a soil fungus that disturbs the life cycle of the insect has stalled the invasion line of gypsy moths to the South. The longevity of this phenomenon is unknown.

Adelgids—balsam wooly and hemlock. The balsam woolly adelgid was introduced in 1908 on European nursery stock in Maine.[45] The vulnerability of North American fir species varies, depending on their ability to withstand abnormal water stress caused by premature heartwood formation that results from adelgid feeding.[46] Mature Fraser fir trees, endemic to the Southern Appalachian highlands, have been largely destroyed by this insect. Remaining younger trees are now being infected. Most mature Fraser fir trees have died.

The hemlock wooly adelgid was introduced into North America from Asia in the 1920s.[47] In the Appalachians, the pest feeds on eastern hemlock, an important species in watersheds that replaced the American chestnut trees at mid-elevations. Although not completely understood, salivary secretions released during feeding are thought to be the cause of tree mortality. Mature trees can die within four years.[48]

Southern pine beetle. This pest is endemic to North America. The fluctuations of its population are considered to be a normal ecological function. These insects invade the bark of mature pine trees, resulting in mortality. In 2001–2, a particularly large population killed an untold number of pine stands throughout the southeastern United States. Some forest ecologists believe that the rate of mortality of pine stands will permanently alter the forest composition in the Southern Appalachians.[49] Many of these pine stands are artifacts of previous land use, such as fire management by Native Americans.[50] So why is this included as a key pest if it is a natural phenomenon and many of the pine stands are cultural artifacts? Because the populations of bark beetle are abnormally large, due to a lack of adequately severe cold spells to control the population. This is a prime example of how one driver of ecological change, global warming, has influenced another, the pine bark beetle.

European wild boar. The European wild boar populations were first introduced as a wild game species in western North Carolina in 1915 and accidentally escaped from a hunting enclosure near Hoopers Bald, North Carolina, which is now part of Nantahala National Forest.[51] Others escaped in the 1920s in the same area. These wild boars interbred with domestic swine, but the wild hogs of today retain many of the European boar's characteristics. There are currently about a thousand wild hogs in Great Smoky Mountains National Park.[52] These animals are quite harmful to the forest understory and soils. Wild hogs are omnivorous, consuming fish, snakes, frogs, salamanders, crayfish, mussels, snails, small mammals, carrion, earthworms, immature and adult insects, and the young and eggs of ground-nesting birds. However, plants usually constitute the bulk of their diet. They root through the soils for roots and insects. Hog rooting in gray beech forests can reduce cover of herbacious understory to less than 5 percent of its expected value.[53] Rare plants have disappeared from the park because of hog rooting and wallowing.[54]

Exotic plants. Of the more than three hundred exotic plant species found in Great Smoky Mountains National Park, twenty-five are considered to pose significant threats to park resources. Among them are kudzu, Japanese grass, Japanese honeysuckle, oriental bittersweet, Japanese barberry, periwinkle, Tree-of-Heaven, privet, multiflora rose, musk thistle, Johnson grass, paulownia, and mimosa. These problem-causing species in the national park are typical of the region.[55] Six species were among the most prevalent exotic trees, shrubs, and vines in forests in the southeastern United States. Many of these species are early invaders following land disturbance. Paulownia is of particular concern because it is distributed widely by lightweight seed pods that can carry in the wind. They are known to invade following forest fire.[56] This is of great concern, since fire is now being frequently used as a resource-management tool in national parks and forests to re-establish "natural" fire regimen. Invasive species can frequently out-compete with native herbaceous plant species, including rare and endemic as well as popular native wildflowers. Again, climate change can greatly influence the spread of exotic plant species.

West Nile virus. This newly emerging disease is getting a lot of attention because it can kill humans as well as birds. Raptors, owls, blue jays, and the common crow are particularly vulnerable. This disease may produce secondary ecological effects, such as the extensive use of insecticides that would likely kill numerous other insect species and could harm amphibians as well. The cumulative loss of large populations of affected species could have adverse effects on biodiversity in the Southern Appalachian region. A significant reduction in the bird populations has the potential to increase the population of such prey species as rodents and insects, creating an environmental imbalance.[57]

Red oak borer. The red oak borer, an inch-long beetle native to forests in the eastern United States, causes most of its damage while in the larval stage of a two-year life cycle. The larva burrows through the bark of the oak, carving out galleries in the cambium and the heartwood of the tree. As the adults emerge from the oval holes, they chew on the bark in odd-numbered years. The red oak borer is usually an insignificant pest that oaks can easily fend off, but since 1999 the density of red oak borer populations has steadily increased, resulting in an explosion of oak borer infestations. Outbreaks once consisted of 5 to 10 borers per tree. Now there are 1,500 per tree. At this level of attack, the insects literally girdle the tree. Tens of thousands of trees have died so far.[58]

Sudden oak death. The USDA Forest Service has indicated that the Southern Appalachian Mountain forests are vulnerable to Phytohthrora, sudden oak death, which is a devastating forest disease attacking live oak species in California and Oregon.[59] It has also infected other species in the region as its invasion continues. This pathogen has now been detected in nursery stock in Florida.

Pests, Pathogens, and Exotic Species

There are numerous programs addressing these complex, intransigent problems. A few examples are offered below to suggest the range of possibilities.

American Chestnut. A new genetic strain of American Chestnut that retains a majority of the native species characteristics, but with a protective gene to combat the disease, has been devised and tested. There are plans to reintroduce this new strain of American Chestnut into the wild soon. This technology will become more prominent in the future as key tool to cope with pests and pathogens.[60]

Hemlock adelgid. Black ladybug beetles that feed exclusively on the hemlock adelgid have been introduced to infected Hemlock groves to combat the disease. The National Park Service has contracted to get a large supply of these insects to introduce into national parks. This approach to combating pests is likely to become more common in the future to address these types of challenges.

European wild boar. Great Smoky Mountains National Park controls the wild hog population by shooting and trapping the animals. The population requires constant vigilance, because it can expand rapidly. Park wildlife biologists, seasonal employees, and summer interns participate in this expensive but effective population-control program.

Bottom line. Solutions for many of these problems are being pursued in national parks and to a lesser degree in national forests and some local communities. However, initiatives on privately owned lands are rare. These problems are just as challenging on a regional scale as climate change is on a global scale.

Pollution Problems

Previous chapters have described point and nonpoint pollution sources and their individual and collective effects on the environment. The following are a few examples of pervasive pollution sources with considerable ecological consequence. There other major sources as well: vehicle exhaust emissions, industrial plant toxic wastes, and, of course, paper mills.

Fossil-fueled power plants. Of the multitude of types of point-source pollution, fossil-fueled power plants arguably have the most pervasive adverse effect on the environment. Their emissions contribute to global warming, acid rain, loss of soil nutrients, and release of toxic aluminum into soil solution, tropospheric ozone, and regional haze. Nearby fossil-fueled power plants are the primary source of air pollution for Great Smoky Mountains National Park, one of the nation's most threatened national parks.[61] The primary environmental threat to that park is air pollution.

Soil sedimentation. There are a plethora of rivers and streams in the region that do not meet state water quality standards. Sediment is the number-one water quality concern in Tennessee.[62] Soil sedimentation can have long-term adverse impact on aquatic habitat for numerous fish species. Due to a lack of resources, regulations to control sedimentation from construction sites and agricultural activity are poorly enforced. By comparison, major point sources of water pollution, though not wholly eliminated, are fairly well regulated.

Impervious surfaces. A second pervasive, non-point source of water and soil pollution is runoff from roads and parking lots, a toxic soup that is for the most part unmitigated. A good example of this problem is manifest in the highly polluted West Prong of the Little Pigeon River that flows through the densely developed tourist corridor in Sevier County, Tennessee.

Impervious surfaces are impenetrable surfaces that prevent water from penetrating the soil where it can then filter back to the water table. Impervious surfaces are constructed of materials like cement, asphalt, and roofing. Intense storms and heavy rains generate large volumes of stormwater, which, as it runs across impervious surfaces collects pollutants along the way—the larger the surface the more the pollution. This problem is most closely associated with urban areas, but rural areas are faced with the same challenges on a smaller scale. Stormwater collects pollutants from car fluids, pathogens for pet and animal wastes, chemicals from fertilizers and herbicides, and litter. Impervious surfaces are a major source of pollution for the West Prong of the Little Pigeon River, running from Gatlinburg, Pigeon forge and Sevierville.

Failed septic tanks. This intransigent problem significantly pollutes groundwater and streams throughout Appalachia. Generally, it remains largely beyond regulatory control. In most places septic systems are allowed in marginal soil types, too near riparian zones, on slopes that are too steep, and too close together. There are no regulations requiring periodic pumping of septic tanks.

Pollution Solutions

There has been remarkable progress made on pollution problems, particularly on air-pollution reduction, in the Southern Appalachian region. There are numerous evolving technologies, policies, and regulations dedicated to reducing and/or eliminating sources of pollution. Some of the innovative programs are listed below.

Southern Appalachian Mountain Initiative. This initiative is a partnership among federal, state, and local agencies, as well as citizen groups, who are committed to understanding air-quality issues in the Southern Appalachian region. The initiative is focused on Class I areas, which include ten Southern Appalachian national parks and wilderness areas. The mission statement reads "Through a cooperative effort, identify and recommend reasonable measures to remedy existing human-induced air pollution on the air quality-related values of the Southern Appalachians, primarily those of Class 1 parks and wilderness areas, weighing the environmental and socioeconomic implications of any recommendations." A key finding of the ten-year research and modeling program was that nearby fossil-fuel power plants are the primary source of pollution contributing to noncompliance in the Class 1 areas in the Southern Appalachia.[63]

North Carolina Clean Smokestacks Law. In June 2002 the North Carolina General Assembly passed Senate Bill 1078, which will improve air quality in the state by imposing limits on the emission of NO_x and SO_2 generated by coal-burning power plants. The law also provides for recovery by electric utilities of the costs of achieving compliance with those limits. The act gives the state discretion to "use all available resources and [lawful] means to induce other states and entities, including the Tennessee Valley Authority, to achieve reductions in emissions of NO_x and SO_2 comparable to those required by Section 1 of the act." In particular, North Carolina would place the economies of states and other entities whose emissions negatively impact air quality at a competitive disadvantage.[64]

Great Smoky Mountains National Park air-quality monitoring and research. The Great Smoky Mountains National Park monitors visibility and gaseous pollutants at its air-quality station at Look Rock, and the data collected are updated at the visitor center display and their website every fifteen minutes. There are several other monitoring stations around the park as well. This landscape is arguably the most highly monitored for air quality of any in the United States. This is the result of a collaborative program involving the National Park Service, Tennessee Valley Authority, Electric Power Research Institute, and the Environmental Protection Agency. Numerous research projects and programs associated with air quality are ongoing, including investigations of vegetation injury from ozone, chemometric programs (acidic deposition, gaseous pollutants), visibility programs, fine particle monitoring, and impacts of air pollution on human health. This monitoring and research, which began in 1980, provides information necessary for shaping policies that will improve the health of the park and the surrounding region.

Scrubbers on TVA power plants near Class 1 areas. The Tennessee Valley Authority has continued its commitment to cleaner air quality for the mountains of eastern Tennessee and western North Carolina. According to TVA emission controls will result in the reduction of sulfur-dioxide pollutants by up to 75 percent by 2005 and even further beyond that. "TVA is the largest single utility buyer of coal in the United States, purchasing almost 45 million tons in 2001." In October 2001 plans were announced to place more scrubbers on major coal-burning plants "as part of a five billion dollar project to cleanup its dirtiest plants." In 2001 TVA emitted 605,000 tons of SO_2, 269,000 tons of NO_x, and 105,347 tons of CO_2 a year and is using emission credits banked in earlier years to exceed the cap set by federal and state regulators. In November 2002 the board approved a $1.5-billion contract to install five scrubbers at four TVA coal-fired plants. Collectively, the five scrubbers are projected to reduce TVA's annual sulfur-dioxide emissions by at least 200,000 tons a year.[65]

There are a number of strategies to control the amount of impervious surface and the amount of stormwater that runs off from given area or to reduce stormwater pollution. In Chattanooga, filters have been installed to capture pollutants before they enter storm drainage pipes. Other communities in the region have developed best management practices by educating the public on the importance of reducing the use of harmful chemicals and taking proper waste-disposal measures. Some have created design guidelines for parking lots, encouraged permeable paving for overflow parking, or enacted open-space ordinances. The more opportunity that stormwater has to penetrate the soil over a larger area, the less it will harm the environment.

Bottom line. Current state and federal regulations, when enforced, satisfactorily control major point sources of pollution dispensed on land, water, and into the air. Nonpoint sources of pollution are another matter. These are much less likely to be abated. Soil sedimentation is the most serious but most ignored river and stream pollutant.

Natural Resource Utilization Problems

Many environmental threats are associated with the utilization of natural resources. A few examples with significant adverse impacts are briefly described below.

Natural Resource Management Policy. The evolution of policy concerning natural resources management on federally owned lands in the last thirty years has been extraordinary. Exotic species were routinely introduced to stabilize soils, enhance agricultural production, or provide new species for fishing and hunting. Rivers were poisoned to "improve" game fish populations. Wetlands were drained for agriculture. In fact, all these activities occurred in Cades Cove in Great Smoky Mountains National Park.[66] Also, exotic species, such as multiflora rose, were introduced to provide food for birds and wildlife by state natural resources agencies throughout the United States. Until recently, forest fires were suppressed in state and federal parks and forests, though fire plays an important role in the ecosystem process. Several species including the Table Mountain Pine are fire-dependent or fire-enhanced. Smoky the Bear was arguably the best-known symbol of environmental stewardship throughout most of the twentieth century. Even today, state fish and wildlife agencies continue to stock non-native rainbow trout in rivers and streams surrounding Great Smoky Mountains National Park.

Mountaintop removal. This relatively new mining technique involves bulldozing overburden above small coal seams into valleys and streams, providing an economically viable means for coal companies to extract small veins of coal. The downside is that this mining technique is environmentally devastating. It reduces biodiversity in terrestrial ecosystems and the resulting acid mine drainage degrades water quality and aquatic biota in rivers and streams. In eastern Kentucky, vast areas of mountainous forest and stream have been reduced to sediment ponds and arid, rocky grasslands that barely hold enough nutrition to sustain a cow. Remarkably, the state of West Virginia has waved environmental scrutiny of this mining procedure.[68]

Chip mills. The timber industry was initially centered in the northeastern United States. After the forests in that region were depleted, the industry shifted to the Pacific Northwest. After the old-growth timber was largely eliminated in that region, the pulp component of the industry shifted again, this time to the recovering forests in the southeastern United States. A recent analysis of the forest resources in the Southeast found that sprawl is the biggest threat to forests.[69] But the proliferation of chip mills in the last decade is of considerable concern as well. A common harvest technique is to clear-cut privately owned forested land. These chip mills employ only a few individuals but supply the majority of material for the pulpwood industry. Hence, economic benefit to the region is minimal.

Water consumption. As the human population increases rapidly, the demand for water consumption exceeds sustainable supply. This problem is systemic throughout the southeastern United States. The situation was exacerbated by a lengthy drought in 2001–2. Potable water has become a scare commodity. Retaining flow rates to sustain aquatic resources is a growing concern, as is preserving groundwater resources.

The greater Atlanta metropolitan area, for example, is a hotspot of inadequate water supply.

Natural Resource Utilization Solutions

National forest management plans. Balance of forest harvest with other land uses is a longstanding multiplier-use mandate of the U.S. Forest Service. According to federal law, the secretary of agriculture is obligated to collect and maintain a current comprehensive inventory and analysis of renewable resources of U.S. forests and range lands. In addition to federally owned lands, the inventory includes analysis of all privately held forest and range lands with owner consent. Forest inventories provide information that can be vital to decision makers. "Information is collected from over 130,000 permanent sample plots selected to assure statistical reliability,"[70] on which vegetation is measured about every ten years. In another attempt to balance forest harvest and other land uses, the USFS created the National Forest System Land and Resource Management Planning Rule. This rule is intended to improve upon the 2000 rule by providing a planning process which is more readily understood, is within the agency's capability to implement, is within anticipated budgets and staffing levels, and recognizes the programmatic nature of planning. A preliminary review of a report evaluating this rule, published in December 2002, affirmed the concepts of sustainability, monitoring, evaluation, collaboration, and use of science. However, it recognized that though the rule was intended to simplify and streamline the development and amendment of land and resource management plans, it is neither straightforward nor easy to implement. It also found that the rule did not clarify the programmatic nature of land and resource management planning.[71]

Ecosystem Management for Sustainability. The Southern Appalachian Man and Biosphere Cooperative is an interagency federal partnership to facilitate collaborative natural resources management activities for the Southern Appalachian International Biosphere Reserve. Thematic areas of concern to the group include environmental monitoring and assessment, sustainable development and sustainable technologies, conservation biology, ecosystem management, environmental education and training, cultural and historical resource conservation, and public information and education. Their most ambitious project was the conduct of the Southern

Appalachian Assessment, a comprehensive assessment of the area's natural, social, and economic resources.[72]

Mountaintop removal. This practice should be conducted within the context of all federal environmental laws. Adherence to environmental laws would reduce the potential environmental impact of the mining site. Beyond the fact that mountaintop removal is aesthetically unappealing, it is also extremely harmful to adjacent waterways and wetlands, which are important for sustaining wildlife and human populations and for protecting economic activities vital to regional sustainability. In May 2002 Federal Court Judge Charles Haden in West Virginia ruled that mountaintop removal mining violates the Clean Water Act and that the Army Corp of Engineers was no longer to give out permits. The judge said the corps issued permits that allowed companies to fill eighty-seven miles of streambeds with mine waste: "Past . . . permit approvals were issued in express disregard of the corps' own regulations and the Clean Water Act."[73]

Chip mills. Other states should follow North Carolina's lead and withhold any new permits for chip mills until a statewide analysis is conducted to assess how the practice impacts the sustainability of the state's forest resources. Active discussion should take place among all stakeholders to define best management practices that would establish guidelines for regulating chip mills while still allowing them to remain in operation. This would go a long way toward minimizing the devastating effect that chip mills have on the environment. Skip Stokes, in a presentation to Missouri's Governor's Council on Chip Mills, offered these a list of best management practices: "sites for skid trails, planting winter wheat, use of water bars, use of streamside zones, training in erosion potential, control sediment runoff, harvesting dying trees, spacing and leaving trees for wild life, and control of visual trails by buffer strips."[74]

Water consumption. The name of the game should be conservation of water resources *first* before tapping new sources. Bathroom fixtures conserving water use are now federally mandated. Restrictions on the use of water should be mandated as well. Water rights to the Tennessee River are the current legal battleground in today's "use first, conserve as a last resort" climate.

Private land development. There is a critical need for a paradigm shift in how developments are designed in order to minimize the human footprint on the environment. Traditional residential developments usually obliterate the natural environment. The Chaffin/Light Associates development demonstrates how a profitable residential development can be designed to protect and enhance the natural environment in the rural landscape. A 4,400-acre forested parcel in the mountains of western North Carolina is the site of this demonstration of sustainable development, which includes 350 two-acre home sites planned, a golf course, a fitness center, riding stables, and a community center. A review board enforces design and landscaping guidelines. Structures are not to be obtrusive, and their placement is of key concern. Disturbance to native trees is to be minimized. Where feasible, native plants are rescued from construction sites and replanted in environmental restoration areas. No security lights are allowed. All residences have cisterns. The use of bear-proof refuse containers is required. Particular attention is paid to minimizing silt and other pollutant runoff during construction and once the structures are complete.

Conservation easements have been placed on 3,300 acres to insure their protection in perpetuity. There is an ongoing inventory and assessment of the environment by a team of scientists and natural resource specialists. This team has identified endemic and rare species, sensitive areas, environmentally degraded areas, and even a few species new to science. Two watersheds on the property have been targeted for restoration. Siltation from logging roads and environmentally damaging forest practices are of particular concern. A series of scientific experiments are underway to restore the environment on these watersheds. Many of the old logging roads will be obliterated. New technology never before applied to mountainous terrain is being tested to filter runoff to remove/retain sediment and to remove sediment from streambeds in order to restore habitat for the reintroduction of native brook trout. Much of the science is being funded through university grants.

The most remarkable component of this project is the establishment of the 2,800-acre Balsam Mountain Preserve, which has been funded principally by real-estate sales and contributions by landowners. There are ambitious plans for the reserve, including the construction of a nature center. A fun dimension of that structure will be a walk-into-the-forest canopy. A fulltime naturalist has been hired to develop and run programs and coordinate volunteer programs. The executive director of the preserve facilitates the science programs conducted in the reserve and oversees the environmental restoration program and outreach to educational institutions.[75]

Nine Counties One Vision. This organization was formed in Knox-ville, Tennessee, by leading private citizens who are concerned about urban sprawl, environmental degradation, and inequity in social services. An extensive visioning initiative was conducted.[76] From this initiative a series of committees were formed and action plans devised concerning, among other things, managed development, public transportation, preser-vation of regional culture, and technology to reduce the human footprint on the landscape. Their goals and actions are targeted to the following:

- Protecting scenic beauty and natural resources

- Preserving farms and open space

- Rehabilitating older areas

- Requiring environmentally responsible and well-planned development and limit sprawl

Land use. Many concerns about the environmental future center on land use policy and practice. Key dimensions are as follows:

- Creating greenway corridors linking protected areas

- Protecting riparian zones along waterways

- Protecting open space via clustering development and conservation easements

- Protecting sensitive areas such as wetlands (no matter their size), habitats of threatened or endangered species, steep slopes and ridgetops

- Protecting prime farmlands

- Protecting forestlands

Strategies to accomplish these goals being practiced in the region include county-wide zoning, land trusts, conservation easements, greenway and park lands, most importantly sustainable development. Collaboration among a variety of organizations is required.

Forested biomes. Key strategies to stabilize forests include the following:

- Utilizing sustainable forestry practices on privately owned forestlands

- Utilizing forest products only if they are sustainably produced

- Monitoring forest health

- Utilizing only native species

- Rehabilitating environmentally degraded areas

- Focusing on the spatial, elevational, moisture, and temperature dimensions of forest biomes

Native Species. Key strategies to protect native species should be expanded. Helpful strategies include:

- Continued reintroduction of extirpated species, such as the river otter and elk

- Removal of exotic plants and replacement with native species

- Designation of habitat for rare, threatened, or endangered species

- Monitoring populations of species at risk

- Landscaping with native species

- Protecting threatened native species genetics by creating protected populations in appropriate safe places, such as nurseries.

- When appropriate, conducting genetic studies to determine within-species genetic viability and its relevance to resistance to pests and/or pathogens

Water resources. Key strategies now being applied should be expanded.

- Complete watershed plans for all watersheds

- Monitor water resources

- Create more watershed associations focusing on stewardship activities

- Promote water conservation activities

- Create water budgets at a watershed and regional scales

Sense of place. This link with sustainability strategies creates social and political resolve.

- Conduct visioning processes at various levels

- Link values to existing conditions

- Devise strategies to fill in the gaps

Bottom line. As natural resources become scarce, policies and regulations are needed to conserve them for the long term on public and privately held lands. The best way to sustain natural ecosystems is using natural resources sparingly. Conservation strategies must be applied comprehensively as demands for all natural resources will invariably increase over time.

The Good News: Environmental Restoration and Technology Advancement

Not all the news about the environment is abysmal. As noted above there are a plethora of solutions to environmental problems. Below are a few additional examples of major accomplishments in righting the environmental ship, sometimes against great odds.

Species reintroduction. There have been a series of programs to reintroduce extirpated native species to the Southern Appalachian landscape. Examples include the Bald Eagle, wild turkey, black bear, deer, beaver, river otter, Peregrine Falcon, southern strain of brook trout, and, most recently, elk. Testing has taken place to reintroduce red wolves to Great Smoky Mountains National Park.[77]

Genetic engineering. The American Chestnut has been genetically altered to incorporate a gene that provides resistance to the Chestnut blight. This new strain of American Chestnut preserves most characteristics of American Chestnut. This strain is being field tested and will likely be reintroduced into the natural landscape in the near future.[78]

Ecosystem restoration. For ninety-five years the Canton Paper Mill discharged waste into the Pigeon River running from western North Carolina to East Tennessee. The river became, in effect, biologically dead.[79] After over seventy-eight years of discharging under weak permits, political intervention from the White House and the sale of the facility to its employees finally led to a more stringent point-source permit and eventual compliance by the paper mill. As a result the river has recovered remarkably, in spite of a heavy dioxin load remaining in the sediment of the riverbed. A population of game fish has been restored, and the Tennessee Wildlife Resources Agency is reintroducing numerous other aquatic organisms.[80]

Another poignant example of ecosystem restoration is the moonscape-type landscape at Copper Basin, Tennessee, an area consisting thousands of acres of contaminated soils and water caused by over a century of copper mining and smelting. Copper Basin has long been a national example of environmental degradation. Reforestation efforts have been underway on the property for several years. A treatment system is under construction for a highly contaminated stream, contaminated electrical equipment has been removed, voids and shafts have been filled, and fencing is being installed to restrict access to more highly contaminated areas. The restoration project will cost tens of millions of dollars to be paid by responsible companies and the federal government.[81]

Environmental Scenarios for the Future

The alternative future environmental scenarios discussed below speculate on the effects of the drivers discussed above on vulnerable biomes, biological communities, and/or species. The intent is to illustrate the risks of inaction as compared to the rewards of following creative leadership. The collective vision of the leaders suggests the possibilities. The alternative futures illustrate how much is at stake in our rapidly changing world. The timeline of the future scenarios is fifty years; therefore, the year of the scenarios is 2053. Today's younger generation will witness this period and will be in a position to judge the accuracy of the predicted alternative futures discussed below. Indicator biomes and species are utilized to illustrate the nature and potential severity of collective influences of the drivers of environmental change. The biome indicators include the spruce-fir forest, northern hardwood forest, oak hickory forest, and rivers and associated riparian zones. Indicator species are black bears, brook trout, wood thrush, and red cheeked salamanders. The human indicators include Appalachian culture, cost of land, and transportation alternatives.

Future Scenario of "Business as Usual"

In this scenario of business as usual, the influence of current drivers of environmental change as described above is assumed to accelerate unabated. As the human population grows and climate changes, the influence of the drivers of environmental change will likely *increase exponentially,* primarily due to their collective adverse effects on the environment.

Anticipated Trends

Global warming. The dramatic warming in the 1990s in the Southern Appalachians will accelerate, routinely setting new records. Extremes in flooding and droughts, late frosts, and warm winters continually grow more frequent. On a global scale, the end of the window of opportunity to reverse the trends in global warming is rapidly approaching. The burden lies squarely on the current generation of world leaders. There is a growing consensus of concern, but many leaders have concluded that it is too costly to reverse the global trend and the best course of action is to "learn to live with it." Unfortunately, the stakes are too high to ignore the likely cataclysmic social, economic, and environmental consequences of inaction. The cost of sea-level rise, increased number and ferocity of storms and hurricanes, flash flooding, and forest fires from drought are likely to slow global and regional economies dramatically and put untold numbers of people at grave risk. Global political unrest and military conflicts are likely to escalate dramatically as a knee-jerk reaction. The premise of conflicts may shift from religious and/or political demagoguery to scarcity of natural resources. Global economic and social inequities will likely escalate dramatically as well. The good old days (for some) at the beginning of the twenty-first century will be long gone.

Human population. Metropolitan areas will continue to sprawl, following four- to six-lane highways that are spatially designed either as spokes of a wheel or periphery beltways. The beltways will be three to four peripheral rings deep. The Interstate Highway I-26 linking Asheville, North

Carolina, and Johnson City, Tennessee, will result in an unprecedented acceleration of sprawl into the Southern Appalachian highlands. Public transportation will be more prevalent but still limited to serving a small minority of the population. The "newcomers" will become the vast majority and evidence of remnant indigenous culture will all but be removed from these areas. The newcomers will have diversified and redefined the regional culture. Cultural ties to nature will be dramatically different.

Land use. The escalation of land-use conversion from forest and agricultural lands to the built environment will continue to be most extreme at the periphery of sprawling metropolitan areas, but boundaries will be more difficult to distinguish. Continuous development corridors along interstate highways will link metropolitan centers.

In 2002 Knoxville, Tennessee, was listed among the top ten hotspots of sprawl in the United States.[82] Other hotspots of land-use conversion in the Southern Appalachian region include densely populated tourist areas, rural areas near interstate highway interchanges, and development along the boundaries of protected areas. This development will constitute formidable barriers, creating an island effect for most protected biomes in the Southern Appalachian highlands. Ultimately, Great Smoky Mountains National Park will become a biological island surrounded by one big, indistinguishable conglomerate of theme parks, campgrounds, bungee jumps, miniature golf courses, shopping malls, rental condos, cabins, buffet restaurants, and seasonally a never-ending string of energy-consuming and night-sky-diminishing Christmas lights. Wears Valley, in this author's opinion the most scenic cove in the Southern Appalachian highlands, will likely be annexed by Pigeon Forge, becoming undifferentiated from what is now the Pigeon Forge, Tennessee, tourist strip. This ever-expanding physical barrier will isolate native species from other natural refugia in nearby national forests and other natural areas, such as those on English and Chilhowee Mountains.

Forested biomes. The greatest driver of change, global warming, is likely to be the greatest threat to the eastern temperate biome. As described below, several climate-sensitive forest types will be in jeopardy. Pest and pathogen infestations will accelerate in frequency and intensity due to global warming and land disturbance. Combinations of perturbations will likely compromise resistance by native vascular plants to these attacks. There will also be an escalation of invasive species that will overtake native species habitat. In addition, forests on privately owned lands will be marginalized by sprawl. The adverse impact of chip mills will likely be dramatically reduced by the decline of viable forests on privately owned lands. Many deciduous forests will have been converted to pine tree plantations.[83]

Native species. Species particularly at risk from this futures scenario include the following long list of risk categories.

- Species likely to experience multiple environmental stressors

- Species whose life cycle includes risks from environmental stressors

- High-elevation endemic, rare, and/or threatened species

- Species that do not readily disperse their seeds

- Co-dependent species

- Fire-dependent or -enhanced species

- Temperature-sensitive species

- Moisture-sensitive species

- Species requiring a large protected habitat

- Species vulnerable to pests and pathogens

- Species vulnerable to invasive species

- Species of commercial interest to humans

- Species of cultural interest to humans

- Species that repulse humans

- Species considered a threat to human health

- Species considered a threat to agriculture

- Species vulnerable to practices employed to control unwanted species

- Species in habitat likely to be of recreational interest to humans

- Species in habitats likely to be converted to development

Water resources. There will be a serious shortage both of surface and groundwater. Water flows in rivers will likely be seriously depleted. Lower flow rates will concentrate pollution levels. Inadequate water supplies are currently an issue throughout the southeastern United States. The Atlanta and Charlotte metropolitan areas are particularly vulnerable. The Southern Appalachian highlands encompass the headwaters of numerous watersheds, not the least of which is the Tennessee River. Water-conservation measures have been short term during drought crisis and not comprehensive or long lasting. One of the most pervasive and insidious pollutants is sedimentation. Outside of headwater streams in protected areas, it will be *very difficult* to find watercourses that meet state water-quality standards for fishable/swimmable waters.

Sense of place. The cultural and ecological mystique of the Southern Appalachian highlands will largely evaporate. The Southern Appalachians will be less distinguishable from burgeoning sprawl throughout the metropolitan areas and interstate corridors in the southeastern United States. Visibility declines from air pollution due to unabated demand from energy will likely make clear views of the mountain landscape a rare experience. The world-renowned native biodiversity and species richness may eventually be reduced to such an extent that the designation of the Southern Appalachian highlands as an International Biosphere Reserve by the United Nations Environmental Social and Cultural Organization may be in jeopardy.

Effects on Biological and Human Indicators

Spruce-fir forest. The highly stressed, high-elevation spruce-fir forests will be well on their way to conversion first to blackberry and later to rhododendron thickets and northern hardwood forests. The multiple environmental stressors of climate change, air pollution in all forms, and pests and pathogens are driving this ecological response. This crown-jewel biome within the crown-jewel Great Smoky Mountains National Park will be lost, greatly reducing the biological distinction of the region. All organisms endemic to these spruce-fir forests that cannot adapt to climate change and/or forest conversion will be lost to the region.

Northern hardwood forest. These forests will invade the high-elevation spruce-fir forest, but over the long term will likely go the way of the spruce-fir forest, disappearing from the landscape due to climate change and various existing and yet unknown pests and pathogens. Beech trees in beech gaps on ridgetops will succumb to the beech bark disease complex.

Oak-hickory forest. The oak-hickory forests will likely be highly stressed by beech bark disease complex, oak disease complex, red oak borer, and sudden oak death syndrome. The greatest threat will likely be land-use conversion to suburban sprawl and pine plantations. A key potential result will likely be forest fragmentation, which would adversely impact species requiring large-scale, contiguous forests.

Cove hardwood forest. This forest type will be either greatly increased or reduced, depending on the amount and timing of changing temperature and moisture patterns.

Black bear. The black bear population will likely vary dramatically due to the unpredictability of their traditional food sources, such as the fall mast crop. There will be constant pressure to reduce the population of bears due to nuisance bear activity, which will be exacerbated due to periodic fall mast crop failure, loss of favorable habitat, and availability of human-source food. Their natural instinct to expand their range as their traditional habitat becomes overpopulated will be impeded by development.

Brook trout. The brook trout population will be reduced as first-order streams become more acidic, second- and third-order stream flow rates become more sporadic, and introduced exotic game fish migrate up from fourth-order streams. Global warming will reduce trout habitat and/or populations in the Southern Appalachians, which is on the extreme margins of the trout's range.[84]

Wood thrush. Forest fragmentation will open up the forest canopy to cowbirds that lay their eggs in wood thrush nests. Acid rain also reduces available calcium, making wood thrush eggshells more fragile. The Great Smoky Mountains National Park is the primary contiguous forest habitat for wood thrush in the southeastern United States. As the population continues to decline dramatically, the delightful call of the wood thrush will unfortunately become a rare event.

Red cheeked salamander. The endemic salamander population may be significantly reduced due to the potential for more extreme fluctuations of the temperature and moisture regimens in their micohabitats. Periodic extreme droughts would increase the potential of hot wildfires.

Appalachian culture. Indigenous European American culture will likely be highly marginalized to a small minority of the population. Stereotypes of Appalachian culture via tourism and marketing distort and exploit the endemic Euro-American culture. Local citizens of modest means will find it difficult to buy land, due to inflated prices. Inequity of educational and economic opportunity, social services, and cultural amenities will persist. In fact, the stark inequity will likely get worse.

Transportation. One of the greatest deterrents to tourist-based economic growth is and will continue to be traffic jams and the lack of transportation alternatives to the personal automobile.

Bottom line. Under this scenario for the future, degradation of the *native* environment is verging on becoming irreversible. The current distribution and species composition of forest types in the landscape are likely to be dramatically changed. The greatest uncertainty is the synergistic effects of a myriad of environmental perturbations on vulnerable species, communities, and biomes, but it is likely to exacerbate risks exponentially. The current condition of the spruce-fir forest is a portent of the future for many other biomes. There is a high likelihood of a drastically altered environment in the Southern Appalachian highlands, with significantly reduced biodiversity and species richness and the ever-growing list of species on the brink of extirpation from the landscape.

Future Scenario Applying Proactive Sustainability Policies

In this scenario the catch phrase is "follow the leaders." As documented above, there are numerous leaders in the Southern Appalachian region practicing the principles of sustainable natural resources management and sustainable development. This is by no means an exhaustive list, but merely a small sample of these leaders and types of contributions being made. This futures scenario is based on the premise that these examples of leadership are applied regionally to greatly increase their collective impact on the future.

Anticipated Practices

Global warming. This dimension is the most problematic but urgent. The challenge is finding a means to collectively embrace this leadership. An important symbolic gesture would be for the National Park Service and

U.S. Forest Service to apply state-of-the-art alternative energy sources and energy-conservation practices. They could also track in their visitor centers their energy savings and the collective impact in reducing greenhouse gas emissions, as well as the contribution of the forests in sequestering greenhouse gasses. They could also urge all gateway communities to follow the lead of Gatlinburg in signing up to purchase green energy. They could seek financial support for these efforts from industries participating in the movement to reduce greenhouse gas emissions. Their educational programs could emphasize the importance of "getting green" on this issue by highlighting at-risk species and biomes. They should also conduct climate-change risk assessments on their natural resources.

Human population. It is highly speculative to estimate the population fifty years into the future. The issue is not how many people there are, but how they collectively impact the environment. The need here is to get others to follow the lead of organizations, such as the Southern Appalachian Man and Biosphere Reserve Cooperative, Nine Counties One Vision, Blount County, Pittman Center, Gatlinburg Gateway Foundation, Foothills Land Conservancy, and the Balsam Mountain Preserve, dedicated to sustainable development described above.

Effects on Biological and Human Indicators

Spruce-fir forest. This high-elevation forest may already be beyond preservation as an ecosystem, but at the least there needs to be a risk assessment to determine which species associated with the ecosystem need special treatment to guard against extinction. For example, a genetic analysis should be conducted among the various isolated populations of Fraser fir, a species endemic to the Southern Appalachians, to determine if there is genetic variability related to resistance to the pest woolly adelgid. A plantation of these trees from various populations should be established in an area not infected with the pest. The population should be large enough to guarantee adequate reproduction. This action now would provide the opportunity to reintroduce the population once the pest has run its course. This type of strategic planning will increase the options in the future. Other endemic species of the ecosystem are of concern as well.

Northern hardwood forest. Due to the uncertainty of climate and pests and pathogens, it is difficult to predict the future viability of this forest. Finding a means to control pests and pathogens in future climate scenarios is key. This forest type is in less jeopardy than the spruce-fir forest, but it ultimately may be lost from the landscape in the Southern Appalachians unless the global-warming trend is reversed.

Oak-hickory forest. The future of the oak-hickory forests is primarily dependent on land use and the impact from disease. If sustainable land-use practices are in place so that sprawl is controlled, pine plantations are not too intrusive, and the influence of chip mills is reduced, these forests should thrive. There are lots of caveats, but there is genuine optimism for this futures scenario.

Cove hardwood forest. The status of this forest type is likely to rest with the future of protected lands in the Southern Appalachian highlands. As indicated in the first scenario, the future is tied to climatic conditions.

Black bear. This scenario is particularly beneficial to the black bear population. Movement corridors for black bears via greenways and conservation easements will help alleviate the risk of bears moving beyond the boundary of protected areas.

Brook trout. Brook trout will also benefit from this scenario. Water-quality improvement and less competition from non-native salmonid species would be the key. The most important benefit from this scenario is the potential to create a continuum of habitat from higher to lower elevations.

Wood thrush. Contiguous forest canopy is more likely to be sustained in this scenario. This is very important for wood thrush. Current populations would likely be sustained.

Red cheeked salamander. This endemic salamander species is vulnerable to land-use conversion and potentially to climate changes. Sustainable land-use policy is a key contributor to maintaining current population size and distribution.

Appalachian culture. The sustainable growth policies in Pittman Center, Tennessee, are centered on highlighting and "preserving our mountain heritage." If other communities follow similar growth-management strategies, the conservation of Appalachian culture would be considerable. This is a good example of the potential benefits of this futures scenario.

Transportation. In fifty years it is very likely that most visitors to Great Smoky Mountains National Park during the peak of the tourist season will utilize public transportation. By that time more people might routinely utilize public transportation and therefore be accepting of its application in the park. Also, the use of alternative energy sources, such as hydrogen-powered fuel cells, might be commonplace. This might be one of the most striking differences from today and one of the most environmentally important.

Bottom line. In this scenario, there is greater potential to sustain and even improve the current environmental condition in the Southern Appalachian highlands. The way to build momentum is to publicize the individual and collective benefits accomplished by todays visionary leaders discussed above and provide incentives for their adoption throughout the region.

Take Away Message: The Future Is Now—
Think Globally, Act Locally

As disturbing as the message of this chapter is, I remain optimistic. Most citizens in the Southern Appalachian highlands are concerned about the environment, particularly the in-migrants in the rural areas.[85] Many individuals and institutions in the region are leading by example. Those mentioned here are just the tip of the iceberg of progressive change. There is a growing momentum that will inevitably create a massive paradigm shift. I encourage more people to shift from discussing the principles of sustainability to acting on them.[86]

Chapter 11

MODELS OF SUSTAINABILITY

Jonathan Scherch, Thomas Fraser, Athena Lee Bradley, and John Nolt

It is not possible in this book to describe detailed solutions to all the ills we have documented. Solutions must be the product of many more hearts, minds, and hands than we, the authors, have to offer. Fortunately, many are already engaged in this vexing and valuable work. In this final chapter, we describe some outstanding examples.

Building Coalitions

One very hopeful sign is the emergence of coalitions of traditionally antagonistic interests.[1] Hunters and anglers are working with hikers, birdwatchers, and biologists to preserve land, further scientific study, and provide for the natural heritage of all Americans.

While hunters were indeed responsible for the decline, and even extinction, of some American game species (see chapter 4), the numbers today suggest that it is high time the preservationists and conservationists bury the hatchet. As game species continue to rebound, hunters and anglers can adjust their sights to include even more ambitious wildlife and fisheries goals, and those typically benefit, by default, nongame animals and plants.

A 1937 inventory of American game species by the United States Fish and Wildlife Service estimated that there were 1,700 white-tailed deer in Tennessee.[2] In 2001 nine times that number of white-tailed deer—157,599—were recorded killed in the state. The estimated population of white-tailed deer in Tennessee today numbers around a million. There were 300 known black bears in Tennessee in 1937. That's just under half the number of black bears recorded killed—159—by Tennessee hunters in

2001.[3] Great Smoky Mountains National Park alone is now home to an estimated 1,500 to 2,000 black bears.[4]

Some game species, such as the ubiquitous white-tailed deer, have enjoyed natural population growth, and much of the wildlife boom can be attributed to a rebound from early overhunting, but these numbers also tell a positive story that often goes unheard in modern environmental circles. The efforts—and dollars—of hunters and anglers and their representative national groups have resulted not only in exploitable populations of game species, but in the preservation of millions of acres of natural land and enhanced stewardship of other public resources.

These "consuming" enjoyers of the natural world could have the most practical argument of all for environmental preservation. Fishers cannot eat contaminated fish. Hunters cannot hunt suburban neighborhoods. Turkey (whose numbers now exceed 30,000, perhaps the most impressive wildlife recovery story in Tennessee)[5] do not breed in suburban backyards. Sauger and trout do not breed in polluted water.

And their arguments carry over to benefit not only "nonconsuming" observers of the natural world, such as birdwatchers and hikers, but nongame wildlife, as well.

A prime example of this cooperation was the purchase of the Cumberland Forest tract in 2002. Multiple organizations and agencies, with seemingly conflicting agendas, including the Nature Conservancy, National Wild Turkey Federation, Tennessee Ornithological Society, Cumberland Trail Conference, Tennessee Wildlife Resources Agency, Conservation Fund, and Rocky Mountain Elk Foundation, contributed to the 2002 effort to secure a 75,000-acre addition to the Royal Blue Wildlife Management Area in Anderson County. Royal Blue is managed by the Tennessee Wildlife Resources Agency, which receives more than half its budget through the sale of hunting and fishing licenses, for hunting.

The wildlife agency was also a major player in the acquisition of the Cumberland Forest land, which, when combined with Frozen Head State Natural Area and Royal Blue, will represent the third-largest public wildland in Tennessee.[6] The Rocky Mountain Elk Foundation would like ultimately to tie the land to a multimillion-acre wildland corridor linking re-established elk herds in Tennessee with those in Kentucky.[7] While the goal of the Rocky Mountain Elk Foundation is to re-establish huntable herds on the Cumberland Plateau, its help with the Cumberland Forest purchase will aid in preservation of migratory songbird populations on the plateau and provide land for other woodland species and efforts to bring black bear back to the plateau. The multiple-use approach to the Cumberland Forest deal—which ostensibly includes provisions for environmentally friendly logging—was a key to its success and represents a relatively new approach to wildland conservation: the merging of hunting and fishing interests with those of strict environmental preservationists.

Some view this as simply a more pragmatic approach to environmental preservation. Some are not in agreement with the approach. Randy Brown, executive director of the Foothills Land Conservancy, received complaints following a successful effort by his group to secure Blount County land that borders Great Smoky Mountains National Park. Four hundred acres of the 6,200-acre tract were donated as an addition to the national park; the balance was used to form the Foothills Wildlife Management Area, which is managed by TWRA for hunting of black bear, boar, wild turkey, squirrels, and other game species. Brown's philosophy, as summed up by his critic, was that the "best way to protect a species is to have it huntable."[8] Ten thousand of the 14,355 acres the Foothills Land Conservancy has preserved since 1992 are open to hunting, though on strict wildlife-management terms set by the Tennessee Wildlife Resources Agency.

It is important to note that the majority of land protected by the Foothills Land Conservancy is not open to the public for liability reasons. Turning management over to the Tennessee Wildlife Resources Agency guarantees some degree of public access, as the managed tracts are open to passive users of the land as well as hunters.

Hunting- and fishing-oriented groups, such as the National Wild Turkey Federation, Quail Unlimited, Trout Unlimited, the Rocky Mountain Elk Foundation, and the Safari Club Foundation and their members, provided almost a quarter of the $3.9 million the Foothills Land Conservancy has raised since 1992 for the preservation of areas such as the Foothills Wildlife Management Area and Kyker Bottoms Wildlife Refuge in Blount County, and the 2,500-acre Yuchi Wildlife Management Area in Rhea County. Members of the groups often match organizational contributions.[9]

Billy Minser, a hunter and wildlife researcher at the University of Tennessee, describes hunters and fishers as the "first and best environmentalists." Over the past twenty years, Minser has seen "an evolution . . . a joining of hands between consumptive and non-consumptive environmentalists" for reasons of both recreation and natural heritage. There are 1,100 verte-brates in Tennessee, and only about 70 are hunted or fished. Nongame species benefit from the habitat protected for fish and game. Many organi-zations give money to purchase land where no hunting is even allowed. In Tennessee hunting and fishing groups, sometimes in the company of organi-zations such as the Nature Conservancy, have, in cooperation with the state and other governmental agencies, helped preserve relatively undeveloped tracts of land that include an addition to Pickett State Forest, Royal Blue and Yuchi wildlife management areas, the Gulf, Scott's Gulf, Wolf River, North Chickamauga Creek, Lick Creek, Star Mountain tracts, and areas of the Ten-nessee River Gorge.[10]

The Rocky Mountain Elk Foundation, Trout Unlimited, the National Wild Turkey Federation, and other organizations also provide staff and money for scientific research and habitat enhancement on area public lands. Volunteers with East Tennessee chapters of Trout Unlimited pro-vide thousands of hours and dollars for brook trout restoration and acid-deposition research in Great Smoky Mountains National Park, and the Rocky Mountain Elk Foundation has been instrumental in the restoration of elk to public lands in the Cataloochee Valley of western North Carolina and the Cumberland Plateau. The synergy of the "hook and bullet" and strict preservationist philosophies were perhaps inevitable.

All types of outdoor enthusiasts, Minser says, are "realizing the resource is finite," and "we can't go it alone." Given the funding structure of the Ten-nnessee Wildlife Resources Agency, he is absolutely right.

The Tennessee Wildlife Resources Agency is responsible for managing all species of wildlife within Tennessee. Its charges include amphibians and endangered species, but it receives the bulk of its funds from the sale of hunting and fishing licenses within the state. It receives no money from the state's general fund. In 2002 $25.5 million of the agency's $50-million budget was derived from the sale of in-state licenses. The balance of the budget consists of revenues from the distribution of a federal excise tax (put into law by the Pittman-Robertson and Dingell-Johnson acts) on fish-ing and hunting equipment and revenue generated by the agency itself, such as profit from timber sales.[11] Hunters and fishers also provide a direct infusion of funds into local economies. The two groups, representing only 30 percent of the public, spend an estimated two billion dollars a year in Tennessee on licenses, tackle, ammunition, and other gear, food, and gas.[12]

The Tennessee Wildlife Resources Agency does considerably more than tend to the needs of hunters and fishers within the state, though its ninety refuges and management areas, which represent some 1.3 million acres, are managed primarily for hunting and fishing. The agency is also responsible, however, for the care of the state's one hundred threatened and endangered species, as well as the dozens of species declared In Need of Management. The agency also manages a cooperative amphibian-monitoring project that aims to document and investigate suspected amphibian declines in the state and elsewhere.

To assist in its goals, the agency in 1999 formed the Tennessee Wildlife Resources Foundation, which funds programs that include habitat-management education for state landowners and farmers, wildlife education for children, cleanups, wildlife restoration, and wildlife and fisheries research. It is also a participating member in the Whooping Crane Eastern Partnership, improving habitat on its lands for the critically endangered bird and ensuring its ability to make unmolested migratory visits to Tennessee. The foundation also provides additional money for land acquisition, such as the 2002 Cumberland Forest purchase and a 51,000-acre East Tennessee wetland. The purchase of the Cumberland Forest by a consortium of groups representing preservationists, birdwatchers, hunters, and hikers, signals a new full-bore approach to environmental sustainability through cooperation.

Healthy Living

But mere conservation and even preservation is not enough; for, despite all the difficult stands being taken by many dedicated organizations, destruction is advancing on many fronts. Moreover, no victory is permanent. The forest saved from clear-cutting this year may fall to roads and suburban development in a decade or two. Yet destruction is never permanent either. A new forest will eventually arise, even if it must force its way through concrete. If we are to witness such rebirths, however, we must practice the arts of healthy living.

Hundreds of such practitioners are already at work, scattered here throughout Southern Appalachia. We can recognize them by their understanding of the interdependence of land and spirit, respect for physical work, recognition of the limits of growth, and willingness to defend what is healthy and heal what is not. They are free spirits, living unconventionally yet responsibly.

To live responsibly, one must perceive the "hidden costs" of one's actions—the far-reaching effects of buying commercial supermarket produce, for example, or using utility electricity, or recruiting a new industry—and act on that perception to enhance the health of the whole. Though this is not easy, the difficulties are often exaggerated. Living responsibly does not, for example, mean living primitively, without hot water, adequate lighting, effective transportation, good food, or free time. Those who live responsibly can still enjoy all these things and something else besides: the peace of mind in knowing that one's own living does not unnecessarily diminish the lives of others.

Learning by Doing More with Less

During the summer of 2002, Jonathan Scherch undertook a bioregional tour, covering some thousand miles of urban and rural territory to explore existing and emerging sustainable-living enterprises. The tour included visits with individuals, community groups, businesses, and ecovillages. Beginning on the campus of Appalachian State University in Boone, North Carolina, in the mountains of western North Carolina, he traveled south, through Knoxville and the Sequatchie Valley to Chattanooga. Here are his observations.[13]

Although it has been almost four decades since E. F. Schumacher proclaimed that "small is beautiful" and "less is more," on the campus of Appalachian State University, the significance of these words has not been forgotten. Schumacher's remarks reflected a new perspective on social development at that time, emphasizing the use of *intermediate technology* as a bridge "between the capital-intensive advanced technologies of the 'West,' driven by large scale production and profit, and the traditional subsistence technologies of developing countries."[14]

Forty years later the Appropriate Technology Program at Appalachian State exemplifies Schumacher's spirit. Under the direction of Dr. Dennis Scanlin, the program engages students in interdisciplinary academic study, emphasizing "smaller scale technologies, that are ecologically and socially benign, affordable, and often powered by renewable energy."[15]

Students research, design, and improve upon various technologies in the service of regional communities and businesses. Considered *appropriate* if they perform with least possible cost, resource use, and impact while honoring the needs and preferences of the user, their technology experiments have included innovative solar kilns for drying lumber, photovoltaic solar panel and windmill installations for generating electricity, passive solar dehydrators for preserving foods, and more.

Scanlin observes that students develop knowledge and experience in such skill areas as "drafting and design, wood and metal working, computers, architecture, construction, graphic arts as well as renewable energy technologies, energy efficient solar building design and construction, waste management, research methods and contemporary technological problems facing society."[16]

Thus, the program's curriculum affords students opportunities to develop valuable creative problem-solving competencies, while sparking the curiosity of fellow students and faculty. Moreover, by bridging theory to practice, the projects seem to nurture self-confidence and self-reliance in ways not typically cultivated on college campuses.

The activities and intentions of Scanlin's classes seem to dovetail with those of the Earthaven ecovillage, an enclave of sustainable living visionaries located due south of Boone, in Black Mountain, North Carolina.

Learning Melds with Living at Earthaven

By definition, ecovillages constitute an "integrated solution to the global social and ecological crisis, and are as appropriate to the industrialized world, both urban and rural, as to the remaining two thirds of the world. Ecovillages are in essence a modern attempt by humankind to live in harmony with nature and with each other. They represent a 'leading edge' in the movement towards developing sustainable human settlements and provide a testing ground for new ideas, techniques and technologies that can then be integrated into the mainstream."[17] Earthaven represents the will and commitment of some fifty residents to change their ways of living in order to honor, restore, and enhance their relationships with each other and the earth.

An intentional community founded in 1995 amid 320 acres of wooded land, the resident "members" of Earthaven collectively grow, harvest, and preserve much of their own organic food on site; generate electricity from solar, wind, and stream-fed micro-hydro sources; design and build beautiful, efficient dwellings of wood, clay, straw, and recycled materials; and mill and dry their own lumber. By way of these and other enterprises, residents generally support each other as if their livelihoods were, indeed, a common interest.

Central to the emergence of the community are the philosophy and principles of permaculture developed by Bill Mollison and David Holmgren in the early 1970s. A design methodology for both urban and rural environs, permaculture (or permanent agriculture) emphasizes efficient, low-maintenance, and optimally productive integration of trees, plants, animals, structures, and human activities. The outcome is ecological stability via interdependent systems that wisely conserve, enhance, and carefully use natural resources to meet existing and future needs. The resulting social, economic, and even political processes of community living thus become instruments for monitoring how human needs and desires may be fulfilled (or not) in keeping with sensitive ecological thresholds.

To this end, Earthaven showcases many innovative appropriate technology, building design, and community development features and offers a variety of hands-on "Culture's Edge" courses designed to teach skills of sustainable living. Augmenting their educational resources with their *Permaculture Activist* and *Communities* magazines and website, their success to date has garnered worldwide interest in their activities.

Sustainable Living as a New Curriculum for the Twenty-first Century

While many educational institutions call our bioregion their home, the University of Tennessee remains a leading institutional resource and influence on our people, place, and culture. Moreover, with UT's history of instructional, research, and community service activities (since 1746), new insights about how best to enlist the university toward a sustainable bioregional future will likely test its educational value and mettle. One such test may well involve the successful integration of campus life and curriculum to form a contiguous, dynamic experience of sustainable living in action. David Orr argues that "more than any other institutions in modern society, colleges and universities have a moral stake in the health, beauty, and integrity of the world their students will inherit. We have an obligation to provide our students with tangible models that calibrate our highest values with our best capabilities—models that they can see, touch, and experience. . . . When the pedagogical abstractions, words, and whole courses do not fit the way the academic campus in fact works, they learn that hope is just wishful thinking, or worse, rank hypocrisy."[18] Orr further asserts that "the challenge of equipping students to participate in the building of a sustainable and decent society is *the* fundamental challenge to educational institutions at all levels."[19] To Brooke Judkins, a Child and Family Studies doctoral student in the College of Human Ecology at the University of Tennessee, the university is well positioned to explore new niches of sustainable living and learning in the spirit of Orr's charge.

Indeed, Judkins is exploring the relational dynamics amid families who attempt to integrate sustainable living activities into their lives. Understanding that activities like recycling, water and energy conservation, or modifying eating habits toward locally grown organics can evoke many types of reactions and learning processes, and Judkins hopes that her study will produce insights into how families adjust to such changes and the inherent lessons therein for us all.

And Judkins does not simply study sustainable living; she endeavors to model her convictions as well, paying attention to her choices and habits of food and energy consumption. While her lifestyle might well be cast as simple, her interests and aspirations, like her doctoral research, are noticeably complex. Indeed, as she reflects on her collegiate experiences to date, Judkins espouses a vision for the University of Tennessee campus that models a sustainable ethic and pedagogy similar to those being introduced on other university campuses.

In particular, she is intrigued by the development of the campus ecovillage project of Berea College, which calls for the construction of new facilities in keeping with principles of ecological design. More specifically, "When completed, the Berea College ecovillage will provide students and their families with lessons on the interaction of humans and nature. The ecovillage will include additional student family housing, a Sustainability and Environmental Studies House and a Child Development Lab, all of which will have significant ecological features imbedded in their designs. The ecovillage will not only meet the needs of the residents who live there, but will provide services and educational opportunities to the community and visitors as well."[20] This type of campus project, in which the content of curriculum is purposefully interwoven with the actual management and occurrence of campus life, represents an exciting and creative adaptation of university learning and living—one that Judkins hopes will take root at the University of Tennessee at Knoxville.

Environmental Education goes Urban, Rural, Communal

Moving from college campus to community commons, Ijams Nature Center, the Narrow Ridge Earth Literacy Center, and the Knoxville Diocese's Office of Justice, Peace and Integrity of Creation come into view. These three groups offer a distinctive though compatible palette of programs and services, whose missions appear to share common features: 1) environmental education for all ages; 2) ecological health and sustainability; 3) building and development that teaches as well as performs. At Ijams Nature Center, the facilities, grounds, and programs provide a compelling example of a neighborhood park with a purpose. Since 1976, Ijams's 150 acres of forest, dale, and riverbank have been home to activities and resources that enliven ecological stewardship for some fifty thousand visitors annually.

Ijams offers an assortment of innovative environmental education events, like its annual river rescue, which to date has removed tons of tires, piles of plastics, and dumpsters full of assorted debris from the shores of the Tennessee River; the Living Clean & Green program, which features twelve learning units that outline current community eco-issues and realistic responses to them; and their popular Earth Flag Program, which encourages recycling and conservation practices at some sixty elementary schools.

These and many other programs have positioned Ijams as a principal urban environmental education resource for Knoxville and vicinity. About an hour due north, in Washburn, Tennessee, is home of what might be envisioned as Ijams rural counterpart—the Narrow Ridge Earth Literacy Center.

Education Links Urban-to-Rural Learning

Similar in mission to Ijams, though unique in design and location, Narrow Ridge Earth Literacy Center is situated within a peaceful valley in Grainger County, Tennessee. The center, directed by Sister Mary Dennis Lentsch, aims "to provide experiential learning of Earth Literacy based on the cornerstones of spirituality, sustainability and community."[21] Moreover, the mission is "one of justice for human beings, animals, and Earth and sustainability in the way we live our lives and the way we create our institutions and technologies."[22]

Since the late 1970s, the center has evolved into a 500-acre rural educational resource featuring solar-powered, strawbale retreat facilities; programs in natural building, gardening, local food preservation, and blacksmithing; a thriving community-supported agriculture program involving some thirty member consumers; and even a cottage publisher, Earth Knows, whose list of titles include *What Have We Done,* the predecessor to this book.

These 500 acres are also home to three land-trust communities, with 100 acres set aside as permanent wilderness area and governed in accord with bylaws intended to preserve the ecological, spiritual, and aesthetic qualities of the land as new development occurs. An agenda of earth literacy, that is, being "knowledgeable about Earth's expertise as economist, educator, healer, lawgiver, and storyteller," permeates the practices in and of place, as the programs, buildings, people, and land coalesce to form an integrated sustainable living and learning experience.

While evoking a sense of eco-spirituality in community seems uniquely well suited to the setting of this rural locale, it is an important feature of yet another Knoxville-based organization.

Going Green with Faith-Based Communities

To enter the Office of Justice, Peace and Integrity of Creation (JPIC), an affiliate of the Knoxville diocese of the Catholic Church, is to step into a hotbed of activity where ecological stewardship and teachings of scripture are united. Such is the intent of JPIC, where Glenda Struss-Keyes and Sister Anne Hablus enlist the participation of Catholic parishioners across the bioregion toward a common eco-ecumenical voice.

While JPIC engages in a variety of public awareness and community education events, its Eco-Church Ministry claims much of its attention. This innovative program bridges spirituality, environmental stewardship, appropriate technologies, and sustainable building practices.

By participating in Eco-Church Ministry programs, JPIC aims to help parishes build earth-friendly church communities in which parishioners lead healthy, spiritually vibrant lives, church buildings are energy efficient and environmentally benign, and landscape designs are habitats for wild-life and parish social life alike. These and other program outcomes are described as embodying biblical significance and liturgical meaning and so are more than simply aesthetic or economic improvements—rather, they are means of practicing what is preached.

With the 1995 publication of *At Home in the Web of Life,* in which Appalachian Catholic Bishops endorsed a pastoral message for sustainable communities and related calls to "examine how we use and share the goods of the earth, what we pass on to future generations, and how we live in harmony with God's creation,"[23] the advent of the Eco-Church seems timely and pertinent. Moreover, as JPIC collaborates with other faith-based communities from around the region, a new spiritual community may well emerge, spanning the bioregion and representing collective will for a sustainable, peaceful future.

Continuing our Bioregional journey southward, we venture to the homestead and living laboratory that is the Sequatchie Valley Institute.

Evoking Sustainable Community in the Southeast

Like Earthaven to the north, the Sequatchie Valley Institute (SVI) offers an impressive example of intentional, full-featured sustainable living. Nestled amid three hundred densely wooded acres at the foot of sheer escarpments defining the westward boundary of the valley, this original homestead of Johnny and Carol Kimmons, commonly known as Moonshadow, has grown since 1971 to become a regional educational resource touting a rich port-folio of practices and services.

With an aim of "working towards a sustainable future through education, example, and land conservation,"[24] the institute affords visitors an experience of applied curriculum, where hands-on learning and doing invoke themes of permaculture design, natural building, alternative energy, integrated food systems, and creative arts.

Complementing their roles as professors at the University of Tennessee at Chattanooga, where the Kimmons teach biology and environmental science, their family and SVI staff employ the skills they teach amid their routines of daily living. Having built by hand and tool remarkable dwellings of straw, clay, lumber, and stone, growing much of their food, and harvesting the sun for electricity and heat, the staff are indeed multitalented and accomplished at their crafts.

Living what they espouse enlivens theory and connects it to practice. Moreover, the appropriate practices that they model seem reasonably accessible to all regardless of age, stage, or gender. As at Earthaven SVI invites and enlists the equal participation of both men and women. Thus, all who participate find opportunity to gain new insights and skills that heretofore might go abashedly untapped—use of precision machinery and hand tools, designing buildings and energy systems, harvesting and preserving food, or creative arts and dance.

Along with its programs and services, the Sequatchie Institute portrays a blend of self-reliant living and interdependent relationships with neighboring people, land, and community. Such richly textured livelihoods— as opposed to the mainstream's dependence on centralized, often anonymous sources for food, fuel, energy, and the like—may well be the template that preserves land-steward tradition and prepares future generations with knowledge and skill sets necessary for competent, low-impact living. Thus, SVI's sustainable-living innovations, always in community with others, seem a healthy and hopeful alternative.

Eco-nomics Emerging for Chattanooga

In Chattanooga, several business ventures are emerging with distinctive eco-entrepreneurial spirit, as they pursue business models and methods that respect and restore, rather than damage and degrade, the bioregion's environment.

Talk with long-time Chattanooga resident and entrepreneur David White, for example, and you will immediately gain a history lesson, economic vision, practical business plan, and eco-entrepreneurial design at once. Such capacity for storytelling, engineering, and economic development are at the forefront of White's attention as he tracks and calculates the evolution of his present venture.

A whole-systems thinker, White has been a part of the "ebb and flow" of Chattanooga's sustainable development efforts over the years, having collaborated with former Councilman David Crockett and others on initial Sustainable Chattanooga events in the early 1990s. At that time Chattanooga was a hotspot for visions and projects of urban sustainability and warmly welcomed innovation to move ahead. Accordingly, the city revitalized its downtown scene for pedestrian and bike-friendly use, introduced its popular electric bus service to shuttle visitors to its Riverwalk district and world-class freshwater aquarium, and transformed a derelict foundry into the now open-air Cricket Pavilion and adjacent community commons for public events and farmers markets, to name a few celebrated achievements.

Yet despite the celebrations, the lingering, tempering effects of the postindustrial era remain apparent in the community, economy, and environment. Left now are vacant factories (their rusting remains once the economic engines delivering products, jobs, and city tax revenues), mountains of residual industrial discards, and the yet to be restored Chattanooga Creek, all which depict a wholly depressing scene. But David White sees opportunity.

White aims to implement an industrial ecology model that would link new and existing businesses, communities, and government to create a "closed-loop" resource-sharing system, where the output "wastes" of one constituent become the valuable input resources of another. In complement to White's intent, Team Associates, Inc., has devised plans for a *SMART* Park—a proposed model for employing Sustainable Manufacturing, Agricultural, and Recycling Technology. These integrated development models would feature systems that share energy, heat, water, and even employee skills and expertise (human resources) in order to optimize efficiency, reduce use of increasingly expensive resources, achieve low- or even zero-emission standards, and improve social, economic, and environmental contexts.

Moreover, White's vision would have Chattanooga create an attractive scenario for present and prospective businesses to enjoy collaboratively benefits of an interdependent economy. He envisions designs for community re-development projects and new or rejuvenated tourist attractions linking the riverbank promenade to the crest of Lookout Mountain.

The medium for jumpstarting the concept is White's own venture—a veritable diamond in the rough—RamRock Compression Formed Masonry, which produces an innovative, low-cost, durable building material (a sort of high-tech adobe) made of spent industrial sand, found piled hundreds of feet high along the banks of the Tennessee River, coal combustion byproducts, millions of tons of which are produced annually from bioregional power plants, and other salvaged composite materials.

Turning waste into wealth, White hopes to harvest these heaps of unappreciated resources and create healthy, affordable dwellings as part of an evolving re-creation of once-vibrant core communities. RamRock has already been successfully applied within a local Habitat for Humanity project, averaging forty-five dollars per finished square foot, with enthusiastically received results.

However, along with its exciting sustainable community development initiatives, Chattanooga must also steward its waters—and the Chattanooga Creek demands attention. A particularly troublesome community feature representing an infamous legacy of use and abuse, the Chattanooga Creek in its present polluted state flows through the local community and carries with it, some might say, the future or fate of local residents.

Restoring the Chattanooga Creek—Naturally

A tributary of the Tennessee River, the Chattanooga Creek's fouled waters have tainted the health, welfare, and civic trust of local residents for decades, and more recently it has become a focal and vocal point for environmental justice concerns. Fortunately, the creek has also served as a living laboratory for an experimental natural water-treatment approach designed by Dr. John Todd—the "living machine."

These devices, also referred to as Restorer Water Treatment Technologies,[25] enlist networked tanks or floating rafts of biologically diverse ecologies, including plants, animals, fungi, and bacteria, to mimic naturally occurring aquatic systems, and so absorb, digest, and process excess nutrients and dangerous substances. More specifically, these systems "support a wide spectrum of treatment processes and the organism's responsible for the treatment. They can substantially reduce organic loading, control algal communities and improve water clarity (lower turbidity). They have the capacity of removing pathogens and priority pollutants. They also can reduce nutrient levels, including nitrogen and phosphorous [and] also enhance the dynamic ecological cycles within the lagoons, ponds, lakes, and canals in which they reside."[26]

In the mid-1990s, Todd established a prototype "machine" for Chattanooga Creek. Using samples of toxic sludge, railroad ties, and creek debris, he observed that such constructed ecosystems are capable of first tolerating, then acclimating to and eventually harmonizing with the troublesome substances. Using sunlight and photosynthesis to produce its water-cleansing alchemy, Todd's remarkable bioaquatic systems have since been installed by colleges, communities, and businesses around the world to restore and resuscitate life between people and their local waters.

Clean Energy Comes to Chattanooga

Equally entrepreneurial and continuing the local theme of innovation and stewardship in Chattanooga, the staff of Big Frog Mountain (BFM) is busy tapping the energies of sun, water, and wind in service of social, economic, and environmental revitalization.

Launched in 1989 BFM provides easy access to a variety of alternative energy products and appliances for homeowners, businesses, and communities. With so-called "plug and play" technology improvements, today's tools for an energy revolution are no longer the domain of tinkering experimenters, isolated homesteaders, or professorial gurus. Now, tens of thousands of people worldwide generate and efficiently use their own reliable, clean energy products, sometimes grid intertied with their local utilities to feed their excess "green power" back for others to use, or as a self-reliant producer acting as their own independent utility company.

In Chattanooga, and throughout the bioregion, many people have innovatively retrofitted their homes and businesses to meet needs for electricity, hot water, and heating with cost-effective and easy-to-use success. Perhaps an offshoot of the so-called home-power movement that emerged in the early 1970s and continues to grow to date, the demand for clean energy is increasing as never before.

To meet the demand, BFM offers a variety of energy-efficient appliances that, if not yet commonly available at local appliance or hardware stores, are now available via the internet for customers to select precision components for tailor-designed systems that meet their particular needs. From solar panels and windmills harvesting seasonal weather events to power-efficient refrigerators, freezers, and washing machines, innovative lighting, heating, and cooling systems, and more, the process of choosing clean energy alternatives is now easier than ever.

Agriculture and Food Security

With the popular emergence of clean energy as a social and environmental movement of sorts, so too have the health, welfare, and stability of our community food systems received increased attention and scrutiny. Indeed, food security has recently become a topic of great interest within and beyond the bioregion, as the origins, quality, and implications of our food choices are critically examined.

Moreover, as smaller farms gradually give way to economic exigencies favoring corporate agribusiness, an unfortunate phenomenon is revealed— the disconnection and dissolution of farming as an enterprise of agriculture. The demise of the family farmstead produces a mosaic of interactive, compounding effects: an aging small farm community is producing less quantity and diversity of truly fresh organic produce, on farms of uncertain land tenure, and in relative isolation. Moreover, small farmers are having to do more with less working capital, essential to buffering crop failures and offsetting increased debt load, while facing complex and impersonal marketplace realities including strong commodity price-point competition with agribusiness producers.

With these and other acute pressures facing small farm communities, their agricultural traditions are also jeopardized. As increasingly fewer young farmer apprentices wait in the eaves, either willing or capable to succeed and continue their predecessors' tenuous ways of farming, the characteristic soil-building, plant-stewarding, personal practices of "old-time" organic agriculture may be fading—practices that may now seem mistakenly unsophisticated, contradictory, and/or perhaps confusing to those with modern "state-of-the-art" agribusiness-oriented training. In short, the future of food as we have known it and expect it would seem to be in peril—but then people like Will and McNair Bailey come along whose example inspires hope.

An Urban Farm Finds a Home

With a passion for food and a personal commitment to small, organic farming, the Baileys launched what has grown to become an impressive urban food resource, Crabtree Farms. With twenty-two acres of zoned agriculture land (dating back two hundred years), the Baileys and their staff have "grafted" a new model of urban agriculture in the heart of east Chattanooga—one that feeds, teaches, stewards, and leads all at once.

Tranquilly situated amid a typical single-family neighborhood, and with only a few years of operating experience, Crabtree is by no means an unnoticed neighbor. Hundreds of Chattanoogans rely directly on their foodstuffs, and thousands enjoy them at restaurants and markets—their relative success evidenced by their growing mailing list of over thirteen hundred.

The farm supports a thirty-six-plot community garden, a farm stand offering a variety of fruits, vegetables, flowers, and herbs, and series of workshops and community-harvest celebrations. Though still fledgling farmers, the Baileys attribute their humble successes to their unwavering focus on learning the business of farming. From their standpoint, small farming suffers in large part from a dearth of farmers with adequate business management training, and, more specifically, business planning skill for growing a successful, diversified farm enterprise.

In response to this dilemma, the Baileys have offered farm-management training workshops and have identified features that, if organized into a business plan, might buffer small farmers from the effects of overdependency on single-element success: a combination of on/off farm employment, adequate variation of field and orchard crops (avoiding overdiversification and featuring specialty "niche" items), valued-added goods like cut flowers, animal husbandry for both farm work and food products, and educational programs to facilitate community connections and partnerships.

While doing all of this offers no guarantee of success (the Baileys themselves are uncertain about whether current socioeconomic conditions bode well for the continuance of the small farm), the equation of management skills, crop and income diversity, and land stewardship seems a compelling alternative to large-scale, centralized, mono-crop models. Indeed, about thirty-five minutes north of Chattanooga, at what might be Crabtree's rural counterpart, Little Sequatchie Cove Farm is taking care of business in a similarly rewarding way.

Cultivating Wisdom on the Farm

More than a place for raising organic vegetables and animals, Little Sequatchie Cove Farm is home to the Keener family, who engage in farming as a practice of stewardship in action. Offering a range of locally produced organic meats, eggs, fruits, and vegetables, the Keeners' manner of farming is as healthy as their products. Their aim is threefold: to produce fresh, nutritious foods; to restore soil fertility; and to re-enliven and sustain agriculture as a life-affirming vocation.

With some three hundred acres of land, sixty devoted for pasture, and involving three generations of family, the Keeners practice farming with a whole systems perspective. Here, the relationships between system elements are as important to system integrity as the elements themselves—healthy soils support healthy plants and animals, which in turn produce healthy people (i.e., satisfied, returning customers) who lead healthy, productive lives.

With an eye on maintaining system integrity, the Keeners have developed a diverse portfolio of crops and animals, including some 50 cows, 100 sheep, 10 pigs, 100 heritage breed Red turkeys, 250 chickens, vegetable varieties, and even native plants for landscaping applications. Moreover, by employing, for example, rotational and open-space grazing routines, the Keeners afford their animals adequate physical and dietary variation while attending to the health of soils and field plants—allowing their family time to do other things. Indeed, their routines of daily living seem to reflect and respect the diverse rhythms of farm and farmer alike: "We are constantly tinkering with things, and our farm is always changing. Our farm's prosperity depends upon the fertility of the soil and the production methods and creative insight of the farmer. It is a big job and no day is ever long enough. Besides the farm work, there are woods, hills, caves, and cliffs that need to be explored, swimming to be done, buildings to build, tractors and farm equipment to maintain, birds and frogs to listen to, and the rope swing to fly on. . . ."[27] Thus, perhaps more than the quality of their foods, the fertility of the fields, and the participation of their family, the most profound feature of Little Sequatchie Cove Farm may well be the intrinsic quality of life that underpins their truly perennial agriculture—what might arguably be *the* most important value added product of them all.

Sustainable Business Opportunities

A number of innovative regional organizations are supporting the development of small sustainable businesses.[28] The Mountain Microenteprise Fund is a nonprofit organization that operates its programs with a staff of ten, from its main office in Asheville, North Carolina, and four satellite offices in Asheville, Hendersonville, Tryon, and Murphy, North Carolina. Its annual budget is $550,000, with the majority of funding coming from foundations and institutions. MMF helps entrepreneurs by providing business-planning courses as well as ways to finance small businesses. Their "Individual Development Account," is a capitalizing fund that matches individual contributions of up to $1,000.[29] MMF assists artisans and entrepreneurs of western North Carolina by helping to develop a viable market for regionally produced goods and services. MMF also provides small business loans. MMF gave out 20 loans, and 78 jobs were created through their efforts. In 2001 more than 22 entrepreneurs completed the business-training course. Of those, 52 percent were low-income individuals/families. MMF disbursed 22 loans for a total of $59,520.[30]

Appalachian Sustainable Development (ASD) is another regional, nonprofit organization that seeks to build ecologically healthy, locally rooted economics in southwest Virginia and northeast Tennessee. ASD supports sustainable development through establishing programs that meet five working principles of sustainable community development:

is locally rooted, diversifying the economy and culture of communities and regions;

fits within the ecosystem, building upon natural assets, honoring limits of absorption and regeneration;

promotes regional self-reliance by building both individual skills and cooperative, innovative networks;

adds value to raw materials and shortens the distance between "producers" and "consumers"; and

lasts indefinitely by building the assets—ecological, human and financial—of particular places.[31]

ASD's service area focuses on the Tennessee counties of Hancock, Hawkins, Sullivan, Washington, and Greene, along with the Virginia counties of Lee, Wise, Scott, Russell, Dickenson, and Washington. The total land-service area is about 3,500 square miles and includes some 400,000 people. ASD's "field-to-table" strategies strive to link entrepreneurs and the land they manage with the marketplace. ASD provides hands-on education for farmers, landowners, and others and sponsors a cooperative marketing system in order to sustain locally based businesses. Currently, ASD's two main primary focus areas are *sustainable agriculture* and *sustainable forestry and wood-product development.*

ASD works to develop "action-oriented" relationships with farmers, loggers, and other small-scale enterprises, or community-based organizations and public agencies involved in sustainable development. The organization has partnered with the Virginia and Tennessee Cooperative Extension, the Dickenson County Department of Economic Development, and the Russell County Industrial Development Authority, the Nature Conservancy, Appalachian Spring Cooperative, and many more. ASD works with members of the Central Appalachian Network on the development of product identities for sustainably produced agriculture products.

ASD's Appalachian Harvest and Sustainable Forestry and Wood Products (SFWP) Program are just two of the locally based, sustainable businesses that have developed in the bioregion in recent years. The goals of the SFWP program are to improve the quality of forest practices on private lands, and to encourage local processing of forest resources in order to add value and create jobs.[32] ASD's market-based "forest-to-table" strategy includes outreach, education, and technical assistance for the conservation and sustainable use of private forestlands, the creation of processing capacity to turn logs into kiln-dried lumber and other value-added forest products and the development of markets for sustainably produced forest products. ASD partners with local wood manufacturers to make quality, kiln-dried lumber products including custom cabinets, trim and molding, hardwood flooring, paneling, and finished lumber. ASD's Sustainable Woods Processing Center in Castlewood, Virginia, makes available kiln-dried lumber, green lumber, cants for pallets and crossties, firewood, sawdust, and timber bridges.

Through ASD and the Russell County, Virginia, Industrial Development Authority, logs are processed using a unique 20,000-board-foot kiln that uses solar power and wood waste to dry the wood to 6- to 8-percent moisture content. The business has also spawned related ventures, including a potting soil company that utilizes leftover sawmill sawdust as its primary component.

ASD also supports an agricultural enterprise, Appalachian Harvest, which has increased small farm production in the region and produced spin-off businesses, including a garlic seed enterprise created to serve the growing number of garlic producers in the area. With assistance from ASD, a network of farmers have come together to grow certified organic produce and make it available wholesale to a number of grocery chains. Selling under the brand name Appalachian Harvest, these farmers are now making their fresh, organic produce available in Virginia, eastern Tennessee, and North Carolina. Sweet peppers, heirloom and slicing tomatoes, cucumbers, eggplant, yellow squash, sauce tomatoes—all organically grown in Virginia and Tennessee—are just some of the items marketed through Appalachian Harvest.

These developments can provide the foundation for a diversified base of economically and ecologically sustainable local businesses, including farmers, loggers, and wood manufacturers. Additionally, nearly one hundred microloans have been made to entrepreneurs from Business Start, an ASD partner. These start-up efforts are promoting ecotourism and other green businesses. Incorporating traditional Appalachian values of frugality, resourcefulness, and community, these businesses represent the sustainable potential of our bioregion.[33]

The history of the Appalachian Mountains is rich with craftspeople and artisans. As modern retailing developed and our culture changed, markets became difficult to reach, and many could no longer afford to live off the crafts they made. Appalmade, a community action agency, is striving to change this. Appalmade is part of People, Incorporated, of Southwest Virginia—a private, nonprofit organization whose mission is to improve social and economic conditions in the Appalachian region. Created by a few community development specialists, Appalmade provides marketing opportunities for 150 Appalachian craftspeople, working from their homes or community centers.

Appalmade began by marketing crafts through a wholesale program funded by a grant from the Levi Strauss Foundation. The wholesale program expanded to catalog sales and sales in gift shops, including one at the Museum of Contemporary Folk Art in New York City. Appalmade's wholesale customer base has grown to include some one hundred retail shops, primarily in the northeastern United States. Appalmade also sells its products through its website.[34]

Located in the beautiful Southern Appalachian Highlands of Hancock County, Tennessee, Appalachian Spring Cooperative is a member-owned, value-added marketing association of family farmers and food and herbal product manufacturers. Member producers utilize the Clinch-Powell Community Kitchens in Treadway, Tennessee, to produce premium-quality specialty foods and herbal products.

Appalachian Spring Cooperative offers a range of member services to support the development of value-added food and farm-product businesses. The cooperative also offers consulting services to nonmembers and community organizations to assist in the development of sustainable farm-based enterprises.

Appalachian Spring Cooperative is partnering with Heifer Project International to place honeybee hives on the farms and lands of cooperative members. Heifer International, a nonprofit based in Little Rock, Arkansas, combats hunger, alleviates poverty, and restores the environment by providing appropriate livestock, training, and related services to small-scale farmers worldwide. Heifer Project helps people utilize livestock as an integral component of sustainable agriculture and holistic development. In exchange for a percentage of the hives' production to support the development of the cooperative, the project provides participants with bee-keeping training, hives, and markets for honey products. Honey-processing equipment at Clinch-Powell Community Kitchens is available for use by all project participants. Participants also enjoy the added benefit of bees for crop pollination and honey for home use or private commercial sales.

With the aid of grant funding from Southern SARE, the Southern Rural Development Council, and USDA, the Appalachian Spring Cooperative is also implementing several programs to promote and improve the economic and environmental sustainability of family farms in the Southern Appalachian region. Utilizing workshops, printed materials, websites, and on-farm field days, the coop's programs promote access to technology and technical assistance for entrepreneurs; market research and development; crop diversification and the planting of high-value specialty crops in the Southern Appalachian region; pesticide use reduction through implementation of organic and integrated pest-management crop production systems; soil-erosion control through use of buffer strips, cover crops, and low-till and no-till practices and more.[35]

There are many farmers growing fresh produce in the mountains of Tennessee and southwest Virginia. Additionally, there is a growing local production of meat and animal products for direct sale to local customers. Saturday Morning Farmers Markets during the growing season include markets in Abingdon, Virginia, Bristol, Tennessee, Kingsport, Tennessee, Johnson City, Tennessee, Lebanon, Virginia, Tazewell, Virginia, Greenville, Tennessee, Morristown, Tennessee. Community Supported Agriculture (CSAs) can also be found in the region. ASD, Rural Resources (in Green County, Tennessee), and Narrow Ridge in Grainger County have worked with or started several CSAs in the area.

These are but a few of the innovative works of sustainability emerging across the bioregion. They are signs of health and hope.

Recommended Resources

Websites

Big Frog Mountain[36]
http://www.bigfrogmountain.com/
Crabtree Farms of Chattanooga
http://www.crabtreefarms.org
Earthaven EcoVillage
http://www.earthaven.org/
Home Power Magazine
http://www.homepower.com
Ijams Nature Center
http://www.ijams.org/
Narrow Ridge Earth Literacy Center
http://www.narrowridge.org/
Ocean Arks International
http://www.oceanarks.org
Sequatchie Valley Institute

http://www.svionline.org/
Sustainable Social Work Mandala
http://www.designtrek.net

Publications

Communities Magazine. *Journal of Cooperative Living.* 138 Twin
Oaks Road, Louisa, Va. 23093.

Hawkin, P. A. Lovins, and L. H. Lovins. *Natural Capitalism:
Creating the Next Industrial Revolution.* Boston, Mass.: Little,
Brown, 1999.

Mollison, B. *Permaculture: A Designers Manual.* Tasmania: Tagari
Publications, 1997.

————. *Introduction to Permaculture.* Tasmania: Tagari
Publications, 2002.

Orr, D. *Earth in Mind: On Education, Environment, and the Human
Prospect.* Washington, D.C.: Island Press, 1994.

The Permaculture Activist. P.O. Box 1209W, Black Mountain, N.C.
28711.

Schimtz-Gunter, T., ed. *Living Spaces: Sustainable Building and
Design.* Cologne: Konemann, 1998.

Todd, J., and N. J. Todd. *From EcoCities to Living Machines:
Principles of Ecological Design.* Berkeley, Calif.: North Atlantic
Books, 1996.

Traina, F., and S. Darley-Hill, eds. *Perspectives in Bioregional
Education.* Rock Spring, Ga.: North American Association
for Environmental Education (NAAEE), 1997.

Appendix 1

Bacteriological Advisories
on Waterways in the Bioregion, 2002

TENNESSEE WATERWAYS

Waterway	Portion	County	Comments
Beaver Creek(Bristol)	Tenn./Va. line to Boone Lake (20.0 miles)	Sullivan	Nonpoint sources in Bristol and Virginia.
Cash Hollow Creek	Mile 0.0 to 1.4	Washington	Septic tank failures.
Coal Creek	Sewage treatment plant to Clinch R. (4.7 miles)	Anderson	Lake City Sewage Treatment Plant (STP).
East Fork Poplar Creek	Mouth to Mile 15.0	Roane	Oak Ridge area.
First Creek	Mile 0.2 to 1.5	Knox	Knoxville urban runoff.
Goose Creek	4.0 miles	Knox	Knoxville urban runoff.
Leadvale Creek	Douglas Lake to headwaters (1.5 miles)	Jefferson	White Pine STP.
Little Pigeon River	Mile 0.0 to 4.6.	Sevier	Improper connections to storm sewers, leaking sewers, and failing septic tanks.
Second Creek	Mile 0.0 to 4.0	Knox	Knoxville urban runoff.
Sinking Creek	Mile 0.0 to 2.8	Washington	Agriculture & urban runoff.
Sinking Creek Embayment of Fort Loudoun Reservoir	1.5 miles from head of embayment to cave	Knox	Knoxville Sinking Creek STP.
Third Creek	Mile 0.0 to 1.4, Mile 3.3	Knox	Knoxville urban runoff.

Waterway	Portion	County	Comments
East Fork of Third Creek	Mile 0.0 to 0.8	Knox	Knoxville urban runoff.
Johns Creek	Downstream portion (5.0 miles)	Cocke	Failing septic tanks.
Baker Creek	Entire stream (4.4 miles)	Cocke	Failing septic tanks.
Turkey Creek	Mile 0.0 to 5.3	Hamblen	Morristown collection system.
West Prong of Little Pigeon River	Mile 0.0 to 17.3	Sevier	Improper connections to storm sewers, leaking sewers, and failing septic tanks.
Beech Branch	Entire stream (1.0 mile)		
King Branch	Entire stream (2.5 miles)		
Gnatty Branch	Entire stream (1.8 miles)		
Holy Branch	Entire stream (1.0 mile)		
Baskins Branch	Entire stream (1.3 miles)		
Roaring Creek	Entire stream (1.5 miles)		
Dudley Creek	Entire stream (5.7 miles)		
Chattanooga Creek	Mouth to Ga. line (7.7 mi.)	Hamilton	Chattanooga collection.
Oostanaula Creek	Mile 28.4–31.2 (2.8 miles)	McMinn	Athens STP and upstream dairies.
Stringers Branch	Mile 0.0 to 5.4	Hamilton	Red Bank collection system.
Citico Creek	Mouth to headwaters (7.3 miles)	Hamilton	Chattanooga urban runoff and collection system.

VIRGINIA WATERWAYS

Waterway	Portion	County	Comments
Straight Creek	Mile 0.0 to 6.66	Lee	The source of the fecal coliform violations is numerous raw sewerage discharges.
Cigarette Hollow	Mile 0.0 to 1.08	Russell	Fecal coliform violations are probably attributable to the failing septic systems and straight pipes along the stream.
Lick Creek	Mile 0.0 to 4.83	Russell	Fecal violations are due to failing septic systems and straight pipes along the stream.
Right Fork	Mile 0.0 to 2.91	Russell	Fecal violations are attributable to the raw sewage discharges from individual homes and failing septic systems along the stream.
Laurel Branch	Mile 0.0 to 4.96	Russell	Fecal violations are attributable to the raw sewage discharges from individual homes and failing septic systems along the stream.
Cedar Creek	Mile 0.0 to 5.24	Washington	Septic tank inputs and agricultural runoff.
Byers Creek	Mile 0.0 to 0.87	Washington	Septic tank inputs and agricultural runoff.
Hall Creek	Mile 0.0 to 5.87	Washington	Septic tank inputs and agricultural runoff.

Waterway	Portion	County	Comments
Hutton Creek	Mile 0.0 4.79	Washington	Septic tank inputs and agricultural runoff.
Middle Fork Holston River	Mile 17.03 to 27.15	Washington	Town of Chilhowie STP is directly above this segment and may be a contributing source to the violations.
Beaver Creek	Mile 15.27 to 28.73	City of Bristol, Washington	Beaver Creek flows through an intense agricultural area as well as being an urban stream as it crosses the state line. Both of these land uses contribute to water quality impacts. DCR ranks the watershed high for overall nonpoint source potential impacts.
Little Creek	Mile 0.26 to 5.78	City of Bristol, Washington	The source is unknown, however the watershed has a high ranking for urban impacts. It is suspected that fecal violations are a combination of urban and agricultural impacts of nonpoint sources.
Crab Orchard Creek	Mile 0.0 to 2.43	Wise	Urban septage disposal is suspected as the source for fecal violations.
Little Toms Creek	Mile 0.0 to 4.37	Wise	Fecal violations are attributable to the raw sewage discharges from individual homes and failing septic systems along the stream.

Waterway	Portion	County	Comments
Sepulcher Creek	Mile 0.0 to 2.60	Wise	The population is dense along the stream banks so that Urban Nonpoint sources are suspected as the reason for the high fecal coliform counts.
Toms Creek	Mile 0.0 to 11.61	Wise	There are many communities and houses along the stream. Although a regional sewage treatment plant, Coeburn Norton Wise STP, has improved sewage treatment for the three towns, inflow and infiltration has not been completely corrected and there are still unsewered communities.

GEORGIA WATERWAYS

Waterway	Portion	County	Comments
Nottely River	U.S. Hwy 19 to Lake Nottely (8 miles)	Union	Fecal Coliform Nonpoint Pollution.

SOURCES: Compiled by Keith Bustos. Information for this table was taken from TDEC, *The Status of Water Quality in Tennessee: Year 2002 305(b) Report* (Nashville, Tenn.: Division of Water Pollution Control, 2002); North Carolina Dept. of Environment and Natural Resources, *Water Quality Assessment and Impaired Waters List (2002 Integrated 305(b) and 303(d) Report)* (Raleigh, N.C.: Division of Water Quality, 2003); Georgia Environmental Protection Division, *2000 303(d) List: Rivers/Streams Not Supporting Designated Uses* (Atlanta: Water Protection Branch, 2000); Virginia Department of Environmental Quality, *2002 Virginia Water Quality Assessment Report* (Virginia: Department of Conservation and Recreation, 2002).

Appendix 2

Fish Tissue Advisories on Waterways
in the Bioregion, 2002

Stream	County	Portion	Pollutant	Comments
Boone Reservoir	Sullivan, Washington	Entirety (4,400 acres)	PCBs, chlordane	Precautionary advisory for carp and catfish.*
Chattanooga Creek	Hamilton	Mouth to Georgia Stateline (11.9 miles)	PCBs, chlordane,	Fish should not be eaten. Also avoid contact with water.
East Fork of Poplar Creek including Poplar Creek embayment	Anderson, Roane	Mile 0.0 – 15.0	Mercury, PCBs	Fish should not be eaten. Also avoid contact with water.
Fort Loudoun Reservoir	Loudon, Knox, Blount	Entirety (14,600 acres)	PCBs	Commercial fishing for catfish prohibited by TWRA. No catfish or large mouth bass over two pounds should be eaten. Do not eat largemouth bass from the Little River embayment.
Melton Hill Reservoir	Knox, Anderson	Entirety (5,690 acres)	PCBs	Catfish should not be eaten.
Nickajack Reservoir	Hamilton, Marion	Entirety (10,370 acres)	PCBs	Precautionary advisory for catfish.*
North Fork Holston River	Sullivan, Hawkins	Mile 0.0–6.2 (6.2 miles)	Mercury	Do not eat the fish. Advisory goes to TN/VA line.
Tellico Reservoir	Loudon	Entirety (16,500 acres)	PCBs	Catfish should not be eaten.

Stream	County	Portion	Pollutant	Comments
Watts Bar Reservoir	Roane, Meigs, Rhea, Loudon	Tennessee River portion (38,000 acres)	PCBs	Catfish, striped bass, & hybrid (striped bass, whitebass) should not be eaten. Precautionary advisory* for whitebass, sauger, carp, smallmouth buffalo and large-mouth bass.
Watts Bar Reservoir	Roane, Anderson	Clinch River arm (1,000 acres)	PCBs	Striped bass should not be eaten. Precautionary advisory* for catfish and sauger.

VIRGINIA WATERWAYS

Stream	County	Portion	Pollutant	Comments
North Fork Holston River	Scott, Smyth, Washington	Mile 3.0 to 85.4	Mercury	The Olin Matheson Plant site is the mercury source.

NORTH CAROLINA WATERWAYS

Stream	County	Portion	Pollutant	Comments
Waterville Lake	Haywood	340 acres	Dioxins	Unknown

NOTE: *Precautionary Advisory: children, pregnant women, and nursing mothers should not consume the fish species named. All other persons should limit consumption of the named species to one meal per month.

SOURCES: Compiled by Keith Bustos. Information for this table was taken from TDEC, *The Status of Water Quality in Tennessee: Year 2002 305(b) Report* (Nashville, Tenn.: Division of Water Pollution Control, 2002); North Carolina Dept. of Environment and Natural Resources, *Water Quality Assessment and Impaired Waters List (2002 Integrated 305(b) and 303(d) Report)* (Raleigh, N.C.: Division of Water Quality, 2003); Georgia Environmental Protection Division, *2000 303(d) List: Rivers/Streams Not Supporting Designated Uses* (Atlanta: Water Protection Branch, 2000); Virginia Department of Environmental Quality, *2002 Virginia Water Quality Assessment Report* (Virginia: Department of Conservation and Recreation, 2002).

Appendix 3

Federal and State Listed Threatened and Endangered Species in the Upper Tennessee Drainage

Scientific Name	Common Name	Federal Status	State Status
BIRDS			
Falco peregrinus	Peregrine falcon	Dlm[a]	End[f]
Gallinula chloropus	Common moorhen	Le[b]	Nmgt[g]
Haliaeetus leucocephalus	Bald eagle	Lt[c]	Nmgt
Lanius ludovicianus	Loggerhead shrike	Spco[d]	End
Picoides borealis	Red-cockaded woodpecker	Le	Exti[h], end
FISHES			
Cyprinella monacha	Spotfin chub	Lt	Exti, end, thr[i]
Erimystax cahni	Slender chub	Lt	End, thr
Etheostoma percnurum	Duskytail darter	Le	End
Notropis albizonatus	Palezone shiner	Le	End
Noturus baileyi	Smoky madtom	Le	End
Noturus flavipinnis	Yellowfin madtom	Lt	Exti, end
Noturus sp	Chucky madtom	C[e]	End
Noturus stanauli	Pygmy madtom	Le	End
Percina tanasi	Snail darter	Lt	End, thr
Phoxinus cumberlandensis	Blackside dace	Lt	

Scientific Name	Common Name	Federal Status	State Status
MAMMALS			
Corynorhinus townsendii virginianus	Virginia big-eared bat	Le	End
Glaucomys sabrinus coloratus	Carolina northern flying squirrel	Le	End
Glaucomys sabrinus fuscus	Virginia northern flying squirrel	Le	End
Lasiurus cinereus	Hoary bat	Ps	Potl[j]
Myotis grisescens	Gray bat	Le	End
Myotis sodalis	Indiana bat	Le	End
Zapus hudsonius	Meadow jumping mouse	Ps	Nmgt
REPTILES			
Clemmys muhlenbergii	Bog turtle	Lt	Thr
ARTHROPODS			
Lirceus usdagalun	Lee county cave isopod	Le	Spco
Glyphopsyche sequatchie	Owen spring limnephilid caddisfly	C	Potl
Pseudanophthalmus holsingeri	Holsinger's cave beetle	C	
Microhexura montivaga	Spurce-fir moss spider	Le	Stun[k]
GASTROPODS			
Athearnia anthonyi	Anthony's river snail	Le	End
Elimia interrupta	Knotty elimia	C	End
Mesodon clarki nantahala	Noonday globe	Lt	Thr
Pyrgulopsis ogmorhaphe	Royal marstonia	Le	End

Scientific Name	Common Name	Federal Status	State Status
MUSSELS			
Alasmidonta raveneliana	Appalachian elktoe	Le	End
Cyprogenia stegaria	Fanshell	Le	End
Dromus dromas	Dromedary pearlymussel	Le	End
Epioblasma brevidens	Cumberland combshell	Le	End
Epioblasma capsaeformis	Oyster mussel	Le	End
Epioblasma florentina walkeri	Tan riffleshell	Le	End
Epioblasma torulosa gubernaculum	Green blossom pearlymussel	Le	End
Epioblasma torulosa torulosa	Tuberculed blossom pearlymussel	Le	End
Epioblasma turgidula	Turgid blossom pearlymussel	Le	End
Fusconaia cor	Shiny pigtoe pearlymussel	Le,	End
Fusconaia cuneolus	Fine-rayed pigtoe	Le	End
Hemistena lata	Cracking pearlymussel	Le	End
Lampsilis abrupta	Pink mucket	Le	End
Lampsilis virescens	Alabama lampmussel	Le	End
Lemiox rimosus	Birdwing pearlymussel	Le	End
Lexingtonia dolabelloides	Slabside pearlymussel	C	Thr
Pegias fabula	Little-wing pearlymussel	Le	Exti, end
Plethobasus cicatricosus	White wartyback Le End		
Plethobasus cooperianus	Orange-foot pimpleback	Le	End
Pleurobema plenum	Rough pigtoe	Le	End
Ptychobranchus subtentum	Fluted kidneyshell		C
Quadrula cylindrica strigillata	Rough rabbitsfoot	Le	End, thr
Quadrula intermedia	Cumberland monkeyface	Le	End

Scientific Name	Common Name	Federal Status	State Status
MUSSELS (CONT.)			
Quadrula sparsa	Appalachian monkeyface	Le	End
Toxolasma cylindrellus	Pale lilliput	Le	End
Villosa perpurpurea	Purple bean	Le	End
Villosa trabalis	Cumberland bean	Le	End
LICHENS			
Gymnoderma lineare	Rock gnome lichen	Le	Thr
PLANTS			
Apios priceana	Price's potato-bean	Lt	End
Betula uber	Virginia round-leaf birch	Lt	End
Conradina verticillata	Cumberland rosemary	Lt	Thr
Geum radiatum	Spreading avens .	Le	End
Hedyotis purpurea var *montana*	Mountain bluet	Le	End
Helianthus eggertii	Eggert's sunflower	Lt	Thr
Helonias bullata	Swamp-pink	Lt	Thr, spco
Isotria medeoloides	Small whorled pogonia	Lt	End
Liatris helleri	Heller's blazing star	Lt	Spco
Marshallia mohrii	Mohr's barbara's buttons	Lt	Thr
Pityopsis ruthii	Ruth's golden aster	Le	End
Platanthera integrilabia	White fringeless orchid	C	End
Sagittaria fasciculata	Bunched arrowhead	Le	End
Sagittaria secundifolia	Little river arrow-head	Lt	Stun
Sarracenia jonesii	Mountain sweet pitcher-plant	Le	Spco
Sarracenia oreophila	Green pitcher plant	Le End,	Spco

Scientific Name	Common Name	Federal Status	State Status
Scutellaria montana	Large-flowered skullcap	Lt	End
Solidago spithamaea	Blue ridge goldenrod	Lt	End
Spiraea virginiana	Virginia spiraea	Lt	End, thr
Asplenium scolopendrium var *americanum*	Hart's-tongue fern	Lt	End

NOTES: Compiled by Stan Guffey. Federal status: [a]Dlm: delisted monitored; [b]Le: listed endangered; [c]Lt: listed threatened; [d]Spco: special concern; [e]C: candidate State status (number of state categories collapsed for this table): [f]End: endangered; [g]Nmgt: need of management; [h]Exti: extirpated; [i]Thr: threatened; [j]Potl: potential listing; [k]Stun: status unknown.

SOURCES: TVA natural heritage project list of taxa of concern: http://www.tva.gov/environment/pdf/class01.pdf.

U.S. Endangered Species List: http://endangered.fws.gov/wildlife.html#Species.

Notes

INTRODUCTION

1. See Wendell Berry, "Health is Membership," in *Another Turn of the Crank: Essays* (Boulder, Colo.: Counterpoint Press, 1995), 86; and Berry, *The Unsettling of America*, 102–4, 222.
2. Such broadened usage is a matter both of public policy and of science. The state of Tennessee's Division of Forestry, for example, has a "Forest Health Management Program," and there is now an academic journal entitled *Ecosystem Health*.
3. This theme is elaborated in W. Berry, "The Gift of Good Land," in *The Gift of Good Land: Further Essays Cultural and Agricultural* (San Francisco: North Point Press, 1981), 267–81.
4. This idea of the expanding moral community originates, so far as I am aware, with Aldo Leopold in 1949. See his *A Sand County Almanac with Essays on Conservation from Round River* (New York: Ballentine Books, 1970), 239–43. The concept has since become pervasive in the environmental ethics literature.
5. Quoted in D. S. Pierce, *The Great Smokies: From Natural Habitat to National Park* (Knoxville: Univ. of Tennessee Press, 2000), 47–48; italics mine.
6. T. Regan, *The Case for Animal Rights* (Berkeley and Los Angeles: Univ. of California Press, 1983), 151–56.
7. For a fuller treatment of the issue of factory farms, see Peter Singer, *Animal Liberation* (New York: Avon Books, 1990), chap. 3.
8. Regan, *The Case for Animal Rights*, 243–48.
9. Singer, *Animal Liberation*.
10. See, for example, B. E. Rollin, *The Unheeded Cry: Animal Consciousness, Animal Pain and Science* (Oxford: Oxford Univ. Press, 1989).
11. The most influential expression of this argument is Peter Singer's in *Animal Liberation*. Readers who find the argument as I have expressed it here too compact or who have objections that I have not addressed are urged to consult the much more adequate treatment in Singer's book.
12. The best introductory account of all these views is J. R. Des Jardins, *Environmental Ethics: An Introduction to Environmental Philosophy*, 3d ed. (Belmont: Calif., Wadsworth, 2001).
13. R. Rees, ed., *The Way Nature Works* (New York: Macmillan, 1992), 231.
14. I borrowed this idea from conversations with John Hardwig, who developed it in a slightly different form.

1. HISTOry

1. W. Bartram, "Travels through North and South Carolina, Georgia, East and West Florida, the Cherokee Country, the Extensive Territories of the Muscogulges or Creek Confederacy, and the Country of the Choctaws," in *Bartram: Travels And Other Writings* (New York: Library of America, 1996), 280 ff.

2. The environmental history of all or part of the upper Tennessee River bioregion are treated in a number of excellent works, including M. L. Brown, *The Wild East: A Biography of the Great Smoky Mountains* (Gainesville: Univ. Press of Florida, 2000); Pierce, *The Great Smokies;* C. Bolgiano, *The Appalachian Forest* (Mechanicsburg, Pa.: Stackpole, 1998); A. Cowdrey, *This Land, This South: An Environmental History* (Lexington: Univ. Press of Kentucky, 1983); and M. Frome, *Strangers in High Places* (New York: Garden City, 1966). D. E. Davis, *Where There Are Mountains: An Environmental History of the Southern Appalachians* (Athens: Univ. of Georgia Press, 2000) is an excellent environmental history of the region, and I have relied on it extensively in constructing my account of the historical period. A good short overview can be found in S. L. Yarnell, *The Southern Appalachians: A History of the Landscape,*General Technical Report SRS-18, Southern Research Station, U. S. Forest Service, Asheville, N.C., 1998. T. Steinberg, *Down to Earth: Nature's Role in American History* (New York: Oxford Univ. Press, 2002), provides a good general treatment of the role of nature in American history, and T. Flannery, *The Eternal Frontier: An Ecological History of North America and Its Peoples* (New York: Atlantic Monthly Press, 2001), is an excellent overview of the environmental history of the North American continent.

3. R. F. Flint, *Glacial and Quaternary Geology* (New York: John Wiley and Sons, 1971).

4. Description of plant communities and climate from Wisconsin last glacial maximum to modern times is based on the enormous amount of research and interpretation by Paul and Hazel Delcourt, especially P. A. Delcourt and H. R. Delcourt, "Vegetation Maps For Eastern North America: 40,000 yr B.P. to the Present," in *Geobotany II. Plenum,* ed. R. C. Romans (New York: Plenum Press, 1981), 123–65; P. A. Delcourt and H. R. Delcourt, "Paleoclimates, Paleovegetation, and Paleofloras During the Late Quaternary," in *Flora of North America North of Mexico, Vol. 1,* ed. N. R. Morin (New York: Oxford Univ. Press, 1993), 71–96; and H. R. Delcourt and P. A. Delcourt, "Eastern Deciduous Forests," in *North American Terrestrial Vegetation,* ed. M. G. Barbour and W. D. Billings (Cambridge: Cambridge Univ. Press, 2000), 357–95.

5. Flannery, *The Eternal Frontier.*

6. Descriptions of late glacial and early post-glacial mammals are based on B. Kurten and E. Anderson, *Pleistocene Mammals of North America* (New York: Columbia Univ. Press, 1980); B. Kurten, *Before the Indians* (New York: Columbia Univ.Press, 1988); and J. E. Guilday, "Appalachia 11,000–12,000 Years Ago," *Archaeology of Eastern North America* 10 (1982):22–25.

7. B. M. Fagan, *Ancient North America* (London: Thames and Hudson, 1995).

8. T. D. Dillehay, *The Settlement of the Americas: A New Prehistory* (New York: Basic Books, 2000).

9. Fagan, *Ancient North America.*

10. D. G. Anderson, "Examining Prehistoric Settlement Distribution In Eastern North America," *Archaeology of Eastern North America* 19 (1991): 1–22; D. G. Anderson and M. K. Faught, "The Distribution of Fluted Paleoindian Projectile Points: Update 1998," *Archaeology of Eastern North America* 26 (1998):163–87.

11. Fagan, *Ancient North America.*

12. Flannery, *The Eternal Frontier.* See also P. S. Martin and R. G. Klein, eds., *Quaternary Extinctions: A Prehistoric Revolution* (Tucson: Univ. of Arizona Press, 1984).

13. S. L. Cox, "A Re-Analysis of the Shoop Site," *Archaeology of Eastern North America* 14 (1986): 101–70; W. M. Gardner, "The Flint Run Paleo-Indian Complex and Its Implications for Eastern North American Prehistory," *Annals of the New York Academy of Sciences* 288 (1977):257–63; Summarized in Fagan, *Ancient North America.*

14. D. Morse and P. Morse, *Archaeology of the Central Mississippi Valley* (New York: Academic Press, 1983).

15. Fagan, *Ancient North America.*

16. Ibid.

17. Description of early and middle Archaic period prehistory in the Little Tennessee River Valley is based on J. Chapman, "The Icehouse Bottom Site, 40MR23," *Reports of Investigations No. 13,* Dept. of Anthropology, Univ. of Tennessee, Knoxville, 1973; J. Chapman, "Archaic Period Research in the Lower Little Tennessee River Valley–1975," *Reports of Investigations No. 19,* Dept. of Anthropology, Univ. of Tennessee, Knoxville, 1977; J. Chapman, "The Bacon Farm Site and a Buried Site Reconnaissance," *Reports of Investigations No. 21,* Dept. of Anthropology, Univ. of Tennessee, Knoxville, 1978; J. Chapman, "The 1979 Archaeological and Geological Investigations in the Tellico Reservoir," *Reports of Investigations No. 29,* Dept. of Anthropology, University of Tennessee, Knoxville, 1980; and as synthesized in J. Chapman, *Tellico Archaeology: 12,000 Years of Native American History* (Knoxville: Univ. of Tennessee Press), 1985.

18. B. D. Smith, "The Archaeology of the Southeastern United States: From Dalton to DeSoto, 10,500 to 500 B.P.," *Advances in World Archaeology* 5 (1986): 1–92; Fagan, *Ancient North America.*

19. Fagan, *Ancient North America.*

20. H. R. Delcourt and P. A. Delcourt, "Pre-Columbian Native American Use of Fire on Southern Appalachian Landscapes," *Conservation Biology* 11 (4) (1997):1010–14; Delcourt and Delcourt, "Eastern Deciduous Forests," 357–95.

21. Fagan, *Ancient North America.*

22. Delcourt and Delcourt, "Eastern Deciduous Forests," 357–95.

23. Fagan, *Ancient North America.*

24. Smith, "Archaeology of the Southeastern United States," 1–92; S. Struever and F. Holton, *Koster: Americans in Search of the Prehistoric Past* (New York: Anchor Press, 1979).

25. Fagan, *Ancient North America;* P. Phillips and J. A. Brown, eds., *Archaic Hunters and Gatherers in the American Midwest* (New York: Academic Press, 1983).

26. Fagan, *Ancient North America.*

27. Delcourt and Delcourt, "Pre-Columbian Native American Use of Fire," 1010–14; Delcourt and Delcourt, "Eastern Deciduous Forests," 357–95.

28. B. D. Smith, *Rivers of Change: Essays on Early Agriculture in Eastern North America* (Washington, D.C.: Smithsonian Institution Press, 1992); B. D. Smith, *The Emergence of Agriculture* (New York: Scientific American Library, 1994).

29. J. Stoltman, "New Radiocarbon Dates for Southeastern Fiber Tempered Pottery," *American Antiquity* 31 (1966): 872–73.

30. *Ancient North America.*

31. W. C. Galinat, "Domestication and Diffusion of Maize," in *Early Food Production in North America,* ed. R. I. Ford, Museum of Anthropology, Univ. of Michigan, Ann Arbor, 1985, 245–82.

32. Smith, *Rivers of Change;* Smith, *Emergence of Agriculture;* G. D. Crites, "Human-Plant Mutualism and Niche Expression in the Paleobotanical Record: A Middle Woodland Example," *American Antiquity* 52 (1987): 725–40.

33. Smith, *Emergence of Agriculture;* Crites, "Human-Plant Mutualism," 725–40; C. B. Heiser, "Some Botanical Considerations of the Early Domesticated Plants North Of Mexico," in *Early Food Production in North America,* ed. R. I. Ford, Museum of Anthropology, Univ. of Michigan, Ann Arbor, 1985, 57–72.

34. Smith, "Archaeology of the Southeastern United States," 1–92; Fagan, *Ancient North America.*

35. Smith, *Emergence of Agriculture.*

36. Fagan, *Ancient North America.*

37. Smith, "Archaeology of the Southeastern United States," 1–92; Fagan, *Ancient North America.*

38. Ibid. See also J. D. Rogers and B. D. Smith, eds., *Mississippian Communities and Households* (Tuscaloosa: Univ. of Alabama Press, 1995), and P. N. Peregrine, *Archaeology of the Mississippian Culture: A Research Guide* (New York: Garland Publishing, 1996), for a guide to the enormous literature concerning the Mississippian tradition.

39. Fagan, *Ancient North America.*

40. Davis, *Where There Are Mountains.*

41. J. P. Nass, Jr. and R. W. Yerkes, "Social Differentiation in Mississippian and Fort Ancient Societies," in *Mississippian Communities and Households,* ed. Rogers and Smith, 58-80; L. P. Sullivan, "Mississippian Household and Community Organization in Eastern Tennessee," in *Mississippian Communities and Households,* ed. Rogers and Smith, 99-123.

42. Fagan, *Ancient North America.*

43. R. S. Dickens, *Cherokee Prehistory: The Pisgah Phase in the Appalachian Summit Region* (Knoxville: Univ. of Tennessee Press, 1976); C. M. Hudson, *The Southeastern Indians* (Knoxville: Univ. of Tennessee Press, 1976). See also B. G. McEwan, ed., *The Indians of the Greater Southeast: Historical Archaeology and Ethnohistory* (Gainesville: Univ. Press of Florida, 2000).

44. This discussion of early Spanish activities is based on A. Taylor, *American Colonies* (New York: Viking, 2001).

45. Ibid., 72.

46. The route of the de Soto expedition remains the subject of considerable speculation. Recent scholarship has clarified much, particularly with respect to the initial route and return, but the route of the bulk of the expedition between remains uncertain. Davis, *Where There Are Mountains,* and Taylor, *American Colonies,* generally concur in

their interpretations and follow C. M. Hudson, "The Hernando de Soto Expedition," in *The Forgotten Centuries: Indians and Europeans in the American South, 1521–1704,* ed. C. M. Hudson and C. C. Tesser (Athens: Univ. of Georgia Press, 1994), 74–122; C. M. Hudson, *Knights of Spain: Warriors of the Sun* (Athens: Univ. of Georgia Press, 1997); and D. J. Weber, *The Spanish Frontier in North America* (New Haven, Conn.: Yale Univ. Press, 1992). However, other researchers have alternative interpretations from readings of the chroniclers and interpretations of contact period archaeology and ethnohistory; see, for example, D. E. Sheppard, "DeSoto's Midwestern Conquest: Starting Point," *Native American Conquest Corp.,* 2003, <http://www.floridahistory.com/inseta.html> (23 July 2003).

47. Quoted in Taylor, *American Colonies,* 73.
48. Davis, *Where There Are Mountains.*
49. Ibid; Smith, "Archaeology of the Southeastern United States," 1–92; Hudson, *The Southeastern Indians;* Cowdrey, *This Land, This South.* See also A. W. Crosby, "Virgin Soil Epidemics as a Factor in the Aboriginal Depopulation in America," *William and Mary Quarterly* (Apr. 1976): 289–99.
50. Hudson, *The Southeastern Indians;* Dickens, *Cherokee Prehistory;* B.C. Keel, *Cherokee Archaeology: A Study of the Appalachian Summit* (Knoxville: Univ. of Tennessee Press, 1976).
51. Bartram, "Travels," 3–425.
52. Davis, *Where There Are Mountains.* See also Hudson and Tesser, eds., *The Forgotten Centuries.*
53. Davis, *Where There Are Mountains.*
54. Ibid.; E. Rostlund, "The Geographic Range of Historic Bison in the Southeast," *Annals of the Association of. American Geographers* 50 (1969): 395–407.
55. Hudson, *The Southeastern Indians.*
56. Taylor, *American Colonies;* P. H. Bergeron, S. V. Ash, and J. Keith, *Tennesseans and Their History* (Knoxville: Univ. of Tennessee Press, 1999).
57. Davis, *Where There Are Mountains.*
58. Ibid; Taylor, *American Colonies.*
59. Davis, *Where There Are Mountains.*
60. Ibid.
61. F. Anderson, *Crucible of War* (New York: Knopf, 2000).
62. Discussion of Euro-American interactions with the Cherokee based on Williams, Samuel Cole, ed. *Lieut. Henry Timberlake's Memoirs.* (Johnson City, Tennessee: Watauga Press, 1927); Bergeron, Ash, and Keith, *Tennesseans and Their History;* Anderson, *Crucible of War.*
63. Williams, *Timberlake's Memoirs;* Bergeron, Ash, and Keith, *Tennesseans and Their History;* H. H. Jackson, *A Century of Dishonor: A Sketch of the United States Government's Dealings With Some of the Indian Tribes* (1885; reprint, Norman: Univ. of Oklahoma Press, 1994).
64. Davis, *Where There Are Mountains.*
65. This discussion of Cherokee and American relations after the Revolutionary War is based on Bergeron, Ash, and Keith, *Tennesseans and Their History;* Davis, *Where There Are Mountains;* and Jackson, *Century of Dishonor.* See also G. Jahoda, *The Trail of Tears* (New York: Holt, Rhinehart, and Winston, 1975); J. Wilson, *The Earth Shall Weep: A History of Native America* (New York: Grove Press, 1998).

66. Jackson, *Century of Dishonor,* 277.

67. This discussion of the history and ecology of nineteenth-century settlement is based on Brown, *The Wild East;* Davis, *Where There Are Mountains;* J. S. Otto, *The Southern Frontiers, 1607–1860: The Agricultural Evolution of the Colonial and Antebellum South* (New York: Greenwood Press, 1989); Bergeron, Ash, and Keith, *Tennesseans and Their History;* and Cowdrey, *This Land, This South.* See also Frome, *Strangers in High Places;* and Yarnell, *The Southern Appalachians.*

68. S. J. Pyne, *Fire in America: A Cultural History of Wildland and Rural Fire* (Seattle: Univ. of Washington Press, 1982).

69. Otto, *The Southern Frontiers.*

70. This discussion of the development of transportation networks is based on Davis, *Where There Are Mountains;* Roads and their importance in the region are discussed in D. C. Hsiung, *Two Worlds in the Tennessee Mountains: Exploring the Origins of Appalachian Stereotypes* (Lexington: Univ. Press of Kentucky, 1997). See also D. C. Hsiung, "How Isolated Was Appalachia? Upper East Tennessee, 1780–1835," *Appalachian Journal* 16 (4) (1989): 336–49.

71. W. Berry, "From a Native Hill," in *At Home on the Earth,* ed. D. L. Barnhill (Berkeley: Univ. of California Press, 1999), 48–49.

72. J. S. Otto, "The Decline of Forest Farming in Southern Appalachia," *Journal of Forest History* (Jan. 1983): 18–27.

73. Davis, *Where There Are Mountains.*

74. This discussion of agriculture is based on Otto, *The Southern Frontiers;* and Davis, *Where There Are Mountains.*

75. A. F. Ganier, "The Wild Life of Tennessee," *Journal of the Tennessee Academy of Science* 3 (3) (1928): 10–22.

76. This discussion of livestock raising and marketing is based on F. McDonald and G. McWhiney, "The Antebellum Southern Herdsman: A Reinterpretation," *Journal of Southern History* 41 (May 1975): 148–68; and Davis, *Where There Are Mountains.* See also R. K. McMaster, "The Cattle Trade in Western Virginia, 1760–1830," in *Appalachian Frontiers: Settlement, Society, and Development in the Pre-Industrial Era,* ed. R. D. Mitchell (Lexington: Univ. Press of Kentucky, 1991).

77. M. J. Lacki and R. A. Lancia, "Effects of Wild Pigs on Beech Growth in Great Smoky Mountains National Park," *Journal of Wildlife Management* 50 (1986): 655–59; L. C. Stegeman, "The European Wild Boar in the Cherokee National Forest, Tennessee," *Journal of Mammalogy* 19 (1938): 279–90.

78. Southern Appalachian grassy balds have been variously thought to be of Native American or Euro-American origin, or natural edaphic/climatic communities; ecological and historical evidence suggests the balds are of historical origin and in part edaphically and climatically maintained. See Brown, *The Wild East,* for a discussion of these various interpretations. See also Davis, *Where There Are Mountains.*

79. Ginseng had been a valuable trade commodity since the colonial period, but expanded transportation networks increased the scale of trade in this and other forest herbs. See Davis, *Where There Are Mountains.*

80. United States Historical Census Browser, 21 February 2003 <http://fisher.lib.virginia.edu/census/> (23 July 2003).

81. Davis, *Where There Are Mountains.*

82. This discussion of industry is based on ibid. See also Yarnell, *The Southern Appalachians.* See H. T. Blethen and C. Wood Jr., "The Antebellum Iron Industry in Western North Carolina," *Journal of the Appalachian Studies Association* 4 (1992): 79–87 for a discussion of early iron smelting. See O. W. Price, "Lumbering in the Southern Appalachians." *Forestry and Irrigation* 11 (1905): 469–76, for a discussion of early logging and sawmilling.

83. Davis, *Where There Are Mountains,* 149–51.

84. This discussion of copper smelting is based on ibid. and on R. E. Barclay, *Ducktown Back in Raht's Time* (Chapel Hill: Univ. of North Carolina Press, 1946). Yarnell, *The Southern Appalachians.*

85. Tennessee Valley Authority, *The First Fifty Years: Changed Land, Changed Lives* (Knoxville: Tennessee Valley Authority, 1983).

2. AIr

1. TVA, *Energy Vision 2020: Integrated Resource Plan Environmental Impact Statement* (1995), vol. 2, T1.32–T1.33, T1.40.

2. D. W. Johnson and S. E. Lindberg, *Atmospheric Deposition and Forest Nutrient Cycling: A Synthesis of the Integrated Forest Study* (New York: Springer Verlag, 1992).

3. Y. Hong, J. Lee, H. Kim, E. Ha, J. Schwartz, and D. C. Christiani, "Effects of Air Pollutants on Acute Stroke Mortality," *Environmental Health Perspective* 110 (2) (2002): 187–91; H. Desqueroux, J. C. Pujet, M. Prosper, F. Squinazi, and I. Momas, "Short-Term Effects of Low-Level Air Pollution on Respiratory Health of Adults Suffering from Moderate to Severe Asthma," *Environmental Research* 89 (1) (2002): 29–37; H. J. Kwon, S.H. Cho, F. Nyberg, and G. Pershagen, "Effects of Ambient Air Pollution on Daily Mortality in a Cohort of Patients with Congestive Heart Failure," *Epidemiology* 12 (2001): 413–19.

4. J. R. Renfro, "Air Quality Monitoring and Research Program at Great Smoky Mountains National Park: An Overview of Results and Findings," Gatlinburg,Tenn., USDI-National Park Service, Great Smoky Mountain National Park, Division of Resource Management and Science, 1996, 69.

5. American Lung Association, "State of the Air 2003," http://www.lungusa.org/air/.

6. J. R. Renfro, air quality specialist, Great Smoky Mountains National Park, personal communication.

7. National Park Service Press Release, Aug. 17, 2000. "Ozone Pollution Advisory Issued for Smokies," formerly at http://www2.nature.nps.gov/ard/gas/advisory/O3Advis60mo.pdf.

8. J. D. Peine, J. C. Randolph, and J. J. Presswood, "Evaluating the Effectiveness of Air Quality Management Within the Class 1 Area of Great Smoky Mountains National Park," *Environmental Management* 19 (1995): 519.

9. S. V. Krupa and W. J. Manning, "Atmospheric Ozone: Formation and Effects on Vegetation," *Environmental Pollution* 50 (1988): 101–37.

10. Peine, Randolph, and Presswood, "Air Quality Management," 524.
11. U.S. Environmental Protection Agency draft documents cited in TVA, *Energy Vision 2020,* vol. 2, T1.81.
12. 1998 Crop data from E. Baum, "Estimate of Impact of Ozone on Crop Value in the Southeast," Clean Air Task Force, Boston, Mass.
13. Southern Alliance for Clean Energy (SACE), *Clearing the Air: Getting the Dirt on TVA's Coal-fired Power Plants,* 2d ed. (May 2000). (Some figures updated using 2002 TVA emission data.)
14. C. A. Pope et al., "Lung Cancer, Cardiopulmonary Mortality, and Long-term Exposure to Fine Particulate Air Pollution," *Journal of the American Medical Association* 287 (9) (2002): 1132–41. Kwon et al., "Ambient Air Pollution," 413–19. R. D. Brook et al., "Inhalation of Fine Particulate Air Pollution and Ozone Causes Acute Arterial Vasoconstriction in Healthy Adults," *Circulation* 105 (2002): 1534–36. Y. C. Hong, J. T. Lee, H. Kim, and H. J. Kwon, "Air Pollution: A New Risk Factor in Ischemic Stroke Mortality," *Stroke* 33 (9) (2002): 2165–69.
15. C. Ballantine, M.D., Presentation to the 4th Annual Governor's Summit, Charlotte, NC, May 2002.
16. Pope et al., "Fine Particulate Air Pollution," 1132–41.
17. D. Hattis et al., "Human Interindividual Variability in Susceptibility to Airborne Particles," *Risk Analysis* 4 (2001): 585–99.
18. M. Lin et al., "The Influence of Ambient Coarse Particulate Matter on Asthma Hospitalization in Children: Case-Crossover and Time-Series Analyses," *Environmental Health Perspectives* 110 (2002): 575–81.
19. Ballantine, Presentation to the 4th Annual Governor's Summit.
20. C. Schneider, *Death, Disease and Dirty Power: Mortality and Health Damage Due to Air Pollution from Power Plants* (Boston: Clean Air Task Force, 2000).
21. Ibid.
22. TVA, *Energy Vision 2020,* vol. 2, T1.19, T1.32–T1.33; Renfro, "Air Quality Monitoring," 31; W. C. Malm, "Visibility and Acid Aerosols at Great Smoky Mountains National Park," in *Proceedings of the SAMAB Forum on Air Quality Management in the Southern Appalachian Class 1 Areas,* Air Quality Division, National Park Service, CIRA-Foothills Campus, Colorado State Univ., Fort Collins, Colo., 1992.
23. EPA, "Table B2: Plant-by-plant Summary Data Organized by State," *Emissions Scorecards,* 4 April 2002, http://www.epa.gov/airmarkets/emissions/score01/index.html (31 July 2003).
24. SACE, *Clearing the Air.*
25. Southern Alliance for Clean Energy calculations based on Environmental Protection Agency Continuous Emissions Monitoring reports for TVA, 2002.
26. Renfro, "Air Quality Monitoring," 22, 67–68.
27. J. R. Renfro, personal communication.
28. Ibid.
29. M. Keating, *Casting Doubt* (Boston: Clean Air Task Force, 2000), 5.
30. A. Hennen, *Mercury in the Upper Midwest* (St. Paul, MN: The Izaak Walton League Of America, 1999). Available at: <http://www.iwla.org/reports/mercury.html>.
31. U.S. Department of the Interior, Fish and Wildlife Service and U.S. Department of Commerce, U.S. Census Bureau, *2001 National Survey of Fishing, Hunting, and*

Wildlife-Associated Recreation (October 2002). T. Kuiken and F. Stadler, *Cycle of Harm: Mercury's Pathway from Rain to Fish in the Environment,* 2d ed. (Ann Arbor, Mich.: National Wildlife Federation, 2003).

32. Kuiken and Stadler, *Cycle of Harm.*
33. Keating, *Casting Doubt.*
34. EPA, *2001 Toxics Release Inventory: Executive Summary,* U.S. Environmental Protection Agency, Washington, D.C., 2003. Accessible at http://www.epa.gov/tri/tridata/tri01/.
35. T. Dutzik et al., *Toxic Releases and Health: A Review of Pollution Data and Current Knowledge on the Health Effects of Toxic Chemicals,* U.S. Public Interest Research Group, Washington, D.C., 2003.
36. EPA, *2000 Toxics Release Inventory.*
37. Dutzik et al., *Toxic Releases and Health.*
38. N.P. Cheremisinoff, J. A. King, and R. Boyko, *Dangerous Properties of Industrial and Consumer Chemicals* (New York: Marcel Dekker, 1994), 130.
39. EPA, "2001 Toxic Release Inventory: Tennessee," *2001 TRI State Fact Sheet,* http://www.epa.gov/triinter/tridata/tri01/state/Tennessee.pdf (30 July 2003).
40. EPA, *2000 Toxics Release Inventory.*
41. J. Harte et al., *Toxics A to Z: A Guide to Everyday Pollution Hazards* (Berkeley: Univ. of California Press, 1991).
42. EPA, "2001Toxics Release Inventory: Tennessee."
43. J. D. Savitz et al., *Dishonorable Discharge: Toxic Pollution of Tennessee Waters,* Environmental Working Group, Washington, D.C., 1996, 7.
44. EPA, *2001 Toxics Realease Inventory.*
45. Save Our Cumberland Mountains (SOCM), *1995 SOCM Report,* Lake City, Tenn., 1996.
46. EPA, "Facility Report."
47. TVA, *Energy Vision 2020,* vol. 2, T1.66–T1.69.
48. T. Widner, "First Results in Assessment of Iodine-131 Emissions," *Oak Ridge Health Studies Bulletin* 5 (2) (1996).
49. Renfro, "Air Quality Monitoring," 33, 68.
50. Centers for Disease Control and Prevention, "Exposure to Environmental Tobacco Smoke and Cotinine Levels: Fact Sheet," *Environmental Tobacco Smoke,* 16 Oct. 2002, http://www.cdc.gov/tobacco/ets.htm (30 July 2003).
51. M. McKinney and R. M. Schoch, *Environmental Science: Systems and Solutions* (Minneapolis: West Publishing Co., 1996), 484–85.
52. P. H. Raven, L. R. Berg, and G. B. Johnson, *Environment* (Fort Worth, Tex.: Saunders College Publishing, 1993), 435–39; McKinney and Schoch, *Environmental Science,* 484–85.
53. McKinney and Schoch, *Environmental Science,* 484.
54. "FY2002 Coal Plant Emissions," TVA 4th Quarter Submittal to EPA, Mar. 2003.
55. IPCC (Intergovernmental Panel on Climate Change), *Climate Change 2001: Synthesis Report* (2001), 8. Available at http://www.ipcc.ch.
56. EPA, "Climate Change and Tennessee," *Global Warming Where You Live: Tennessee,* 30 Sept. 2002, http://yosemite.epa.gov/oar/globalwarming.nsf/content/us-tennessee.html?OpenDocument&Flash=yes (31 July 2003).
57. Information for this and preceding paragraphs is from McKinney and Schoch, *Environmental Science,* 504–19, and from Hara et al., *Regional Assessment of Climate*

Change Impacts in the Southeast Summary of Workshop I—Vulnerable Resources and Predictive Capabilities, Center for Global Environmental Studies and Environmental Science Division, Oak Ridge National Laboratory (undated).

58. Physicians for Social Responsibility (PSR), *Death by Degrees: The Emerging Health Crisis of Climate Change in Georgia* (Washington, D.C.: Physicians for Social Responsibility, 2000). Available at http://www.psr.org/home.cfm?id=enviro_resources.

59. World Meteorological Organization (WMO), *Scientific Assessment of Ozone Depletion: 2002,* Global Ozone Research and Monitoring Project—Report No. 47, 498 pp., Geneva, 2003. Available at http://www.wmo.ch/web/arep/ozone.html.

60. L. Bjorn et al., eds., *Environmental UV Photobiology* (New York: Plenum, 1993).

61. J. Flynn, "Forest Death in Coal River Valley," *Earth Island Journal* (Fall 1995).

62. H. Ayers, J. Hager, and C. E. Little, *An Appalachian Tragedy: Air Pollution and Tree Death in the Eastern Forests of North America* (San Francisco, Calif.: Sierra Club Books, 1998).

63. McKinney and Schoch, *Environmental Science,* 487–89; S. Shimek, "Your Home May Hide a Silent Killer," *Tennessee Conservationist* 60 (1) (1994): 15–22.

64. McKinney and Schoch, *Environmental Science,* 507..

3. WATER

1. D. Davidson, *The Tennessee,* vol. 1: *The Old River, Frontier to Secession* (New York: Reinhart & Company, 1946), 38–39.

2. National Park Service (NPS), Great Smoky Mountains National Park, "Briefing Statement," December 2001. Accessible at http://data2.itc.nps.gov/parks/grsm/ppdocuments/ACF7B1.doc.

3. D. W. Johnson and S. E. Lindberg, *Atmospheric Deposition and Forest Nutrient Cycling: A Synthesis of the Integrated Forest Study* (New York: Springer Verlag, 1992).

4. E. LeQuire, "The Acid Test: Acid Deposition May Be Damaging the Smokies' Soil, Water, Plants and Animals," *Sightline* 2 (1) (2001). Accessible at http://eerc.ra.utk.edu/sightline/SightlineV2N1.html#anchor278832.

5. Ayers, Hager, and Little, *An Appalachian Tragedy,* 59.

6. NPS, "Briefing Statement."

7. Johnson and Lindberg, *Atmospheric Deposition.*

8. NPS, "Briefing Statement."

9. E. Baum, "Unfinished Business: Why the Acid Rain Problem is Not Solved," Clean Air Task Force, Oct. 2001. Accessible at http://www.catf.us/publications/reports/acid_rain_report.php (21 May 2004).

10. Ibid.

11. Johnson and Lindberg, *Atmospheric Deposition.*

12. LeQuire, "The Acid Test."

13. Ibid.

14. Ayers, Hager, and Little, *An Appalachian Tragedy,* 2.

15. Ibid. provides a lavishly-illustrated account of the decline of the spruce-fir forest, including an excellent bibliography.

16. Tennessee Valley Authority (TVA), *RiverPulse: A Report on the Condition of the Tennessee River and Its Tributaries in 1994* (annual report, 1995), 5.

17. Tennessee Department of Environment and Conservation (TDEC), *The Status of Water Quality in Tennessee: 1994 305(b) Report* Division of Water Pollution Control, Nashville, 1994, 51.

18. R. Dosser, "State Officials Probe Possible Waste Spill at Knox Firm," *Knoxville Journal,* 24 Mar. 1989, sec. 3A.

19. P. Park, "Knoxville Has Troubled Waters: Sewage, Trash, Oil Run High," *Knoxville News-Sentinel,* 28 May 1992, sec. A1.

20. G. M. Brune, "Trap Efficiency of Reservoirs," *Transactions of the American Geophysical Union* 34 (1953): 407–18.

21. Information on the dam comes from Harold N. Mullican et al., "A Survey of Aquatic Biota of the Nolichucky River," Pollution Control Board, Tennessee Department of Public Health, Nashville, Tenn., 1960, 1–2. The closure date is given in TVA, *The First Fifty Years,* 70.

22. A. Watson, "The Greening of Copper Basin," *Tennessee Conservationist* 60 (3) (1994): 21–26.

23. T. W. Mermel, ed., *Register of Dams in the United States* (New York: McGraw-Hill, 1958), 150, and TDEC, *Status of Water Quality in Tennessee* (1994), 55.

24. North Carolina Division of Environment Management (NCDEM), *French Broad River Basinwide Water Quality Management Plan* (draft), Water Quality Section, Raleigh, N.C., 1995, 4-16–4-17.

25. TVA, *TVA Handbook,* Deborah D. Mills, ed., TVA/OCS/PS-87/8, 1987.

26. C. C. Amundsen, "Reservoir Riparian Zone Characteristics in the Upper Tennessee River Valley," *Water, Air and Soil Pollution* 77 (1994): 469–93.

27. Observation by Leaf Myczack, riverkeeper.

28. Tennessee Department of Environment and Conservation (TDEC), *The Status of Water Quality in Tennessee: Year 2002 305(b) Report,* Division of Water Pollution Control, Nashville, 2002, 43–44.

29. TVA, "Fort Loudon Reservoir," http://www.tva.com/environment/ecohealth/fortloudoun.htm (9 July 2003).

30. TDEC, *Status of Water Quality in Tennessee* (2002), 76–77.

31. Letter from A. David McKinney, State of Tennessee Department of Health and Environment, to Stan Stieff, Oak Ridge Operations, Oct. 26, 1983.

32. Oak Ridge Education Project (OREP), "A Citizen's Guide to Oak Ridge," Foundation for Global Sustainability, Knoxville, Tenn., 1992, 26–27.

33. F. C. Kornegay et al., *Oak Ridge Reservation Environmental Report for 1990,* ES/ESH-18/V1, Dept. of Energy, Oak Ridge, Tenn., 1991, 147.

34. TVA, *Energy Vision 2020,* vol. 2: T1.106.

35. McKinney and Schoch, *Environmental Science,* 418–19.

36. B. Kauffman, "Ridge Scientists to Test Fish," *Knoxville News-Sentinel,* 17 July 1988, sec. A2

37. TDEC, *Status of Water Quality in Tennessee* (2002), 73, and North Carolina Department of Environment and Natural Resources, *Water Quality Assessment and*

Impaired Waters List (2002 Integrated 305 (b) and 303 (d) Report), Division of Water Quality, Raleigh, N.C. 2003.

38. Richard Caplan, *Permit to Pollute: How the Government's Lax Enforcement of the Clean Water Act is Poisoning Our Waters* (Washington, D.C., U.S. Public Interest Research Group Education Fund, 2002), 6. Accessible at http://uspirg.org/reports/permittopollute2002.pdf.

39. Ibid.

40. Tennessee Environmental Council, "Report Ranks Tennessee 3rd in Nation for Facilities Violating Clean Water Permits," Press Release, 6 Aug. 2002, accessible at http://www.tectn.org/tectnhome.html (30 July 2003).

41. TDEC, *The Status of Water Quality in Tennessee: Year 2000 305 (b) Report,* Division of Water Pollution Control, Nashville, 2000, 16, 18.

42. TDEC, *Status of Water Quality in Tennessee* (2002), 2.

43. Ibid., 67–70.

44. Ibid., 3.

45. Ibid., 70.

46. Ibid., 30.

47. Kristen Hebestreet, "A Case of Nature's Resilience," *Tennessee Wildlife* (Mar./Apr. 2002): 25–28.

48. TVA, *Energy Vision 2020,* vol. 2: T1.96.

49. EPA, "TRI Explorer," accessible at http://www.epa.gov/triexplorer/facility.htm (09 July 2003).

50. TDEC, *Status of Water Quality in Tennessee* (2002), 67.

51. Ibid., 68–69.

52. TDEC, *The Status of Water Quality in Tennessee: Year 1994 305 (b) Report,* Division of Water Pollution Control, Nashville, 12.

53. "Robertshaw Controls Ordered to Stop Dumping Oil into Creek," *Knoxville News-Sentinel,* 7 Dec. 1989, sec. A10.

54. Information for the last three paragraphs was compiled from TVA, *RiverPulse;* Broadened Horizons Clean Water Project (BHCWP), *A Citizen's Guide to Pollution in the Tennessee River,* Whitwell, Tenn., 1992, and direct observation.

55. BHCWP, *A Citizen's Guide.*

56. U.S. Army Corps of Engineers Public Notice #91-132, Sept. 1991.

57. T. Tinker, G. Moore, and C. Lewis-Younger, "Chattanooga Creek," *Tennessee Conservationist* 62 (1) (1996): 14–18.

58. BHCWP, *Broadened Horizons Clean Water Project Newsletter,* Fall 1991.

59. TDEC, *Status of Water Quality in Tennessee* (2002), 64, 77; TVA, "Nickajack Reservoir," accessible at http://www.tva.gov/environment/ecohealth/nickajack.htm (08 July 2003).

60. TVA, *Energy Vision 2020,* vol. 2: T1.98–T1.100.

61. TVA, "Douglas Reservoir," http://www.tva.com/environment/ecohealth/douglas.htm (09 July 2003).

62. TVA, "Boosting Oxygen Levels," http://www.tva.com/environment/water/rri_oxy.htm (09 July 2003).

63. TVA, *RiverPulse,* 6.

64. TVA, "Reservoir Ratings," http://www.tva.com/environment/ecohealth/ (09 July 2003).

65. TVA, *Energy Vision 2020,* vol. 2: T1.95–T1.96.
66. TVA, *2001 Annual Environmental Report,* 4, and John Shipp, TVA, personal communication.
67. TVA, *RiverPulse,* 2.
68. J. Shipp, TVA, personal communication.
69. TDEC, *Status of Water Quality in Tennessee* (2002), 41.
70. N. A. Murdock, "Rare and Endangered Plants and Animals of Southern Appalachian Wetlands," *Water, Air and Soil Pollution* 77 (1994): 385–405.
71. TVA, U.S. Army Corps of Engineers, and U.S. Fish and Wildlife Service, *Final Environmental Impact Statement: Chip Mill Terminals on the Tennessee River* (1993), vol. 1: 111.
72. K. Morgan and T. Roberts, *An Assessment of Wetland Mitigation in Tennessee,* Tennessee Technological Univ., Cookeville, Tenn., 1999.
73. T. Moss, "Sinking Ground: Karst Topography is Mother Nature's Plumbing System," *Tennessee Conservationist* 59 (1) (1993): 6–9.
74. K/KCMPC (Knoxville/Knox County Metropolitan Planning Commission), *Development Impacts on Drainage Basins* (1994).
75. Raven, Berg, and Johnson, *Environment,* 404.
76. Save Our Cumberland Mountains (SOCM), "Hamblen County Residents Win Water after 15 Month Fight," *The SOCM Sentinel* (Nov./Dec. 1994).
77. Raven, Berg, and Johnson, *Environment,* 472.
78. Information in this and the previous paragraph comes from OREP, "Citizen's Guide to Oak Ridge," and Roger Clapp, Oak Ridge National Laboratory, personal communication.
79. Martin Marietta Energy Systems (MMES), *Environmental Update,* Martin Marietta Energy Systems, Oak Ridge, 1995.
80. TDEC, *Status of Water Quality in Tennessee* (2002), 16.
81. H. Komulainen et al., "Carcinogenicity of the Drinking Water Mutagen 3-chloro-4- (dichloromethyl)-5-hydroxy-2(5H)-furanone in the Rat," *Journal of the National Cancer Institute* 89 (12) (1997): 848–56.
82. D. L. Feldman, "Going Thirsty?" *Now & Then* (Spring 2001): 3–8, and D. Feldman, UT/EERC, personal communication, July 2003.
83. Feldman, "Going Thirsty?"
84. Ibid.
85. TVA, "Could We Run Out Of Water?" *TVA Kids,* accessible at http://www.tvakids.com/river/runout.htm (30 July 2003).

4. BIOTA

1. The role of humans in the extinction of the Pleistocene megafauna of the Americas has been a contentious area of research for over fifty years. Recent simulation studies, e.g., J. Alroy, "A Multispecies Overkill Simulation of the End-Pleistocene Megafaunal Mass Extinction," *Science* 292 (2001): 1893–1896, suggest that hunting

was indeed a significant factor; see also Martin and Klein, *Quaternary Extinctions;* S. Krech III, *The Ecological Indian* (New York: W. W. Norton, 1999); and Flannery, *The Eternal Frontier.*

2. A. W. Schorger, *The Passenger Pigeon: Its Natural History and Extinction* (Norman: Univ. of Oklahoma Press, 1955); J. C. Greenway Jr., *Extinct and Vanishing Birds of the World* (New York: Dover, 1967).

3. Several species in the mussel genus *Epioblasma* probably began their decline toward extinction in the late nineteenth century. The last collection of the round comb-shell, *E. personata,* was in the nineteenth century, and the leafshell, *E. flexuosa,* was last collected in 1900. See P. W. Parmalee and A. E. Bogan, *The Freshwater Mussels of Tennessee* (Knoxville: Univ. of Tennessee Press, 1998).

4. The harelip sucker went extinct throughout its range some time after about 1893. The species had once been widespread in the southcentral United States and was used extensively for food by Native Americans and Euro-Americans in the upper Tennessee valley. See D. A. Etnier and W. C. Starnes, *The Fishes of Tennessee* (Knoxville: Univ. of Tennessee Press, 1993).

5. J. P. Prestemon and R. C. Abt, "Timber Products Supply and Demand," in *Southern Forest Resource Assessment,* ed. D. N. Wear and J. G. Greis, Gen. Tech. Rpt. SRS-53, U.S. Dept. of Agriculture, Forest Service, Southern Research Station, Asheville, N.C., 2002, 299, 306.

6. R. S. Lambert, "Logging in The Great Smoky Mountains National Park, A Report To The Superintendent," Manuscript in Great Smoky Mountains National Park Archives, Sugarlands Headquarters, Gatlinburg, Tenn. (1958). Brown, *The Wild East,* considers this a conservative estimate because Lambert "did not include timber shipped in the log, timber cut for tanbark, timber destroyed by machinery and fire, and timber used in construction and for fuel." In addition to Brown, excellent and informed discussions of the logging of the Smokies are found in Frome, *Strangers in High Places;* R. Houk, *Great Smoky Mountains National Park: A Natural History Guide* (Boston: Houghton Mifflin, 1993); and Pierce, *The Great Smokies.* For the logging of the Southern Appalachians more generally, see C. Bolgiano, *The Appalachian Forest* (Mechanicsburg, Pa.: Stackpole, 1998).

7. Wear and Greis, *Southern Forest Resource Assessment,* Executive Summary, 1–2. We do not count the thirty-two million acres of pine plantations, as these are clearly not natural—nor, for reasons described below, should they even be counted as *forests.*

8. M. K. Trani (Greip), "Terrestrial Ecosystems," in *Southern Forest Resource Assessment,* ed. Wear and Greis, 20.

9. V. A. Rudis, "Refional Forest Resource Assessment in an Ecological Framework: The Southern United States," *Natural Areas Journal* 18 (1998): 319–32; H. Irwin, S. Andrew, and T. Bouts, *Return of the Great Forest: A Conservation Vision for the Southern Appalachian Region* (Asheville, N.C.: Southern Appalachian Forest Coalition, 2002).

10. TVA, U.S. Army Corps of Engineers, and U.S. Fish and Wildlife Service, *Final Environmental Impact Statement,* vol. 1: 214–15.

11. Wear and Greis, *Southern Forest Resource Assessment,* 52.

12. Prestemon and Abt, "Timber Products Supply and Demand," 320–21; R. W. Haynes, *An Analysis of the Timber Situation in the United States: 1952 to 2050,* Gen. Tech. Rpt.

PNW-GTR-560, U.S. Dept. of Agriculture, Forest Service, Pacific Northwest Research Station, Portland, Ore., 2003, 173, 185.

13. Most pine plantations are established on previously forested land. Data from the 1980s and 1990s show that 47 percent were established on what had been hardwood or oak-pine forest and 28 percent on what had been natural pine forest. The remaining 25 percent displaced agricultural land. See Prestemon and Abt, "Timber Products Supply and Demand," 46.

14. Wear and Greis, *Southern Forest Resource Assessment,* 14, 32–4; Haynes, *Analysis of the Timber Situation,* 173.

15. TVA, U.S. Army Corps of Engineers, and U.S. Fish and Wildlife Service, *Final Environmental Impact Statement,* vol. 1: 231–33; W. T. Swank and D. A. Crossley Jr., eds., *Forest Hydrology and Ecology at Coweeta* (New York: Springer-Verlag, 1988); D. W. Johnson, G. S. Henderson, and D. E. Todd, "Changes in Nutrient Distribution in Forests and Soils of Walker Branch Watershed, Tennessee, Over An Eleven Year Period," *Biogeochemistry* 8 (1988): 275–93; D. W. Johnson and R. I. Van Hook, *Analysis of Biogeochemical Cycling Processes in Walker Branch Watershed* (New York: Springer-Verlag, 1989); J. D. Knoepp and W. T. Swank, "Long-Term Soil Chemistry Changes in Aggrading Forest Ecosystems," *Soil Science Society of America Journal* 58 (1994): 325–31.

16. G. E. Likens, F. H. Borman, R. S. Pierce, and W. A. Reiners, "Recovery of a Deforested Ecosystem," *Science* 88 (1978): 492–95.

17. L. R. Boring, W. T. Swank, and C. D. Monk, "Dynamics of Early Successional Forest Structure and Processes in the Coweeta Basin," in *Forest Hydrology and Ecology at Coweeta,* ed. Swank and Crossley Jr., 161–79. See also TVA, U.S. Army Corps of Engineers, and U.S. Fish and Wildlife Service, *Final Environmental Impact Statement,* vol. 1: 233–37.

18. D. C. Duffy and A. J. Meier, "Do Appalachian Herbaceous Understories Ever Recover from Clearcutting?" *Conservation Biology* 6 (1992): 196–201. The Duffy and Meier paper received a great deal of media attention and generated a great deal of criticism and comment, mostly with respect to research methodology and data analysis. See A. S. Johnson, W. M. Ford, and P. E. Hale, "The Effects of Clearcutting on Herbaceous Understories Are Still Not Fully Known," *Conservation Biology* 7 (1993): 433–35; and D. C. Duffy, "Seeing the Forest for the Trees: Response to Johnson et al.," *Conservation Biology* 7 (1993): 436–39. One participant in the discussion, Glenn Matlack of the Harvard Forest, found the criticism surprising, because Duffy and Meier's results were "not very exceptional" in light of the fact that "similar historical effects have been reported for a variety of European and American forest communities" (G. Matlack, "Plant Demography, Land-Use History, and the Commercial Use of Forests," *Conservation Biology* 7 [1994]: 298–99). In a subsequent paper, Meier et al. discussed mechanisms of loss and slow recovery of herbaceous species in logged deciduous forests, mechanisms to which some species are more susceptible than others. All forms of logging—especially clear-cuts but also leave-tree and shelterwood schemes—change forest floor microclimates, resulting in the loss of species susceptible to desiccation. Factors causing a relaxation of diversity during forest recovery and influencing the re-establishment of extirpated species include the following: continued physiological stress and competition with weedy or invasive species during the early stages of forest recovery; low growth and

reproductive rates of forest-floor herbaceous species; and low rates of dispersal (Meier et al., "Possible Ecological Mechanisms for Loss of Vernal-Herb Diversity in Logged Eastern Deciduous Forests," *Ecological Applications* 5 [1995]: 935–46). See also S. M. Hood, "Vegetation Responses to Seven Silvicultural Methods in the Southern Appalachians One Year after Harvesting," M.S. thesis, Virginia Polytechnic Institute and State Univ., Blacksburg, 2001. Accessible at http://scholar.lib.vt.edu/theses/available/etd-05292001-140004/.

19. A. N. Ash, "Disappearance of Salamanders from Clearcut Plots," *Journal of the Elisha Mitchell Scientific Society* 104 (1988): 116–22; A. N. Ash, "Disappearance and Return of Plethodontid Salamanders to Clearcut Plots in the Southern Blue Ridge Mountains," *Conservation Biology* 11 (1997): 983–89; J. W. Petranka, M. E. Eldridge, and K. E. Haley, "Effects of Timber Harvesting on Southern Appalachian Salamanders," *Conservation Biology* 7(1993): 363–70; J. W. Petranka, M. P. Brannon, M. E. Hopey, and C. K. Smith, "Effects of Timber Harvesting on Low Elevation Populations of Southern Appalachian Salamanders," *Forest Ecology Management* 67 (1994): 135–47. See also D. H. Bennett, J.W. Gibbons, and J. Glanville, "Terrestrial Activity, Abundance, and Diversity of Amphibians in Differently Managed Forest Types," *American Midlands Naturalist* 103 (1980): 412–16; P. D. N. Hebert, B. W. Muncaster, and G. L Mackie, "Ecological and Genetic Studies on *Dreissena polymorpha* (Pallas): A New Mollusc in the Great Lakes," *Canadian Journal of Fisheries and Aquatic Science* 46 (1989): 1587–91.

20. A. N. Ash and R. C. Bruce, "Impacts of Timber Harvesting on Salamanders," *Conservation Biology* 8 (1994): 300–301; J. W. Petranka, "Recovery of Salamanders after Clearcutting in the Southern Appalachians: A Critique of Ash's Estimates," *Conservation Biology* 13 (1999): 203–5.

21. R. H. Yahner, "Dynamics of a Small Mammal Community in a Fragmented Forest," *American Midlands Naturalist* 127 (1992): 381–91.

22. Hood, "Vegetation Responses."

23. R. F. Noss and A. Y. Cooperrider, *Saving Nature's Legacy: Protecting and Restoring Biodiversity* (Washington, D.C.: Island Press, 1994), 192–97; Irwin, Andrew, and Bouts, *Return of the Great Forest;* J. C. Baker and W. C. Hunter, "Effects of Forest Management on Terrestrial Ecosystems," in *Southern Forest Resource Assessment,* ed. Wear and Greis, 99–101.

24. Wear and Greis, *Southern Forest Resource Assessment,* 46.

25. Prestemon and Abt, "Timber Products Supply And Demand," 299, 316.

26. J. P. Siry, "Intensive Timber Management Practices," in *Southern Forest Resource Assessment,* ed. Wear and Greis, 327–40.

27. C. Camuto, "Flying is Believing," *Blue Ridge Press,* 30 June 2003.

28. J. D. Ward and P. M. Mistretta, "Impact of Pests on Forest Health," in *Southern Forest Resource Assessment,* ed. Wear and Greis, 410–12.

29. For a discussion of this estimate for the sourcing area, see TVA, U.S. Army Corps of Engineers, and U.S. Fish and Wildlife Service, *Final Environmental Impact Statement,* vol. 1: 53.

30. D. Murray (ForestWatch) and C. Sand (Dogwood Alliance), personal communication.

31. TVA, U.S. Army Corps of Engineers, and U.S. Fish and Wildlife Service, *Final Environmental Impact Statement,* vol. 1: 51.

32. Forest Watch, "Forest Watch," accessible at http://www.forestwatch.net./ (21 July 2003).

33. TVA, U.S. Army Corps of Engineers, and U.S. Fish and Wildlife Service, *Final Environmental Impact Statement,* vol. 1: 51.

34. Southern Appalachian Man and the Biosphere Cooperative (SAMAB), "Southern Appalachian Assessment: Social/Cultural/Economic Technical Report," United States Dept. of Agriculture, Washington (1996), 88.

35. H. Irwin, *Tennessee's Mountain Treasures: the Unprotected Wildlands of the Cherokee National Forest* (Washington, D.C.: The Wilderness Society, 1996), 6.

36. Z. J. Cornett, "Birch Seeds, Leadership, and a Relationship with the Land," *Journal of Forestry* 93, (9) (1995): 6–11. This trend has become still more evident in recent years. See G. Wicker, "Motivation for Private Forest Landowners," in *Southern Forest Resource Assessment,* ed. Wear and Greis, 225–37.

37. Wicker, "Motivation for Private Forest Landowners," 225-237.

38. D. N. Wear, "Land Use," in *Southern Forest Resource Assessment,* ed. Wear and Greis, 153.

39. Except for the 465,000 designated wilderness acres of Great Smoky Mountains National Park, roadless areas are few and far between, and small. The Wilderness Act of 1964 defines a roadless area as an area with no more than one-half mile of improved road for each 1,000 acres. In Cherokee, Nantahala, and Pisgah National Forests, there is a total of 239,183 roadless acres, ranging in size from 335 acres to 13,791 acres. See SAMAB, "Southern Appalachian Assessment," 180–82; see also I. F. Spellerberg, "Ecological Effects of Roads and Traffic: A Literature Review," *Global Ecology and Biogeography Letters* 7 (1998): 317–33, for a global review of the ecological consequences of roads and road construction.

40. Wear, "Land Use," 153–73; R. C. Conner and A. J. Hartsell, "Forest Area and Conditions," in *Southern Forest Resource Assessment,* ed. Wear and Greis, 357–402; Irwin, Andrew, and Bouts, *Return of the Great Forest.* See also D. S. Wilcove, D. Rothstein, J. Dubow, A. Philips, and E. Losos, "Leading Threats to Biodiversity: What's Imperiling U.S. Species?" in *Precious Heritage: The Status of Biodiversity in the United States,* ed. B. A. Stein, L. S. Kutner, and J. S. Adams (New York: Oxford Univ. Press, 2000), 239–54.

41. Warren Parker and Laura Dixon, *Endangered and Threatened Wildlife of Kentucky, North Carolina, South Carolina, and Tennessee,* North Carolina Agricultural Extension Service (1980).

42. Amundsen, "Reservoir Riparian Zone Characteristics," 469–93.

43. J. A. Stanturf, D. D. Wade, T. A. Waldrop, D. K. Kennard, and G. L. Achtemeier, "Background Paper: Fire in the Southern Forest Landscapes," in *Southern Forest Resource Assessment,* ed. Wear and Greis, 607–30; E. R. Buckner and N. L. Turrill, "Fire Management," in *Ecosystem Management for Sustainability: Principles and Practices Illustrated by a Regional Biosphere Reserve Cooperative,* ed. J. D. Peine (Boca Raton, Fla.: CRC Press, 1998), 329–47; Delcourt and Delcourt, "Pre-Columbian Native American Use of Fire," 1010–14; D. H. Van Lear and T. A. Waldorp, "History, Use, and Effects of Fire in the Appalachians," Tech. Rpt. SE-54, U.S. Dept. of Agriculture, Forest Service, Southern Research Station, Asheville, N.C., 1989; P. A. Delcourt, H. R. Delcourt, P. A. Cridlebangh, and J. Chapman, "Holocene Ethnobotanical and Paleoecological Record of Human Impact on Vegetation in the Little Tennessee River Valley, Tennessee," *Quaternary Research* 25 (1986): 330–44; M. E. Harmon, "Fire History of the Westernmost Portion of The Great Smoky

Mountains National Park," *Bulletin of the Torrey Botanical Club* 109 (1982): 74-79; see also Pyne, *Fire in America.*

44. Pyne, *Fire in America.*
45. L. S. Barden and F. W. Woods, "Effects of Fire on Pine and Pine-Hardwood Forests in the Southern Appalachians," *Forest Science* 22 (1976): 399–403; L. S. Barden, "Self-Maintaining Populations of *Pinus pungens* Lam. in the Southern Appalachian Mountains," *Castanea* 42 (1977): 316–23; G. L. Sanders, "The Role of Fire in the Regeneration of Table Mountain Pine in the Southern Appalachian Mountains," M.S. thesis, Univ. of Tennessee, Knoxville, 1992.
46. Stanturf et al., "Background Paper," 607–30; N. T. Welch and T. A. Waldrop, "Restoring Table Mountain Pine (*Pinus pungens* Lamb.) Communities with Prescribed Fire: An Overview of Current Research," *Castanea* 66 (2001): 42–49; P. H. Brose and T. A. Waldrop, "Using Prescribed Fire to Regenerate Table Mountain Pine in the Southern Appalachian Mountains," in *Fire And Forest Ecology: Innovative Silviculture and Vegetation Management,* ed. W. K. Moser and C. F. Moser, Tall Timbers Fire Ecology Conference Proceedings, No. 21, Tallahassee, Fla., 2000, 191–96; Buckner and Turrill, "Fire Management," 329–47.
47. Barden and Woods, "Effects of Fire," 399–403; M. D. Abrams, "Fire and the Development of Oak Forests," *BioScience* 42 (1992): 346–53; Delcourt and Delcourt, "Pre-Columbian Native American Use of Fire," 1010–14; P. A. Delcourt and H. R. Delcourt, "The Influence of Prehistoric Human-Set Fires on Oak-Chestnut Forests in the Southern Appalachians," *Castanea* 63 (1998): 337–45.
48. TVA, U.S. Army Corps of Engineers, and U.S. Fish and Wildlife Service, *Final Environmental Impact Statement,* vol. 1: 55.
49. Hudson, *The Southeastern Indians,* 41.
50. J. Mooney, *Myths of the Cherokee and Sacred Formulas of the Cherokees* (Nashville, Tenn.: C. Elder—Bookseller, 1972), 250–52; C. Hudson, "The Cherokee Concept of Natural Balance," *Indian Historian* 3 (1970): 51–54; Hudson, *The Southeastern Indians,* 158, 346.
51. Hudson, *The Southeastern Indians,* 427–43.
52. Ibid., 173.
53. Such thinking seems to have been common among Native American tribes elsewhere. See C. Martin, *Keepers of the Game: Indian–Animal Relationships and the Fur Trade* (Berkeley: Univ. of California Press, 1978).
54. J. W. Ross, "Tennessee Wildlife on the Comeback Trail," *Tennessee Conservationist* 59 (5) (1993): 24–28.
55. Noss and Cooperrider, *Saving Nature's Legacy,* 204.
56. P. A. Morton, "Charting a New Course: National Forests in the Southern Appalachians," *The Living Landscape* 5 (1994): 58.
57. S. G. Miller, S. P. Bratton, and J. Hadidian, "Impacts of White-Tailed Deer on Endangered and Threatened Plants," *Natural Areas Journal* 12 (1992): 67–74.
58. J. W. Ross, "Tennessee Vanquished and Vanished Wildlife," *Tennessee Conservationist* 59 (3) (1993): 8–11.
59. Ibid.
60. L. D. Bryant and C. Maseer, "Classification and Distribution," in *Elk of North America: Ecology and Management,* ed. J. W. Thomas and D. E. Toweill (Harrisburg, Pa.: Stackpole, 1982), 1–60.
61. A. Leopold, *A Sand County Almanac, with Essays from Round River* (New York: Ballentine, 1966), 139–40.

62. Ross, "Tennessee Vanquished and Vanished Wildlife," 8-11.
63. J. L. Gittleman and S. L. Pimm, "Crying Wolf in North America," *Nature* 351 (1991): 524–25.
64. A. McDade, "Cry of the Red Wolf," *Tennessee Conservationist* 59 (2) (1993): 18–22; and National Park Service (NPS), "Smokies Guide" (Summer Issue 1995), Gatlinburg, Tenn., 6.
65. U.S. Fish and Wildlife Service Southeast Region, "U.S. Fish and Wildlife Service, National Park Service Ends Effort to Establish Endangered Red Wolves in Great Smoky Mountains National Park," press release, Oct. 8, 1998. Accessible at http://southeast.fws.gov/news/1998/r98-091a.html.
66. D. W. Linzey, *Mammals of Great Smoky Mountains National Park* (Blacksburg, Va.: McDonald and Woodward, 1995).
67. A. F. Gainer, "The Wild Life of Tennessee," *Journal of the Tennessee Academy of Science* 3 (3) (1928): 10–22.
68. Ross, "Tennessee Vanquished and Vanished Wildlife," 8–11.
69. National Park Service, "Great Smoky Mountains National Park," http://www.nps.gov/grsm/pphtml/highlights383.html (7 July 2003).
70. C. Toops, "Baiting the Bears," *National Parks* (Nov./Dec. 1992): 39–42.
71. B. Hatcher, "Endangered Species: We CAN Make a Difference," *Tennessee Wildlife* 19 (6) (1996): 25–27; Ross, "Tennessee Wildlife on the Comeback Trail," 24-28.
72. D. C. Eager and R. M. Hatcher, "Tennessee's Rare Wildlife," in *Vol I: The Vertebrates* (Nashville: Tennessee Wildlife Resources Agency, 1980), C11–C12.
73. G. A. Himebaugh, "Battling for Survival: Tennessee's Flying Mammals Face Long Odds," *Tennessee Conservationist* 58 (6) (1992): 23–28; A. R. Richter, S. R. Humphrey, J. B. Cope, and V. Black, Jr., "Modified Cave Entrances: Thermal Effect on Body Mass and Resulting Decline in Endangered Indiana Bats *(Myotis sodalis),*" *Conservation Biology* 7 (1993): 407–15.
74. Y. Baskin, "Winners and Losers in a Changing World," *BioScience* 48 (1998): 788–92; M. L. McKinney and J. L. Lockwood, "Biotic Homogenization: A Few Winners Replacing Many Losers in the Next Mass Extinction," *Trends in Ecology and Evolution* 14 (1999): 450–53.
75. Ross, "Tennessee Vanquished and Vanished Wildlife," 8–11.
76. G. Howard, "The Carolina Parakeet," accessible at http://www.georgehoward.net/parakeet.htm (18 Feb. 2003), citing G. Laycock "The Last Carolina Parakeet," *Audubon Magazine* (Mar. 1969).
77. J. Terborgh, "Why American Songbirds Are Vanishing," *Scientific American* (May 1992): 98–104.
78. J. W. Ross, "Tennessee Vanquished and Vanished Wildlife," *The Tennessee Conservationist* 59, no. 3 (1993): 8-11.
79. Schorger, *The Passenger Pigeon.*
80. Terborgh, "Why American Songbirds Are Vanishing," 98–104.
81. Hatcher, "Endangered Species," 25–27; Ross, "Tennessee Wildlife on the Comeback Trail," 24–28.
82. C. Nicholson, TVA environmental scientist, personal communication.
83. TVA, *RiverPulse: A Report on the Condition of the Tennessee River and Its Tributaries in 1993* (1994), 3.
84. Ross, "Tennessee Wildlife on the Comeback Trail," 24–28.
85. Hatcher, "Endangered Species," 25–27.

86. Ibid.

87. SAMAB, "Southern Appalachian Assessment," 66–67.

88. N. Collar, M. Crosby, and A. Stattersfield, *Birds to Watch 2* (Washington, D.C.: Smithsonian Institution Press, 1994); B. G. Peterjohn, J. R. Sauer, and C. S. Robbins, "Population Trends from the North American Breeding Bird Survey," in *Ecology and Management of Neotropical Migratory Birds,* ed. T. E. Martin and D. M. Finch (New York: Oxford Univ. Press, 1995), 3–39.

89. D. R. Petit, J. F. Lynch, R. L. Hutto, J. G. Blake, and R. B. Waide, "Habitat Use and Conservation in the Tropics," in *Ecology and Management,* ed. Martin and Finch, 145–97; T. W. Sherry and R. T. Holmes, "Winter Habitat Quality, Population Limitation, and Conservation of Neotropical-Nearctic Migrant Birds," *Ecology* 77 (1996): 36–48.

90. Terborgh, "Why American Songbirds Are Vanishing," 98–104.

91. Trani (Greip), "Terrestrial Ecosystems," 18.

92. Martin and Finch, *Ecology and Management.*

93. Terborgh, "Why American Songbirds Are Vanishing," 98–104. See also K. Böhning-Gaese, M. L. Taper, and J. H. Brown, "Are Declines in North American Insectivorous Songbirds Due to Causes on the Breeding Range?" *Conservation Biology* 7 (1993): 76–86; P. W. C. Paton, "The Effect of Edge on Avian Nest Success: How Strong Is the Evidence?" *Conservation Biology* 8 (1994): 17–26; S. K. Robinson, S. I. Rothstein, M. C. Brittingham, L. J. Petit, and J. A. Grzybowski, "Ecology and Behavior of Cowbirds and Their Impact on Host Populations," in *Ecology and Management,* ed. Martin and Finch, 428–60.

94. D. S. Wilcove, "Nest Predation in Forest Tracts and the Decline of Migratory Songbirds," *Ecology* 66 (1985): 1211–14.

95. L. L. Master, B. A. Stein, L. S. Kutner, and G. A. Hammerson, "Vanishing Assets: Conservation Status of U.S. Species," in *Precious Heritage,* ed. Stein, Kutner, and Adams, 93–118; K. O'Neal, "Effects of Global Warming on Trout and Salmon in U.S. Streams," report, Defenders of Wildlife, Washington, D.C., 2002; Eager and Hatcher, "Tennessee's Rare Wildlife"; and Chuck Nicholson, personal communication.

96. Hatcher, "Endangered Species," 25–27.

97. O'Neal, "Effects of Global Warming"; Eager and Hatcher, "Tennessee's Rare Wildlife"; and Chuck Nicholson, personal communication.

98. Terborgh, "Why American Songbirds Are Vanishing," 98–104.

99. K. N. Rabenold, P. T. Fauth, B. W. Goodner, J. A. Sadowski, and P. G. Parker, "Response of Avian Communities to Disturbance by an Exotic Insect Spruce-Fir Forests of the Southern Appalachians," *Conservation Biology* 12 (1) (Feb. 1998): 182.

100. Eager and Hatcher, "Tennessee's Rare Wildlife."

101. M. Barinaga, "Where Have All The Froggies Gone?" *Science* 247 (1990): 1033–34; A. R. Blaustein and D. B. Wake, "Declining Amphibian Populations: A Global Phenomenon?" *Trends in Ecology and Evolution* 5 (1990): 203-204; K. Phillips, "Where Have All the Frogs and Toads Gone?" *Bioscience* 40 (1990): 422–24; A. R. Blaustein and D. H. Olson, "Declining Amphibians," *Science* 253 (1991): 1467.

102. L. E. Long, L. S. Saylor, and M. E. Soulé, "A pH/UV-B Synergism in Amphibians," *Conservation Biology* 9 (1995): 1301–3.

103. United States Geological Survey (USGS), "USGS Diagnoses Causes of Many U.S. Amphibian Die-Offs," News Release, 8 Aug. 2000. Accessible at

http://www.usgs.gov/public/press/public_affairs/press_releases/pr1272m.html (07 Mar. 2003). See also National Biological Information Infrastructure, "FrogWeb: Amphibian Declines and Deformities," *Current Biological Issues,* accessible at http://frogweb.nbii.gov/ (07 Mar. 2003).

104. C. K. Dodd Jr., "Imperiled Amphibians: A Historical Perspective," in *Aquatic Fauna in Peril: The Southeastern Perspective,* ed. G. W. Benz and D. E. Collins (Decatur, Ga.: Lenz Design & Communication, 1999), 165–200; A. R. Blaustein, D. B. Wake, and W. P. Sousa, "Amphibian Declines: Judging Stability, Persistence, and Susceptibility of Populations to Local and Global Extinctions," *Conservation Biology* 8 (1994): 60–71; J. H. K. Pechmann, D. E. Scott, R. D. Semlitsch, J. P. Caldwell, L. J. Vitt, and J. W. Gibbons, "Declining Amphibian Populations: The Problem of Separating Human Impacts from Natural Fluctuations," *Science* 253 (1991): 892–95; Blaustein and Wake, "Declining Amphibian Populations," 203–4.

105. NPS, "Smokies Guide" (Summer Issue 1995).

106. Dodd, "Imperiled Amphibians," 165–200.

107. Ibid.

108. J. W. Huckabee, C. P. Goodyear, and R. D. Jones, "Acid Rock in the Great Smokies: Unanticipated Impact on Aquatic Biota of Road Construction in Regions of Sulfide Mineralization," *Transactions of the American Fisheries Society* 104 (1975): 677–84; R. C. Matthews Jr. and E. L. Morgan, "Toxicity of Anakeesta Formation Leachates to Shovel-Nosed Salamander, Great Smoky Mountains National Park," *Journal of Environmental Quality* 11 (1982): 102–6.

109. J. T. Collins recommended that this previously defined, diagnosed, and allopatric taxon be recognized as a distinct species (*Herpetological Review* 22 [2] [1991]: 42–43). Collins presented this proposed change to his salamander systematist group, composed of D. A. Good, R. Highton, D. M. Sever, H. B. Shaffer, S. G. Tilley, T. A. Titus, D. B. Wake, and A. H. Wynn, and the majority of those individuals responding concurred that *gulolineatus* be so recognized (*Herpetological Circular* 25: 1–40). Collins followed that recommendation. R. Powell, J. T. Collins, and E. D. Hooper recognized this lineage as a species distinct from *G. palleucus* (*A Key to the Amphibians and Reptiles of the Continental United States and Canada* [Lawrence: Univ. Press of Kansas]). W. E. Duellman and S. S. Sweet recognized *G. gulolineatus* as a distinct species, as proposed by Collins (Duellman, W. E., ed., *Patterns of Distribution of Amphibians. A Global Perspective* [Baltimore: Johns Hopkins Univ. Press, 1999], 31–109). See also B. I. Crother, "Scientific and Standard English Names of Amphibians and Reptiles of North America North of Mexico, with Comments Regarding Confidence in Our Understanding," *Society for the Study of Amphibians and Reptiles,* Herpetological Circular No. 29 (2000).

110. Much of the information in this paragraph was gleaned from notes taken by John Nolt at a U.S. Fish and Wildlife Service meeting on the status of the Tennessee and Berry Cave Salamanders, Dec. 10, 2002, in Cookeville, Tenn., and from personal communication with Addison Wynn of the National Museum of Natural History.

111. Letter dated 24 Feb. 2003 to John Nolt from Sam D. Hamilton, acting regional director of the U.S. Fish and Wildlife Service, Atlanta, Ga.

112. Etnier and Starnes, *The Fishes of Tennessee;* Eager and Hatcher, "Tennessee's Rare Wildlife."

113. TVA, "TVA River Neighbors," September Resource Group, Norris, Tenn. (1996), 4.

114. D. A. Etnier, "Our Southeastern Fishes—What Have We Lost and What Are We Likely to Lose," *Southeastern Fishes Council Proceedings* 29 (1994): 5–9; Eager and Hatcher, "Tennessee's Rare Wildlife," B101–B102.

115. D. A. Etnier, personal communication; as of June 2002 repeated efforts to locate slender chubs had still failed. See Conservation Fisheries, Inc., "The Search Goes On," *Newsletter #12* (June 2002), accessible at http://www.conservationfisheries.org/Newsletters/newsletter12_june_2002.htm (21 July 2003).

116. G. R. Dinkins and P. W. Shute, "Life History of *Noturus baileyi* and *N. flavipinnis* (Pisces: Ictaluridae), Two Rare Madtom Catfishes in Citico Creek, Monroe County, Tennessee," *Bulletin of the Alabama Museum of Natural History* 18 (1996): 43–69.

117. D. A. Etnier, "Jeopardized Southeastern Freshwater Fishes: A Search for Causes," in *Aquatic Fauna in Peril,* ed. Benz and Collins, 87–104; Etnier, "Our Southeastern Fishes," 5–9; R. E. Jenkins and N. M. Burkhead, *Freshwater Fishes of Virginia* (Bethesda, Md.: American Fisheries Society, 1993).

118. P. Shute, personal communication.

119. R. E. Jenkins and N. M. Burkhead, "Description, Biology, and Distribution of the Spotfin Chub *Hybopsis monacha,* a Threatened Cyprinid Fish of the Tennessee River Drainage," *Buletin of the. Alabama Museum of Naural. History* 8 (1984): 1–30. Most of the information in the previous several paragraphs is from Etnier, "Our Southeastern Fishes," 5–9, and Eager and Hatcher, "Tennessee's Rare Wildlife"; see also J. Herrig and P. Shute, "Aquatic Animals and Their Habitats," in *Southern Forest Resource Assessment,* ed. Wear and Greis, 537–80; Etnier, "Jeopardized Southeastern Freshwater Fishes," 87–104.

120. R. E. Lennon and P. S. Parker, "The Reclamation of Indian and Abrams Creeks, Great Smoky Mountains National Park," U.S. Fish and Wildlife Service, Scientific Report 306, Washington, D.C., 1959.

121. Etnier and Starnes, *The Fishes of Tennessee,* 587–90, and P. Shute, personal communication.

122. Etnier and Starnes, *The Fishes of Tennessee,* 161–63.

123. S. Z. Guffey, G. F. McCracken, S. E. Moore, and C. R. Parker, "Management of Isolated Populations: Southern Strain Brook Trout," in *Ecosystem Management for Sustainability,* ed. Peine, 247–65.

124. Ibid., 247–65; Etnier and Starnes, *The Fishes of Tennessee.*

125. Huckabee, Goodyear, and Jones, "Acid Rock in the Great Smokies," 677–84.

126. J. D. Meisner, "Effect of Climate Warming on the Southern Margins of the Native Range of Brook Trout, *Salvilinus fontinalis,*" *Canadian Journal Fisheries and Aquatic Science* 47 (1990): 1065–70; O'Neal, "Effects of Global Warming."

127. Etnier and Starnes, *The Fishes of Tennessee.*

128. Parmalee and Bogan, *The Freshwater Mussels of Tennessee;* see also J. D. Williams, M. L. Warren, Jr., K. S. Cummings, J. L. Harris, and R. J. Neves, "Conservation Status of Freshwater Mussels of the United States and Canada," *Fisheries* 18 (9) (1993): 6–22; R. J. Neves, A. F. Bogan, J. D. Williams, A. A. Ahlstedr, and P. W. Hartfield, "Status of Aquatic Mollusks in the Southeastern United States: A Downward Spiral of Diversity," in *Aquatic Fauna in Peril,* ed. Benz and Collins, 43–86.

129. TVA, "Tennessee Valley Authority Natural Heritage Project," 22 Oct. 2001, accessible at http://www.tva.gov/environment/pdf/class01.pdf (21 July 2003).

130. Hatcher, "Endangered Species," 25–27.

131. A. D. McClure, "Zebras 'Mussel' Their Way into the Tennessee Valley," *EnviroLink*, Chattanooga (1995) 9–10.

132. J. M. Bates and S. D. Dennis, Final Report, Mussel Resource Survey; State of Tennessee, Tennessee Wildlife Resources Agency (1985): 48–61.

133. Discussed in M. L. Shaffer and B. A. Stein, "Safeguarding Our Precious Heritage," in *Precious Heritage*, ed. B. A. Stein et al., 301–21.

134. O'Neal, "Effects of Global Warming"; A. E. Bogan and P. W. Parmalee, *Tennessee's Rare Wildlife, Volume II: The Mollusks* (Nashville: Tennessee Wildlife Resources Agency, 1983).

135. L. Roberts, "Zebra Mussel Invasion Threatens U.S. Water," *Science* 249 (1990): 1370–72; Hebert, Muncaster, and Mackie, "Studies on *Dreissena polymorpha* (Pallas)," 1587–91; P. D. N. Hebert, C. C. Wilson, M. H. Murdoch, and R. Lazer, "Demography and Ecological Impacts of the Invading Mollusc *Dreissena polymorpha*," *Canadian Journal of Zoology* 69 (1991): 405–9; M. L. Ludyanskiy, D. McDonald, and D. MacNeil, "Impact of the Zebra Mussel, a Bivalve Invader," *BioScience* 43 (1993): 533–41; L. E. Johnson and D. K. Padilla, "Geographic Spread of Exotic Species: Ecological Lessons and Opportunities from the Invasion of the Zebra Mussel *Dreissena polymorpha*," *Conservation Biology* 78 (1996): 23–33; H. J. MacIsaac, "Potential Abiotic and Biotic Impacts of Zebra Mussels on the Inland Waters of North Anerica," *American Zoologist* 36 (1996): 287–99; see also H. J. Cathay, "Zebras in our Back Yard!" *Tennessee Conservationist* 58 (5) (1992): 23–26; F. Kuznik, "America's Aching Mussels," *National Wildlife* (Oct.–Nov. 1993): 34–38; McKinney and Schoch, *Environmental Science*; J. Ross, "An Aquatic Invader Is Running Amok in U.S. Waterways," *Smithsonian* (Feb. 1994): 41–50; TVA, *RiverPulse* (1994).

136. McKinney and Schoch, *Environmental Science,* 72.

137. J. Opie, "Where American History Began: Appalachia and the Small Independent Family Farm," *Proceedings of the 1980 Appalachian Studies Conference* (Boone, N.C.: Appalachia Consortium Press, 1981), 61.

138. NPS, "Smokies Guide" (Summer Issue 1995), 7.

139. S. E. Schlarbaum, Dept. of Forestry, Wildlife and Fisheries, Univ. of Tennesssee, Knoxville, personal communication, Aug. 2003.

140. B. Bowen, "The Green Scourge: Invasive Alien Plants," *Tennessee Conservationist* 59 (5) (1993): 29–32; Southern Appalachian Man in the Biosphere Cooperative (SAMAB), "Southern Appalachian Assessment: Social/Cultural/Economic Technical Report," U.S. Dept. of Agriculture, Washington, D.C., 1996, 71; SAMAB, "Southern Appalachian Assessment: Terrestrial Technical Report," U.S. Dept. of Agriculture, Washington, D.C., 1996, 113; F. T. Campbell and S. E. Schlarbaum, *Fading Forests II: Trading Away North America's Natural Heritage* (N.p.: Healing Stones Foundation, 2002), 71–72.

141. Campbell and Schlarbaum, *Fading Forests II,* 71.

142. Ibid., 72.

143. Ibid., 73.

144. Schlarbaum, personal communication.

145. NPS, "Smokies Guide" (Summer Issue 1996).

146. D. Searfoss, "Mountains Mourning: Pest Invaders Threaten Tennessee's Trees," *Tennessee Green* 9 (Summer 1995); SAMAB, "Southern Appalachian Assessment: Terrestrial Technical Report," 109–11; Campbell and Schlarbaum, *Fading Forests II,* 78–79.

147. NPS, "Smokies Guide" (Summer Issue 1996).

148. SAMAB, "Southern Appalachian Assessment: Terrestrial Technical Report," 111–13, and NPS, "Smokies Guide" (Summer Issue 1996).

149. Schlarbaum, personal communication; SAMAB, "Southern Appalachian Assessment: Terrestrial Technical Report," 103–8.

150. SAMAB, "Southern Appalachian Assessment: Terrestrial Technical Report," 114-17. The two-decade estimate is from S. E. Schlarbaum.

151. NPS, "Smokies Guide" (Summer Issue 1996).

152. SAMAB, "Southern Appalachian Assessment: Terrestrial Technical Report," 120–21.

153. Schlarbaum, personal communication.

154. Bowen, "The Green Scourge," 29–32; SAMAB, "Southern Appalachian Assessment: Terrestrial Technical Report," 117-18; Campbell and Schlarbaum, *Fading Forests II,* 75-76; NPS, "Smokies Guide" (Summer Issue 1996).

155. NPS, "Smokies Guide" (Summer Issue 1995).

156. TVA, *Energy Vision 2020,* vol. 2: T1.129.

5. popuLaTion anD urBanizaTion

1. Roderick Nash, *Wilderness and the American Mind,* rev. ed., (New Haven: Yale Univ. Press, 1973).

2. Based on U.S. Census Bureau data. Percentages do not total 100 due to rounding.

3. The definition used by the U.S. Census Bureau is considerably more complicated. See www.census.gov/geo/www/garm.html.

4. As defined by the U.S. Census Bureau, metropolitan areas include metropolitan statistical areas (MSAs), consolidated metropolitan statistical areas, and primary metropolitan statistical areas. In general, a MSA must include at least one city with 50,000 or more inhabitants and a total metropolitan area of at least 100,000. Qualification of outlying counties in the MSA depends on meeting requirements such as population density and percentage of residents commuting to the central county(ies). U.S. Census Bureau, *Statistical Abstract of the United States: 2001,* Appendix II, Metropolitan Areas: Concepts, Components, and Population. R. Ewing, R. Pendall, and D Chen, *Measuring Sprawl and its Impacts* (Smart Growth America, 2002) ranked the Knoxville MSA as the eighth most sprawling of 83 metropolitan areas evaluated nationally. It reached this conclusion in part because of the Knoxville MSA's low residential density. The Knoxville MSA's sprawl is bad, but perhaps not as bad as might be assumed from reading

this report. Whereas the five outlying counties in the MSA had average densities of from 80 (Union County) to 211 (Anderson County) in 2000, Knox County's average density was 751.

5. In the bioregion, the number of housing units increased nearly 16 percent between 1990 and 2000, while the number of units per square mile increased only 3 percent. This is another indication that housing and related land uses are causing large tracts of open land to be converted to low-density development.

6. M. R. English, S. T. Huss, and J. R. Hoffman, *Growth Patterns in Tennessee: State-wide Impacts,* White Paper prepared for the Tennessee Advisory Commission on Intergovernmental Relations, June 2002, 5.

7. For a discussion of demographic changes in "Sunbelt" suburbs, see W. H. Frey, "Boomers and Seniors in the Suburbs: Aging Patterns in Census 2000," *Brookings Institution, Living Cities Census Series* (New York, N.Y.: Center on Urban and Metropolitan Policy, 2003).

8. Partly because of changes in the bioregion's age composition, its average household size dropped 14 percent between 1980 and 2000, from 2.8 to 2.4 persons. In the United States in 2000, the average household size was about 2.6 persons.

9. E. L. Andrews, "Economic Inequality Grew in 90's Boom, Fed Reports," *New York Times,* 23 Jan 2003). According to this article, the wealth of those in the top 10 percent income bracket grew much more than the wealth of any other group during the 1990s, citing a report released by the U.S. Federal Reserve.

10. In contrast, the Virginia portion of the bioregion, which still has a largely rural population, showed a decline in poverty of only 0.6 percent, from 16.6 percent in 1979 to 16.0 percent in 1999.

11. H. K. Cordell and C. Overdevest, *Footprints on the Land: An Assessment of Demographic Trends and the Future of Natural Lands in the United States* (Champaign, Ill.: Sagamore, 2001).

12. Ibid.

13. Based on U.S. Census of Agriculture data.

14. Tennessee, for example, has a law establishing a rebuttable presumption that a farm is not a nuisance if it conforms to generally accepted management practices, or if it existed before a nearby change in land use, and if, prior to the change, it would not have been a nuisance. See TCA 43-26-101 to -104.

15. According to 1997 Census of Agriculture data, more than 47 percent of full-time farmers in Tennessee were age 65 and over; 22 percent were age 55–64.

16. The U.S. Department of Agriculture defines prime farmland as "land that has the best combination of physical and chemical characteristics for producing food, feed, forage, fiber, and oilseed crops and is also available for these uses. . . . In general, prime farmlands have an adequate and dependable water supply from precipitation or irrigation, a favorable temperature and growing season, acceptable acidity or alkalinity, acceptable salt and sodium content, and few or no rocks. They are permeable to water and air. Prime farmlands are not excessively erodible or saturated with water for a long period of time, and they either do not flood frequently or are protected from flooding" (*USDA Handbook* No. 18, Oct. 1993).

17. Based on sampling data from the National Resources Inventory conducted by the U.S. Department of Agriculture.

18. English, Huss, and Hoffman, *Growth Patterns in Tennessee,* 18.

19. S. Postell and S. Carpenter, "Freshwater Ecosystem Services," in *Nature's Services: Societal Dependence on Natural Ecosystems,* ed. G. C. Dailey (Washington, D.C.: Island Press, 1997), 195–214. See Cordell and Overdevest, *Footprints on the Land,* 271.

20. Cordell and Overdevest, *Footprints on the Land.*

21. Ibid., 229–84.

22. E.g., see W. U. Chandler, *The Myth of TVA: Conservation and Development in the Tennessee Valley, 1933–1983* (Cambridge, Mass.: Ballinger, 1984).

23. Ibid., 161.

24. Natalie Pawelski, "Paradise Drowned: Endangered Species Act Blamed for Almost Stopping Progress," *CNN.com,* 23 Dec. 1998, accessible at http://www.cnn.com/TECH/science/9812/23/dam.vs.fish/index.html (31 July 2003).

25. Chandler, *The Myth of TVA,* 162.

26. "Tellico Reservoir Development Agency," accessible at www.tellico.com (31 July 2003).

27. TVA, "Notice of Intent, Environmental Assessment or Environmental Impact Statement, Proposed Commercial Recreational and Residential Developments on Tellico Reservoir, Loudon County, Tennessee," *Federal Register* (Knoxville, Tenn.: 2002).

28. "Tellico Reservoir Development Agency."

29. TVA, Public Comments Sought on Proposed Rarity Pointe Development, Knoxville, Tenn., 8 July 2002.

30. R. Ferrar, "Rarity Pointe to Get Close Look," *Knoxville News-Sentinel,* 16 Nov. 2002.

31. TVA, "Shoreline Management Initiative: An Assessment of Residential Shoreline Development Impacts in the Tennessee Valley," *Abstract of Final Environmental Impact Statement,* Nov. 1998. Accessible at http://www.tva.gov. According to an article in *TVA River Neighbors,* May 2002, TVA completed its first total shoreline assessment in 2001.

32. In March 2000 TVA established a Regional Resource Stewardship Council, with twenty political, civic, and environmental leaders as members. The council makes recommendations on environmental and land-use issues to the three-member TVA board. In the fall of 2002, the council was urged by citizens—including citizens objecting to the Rarity Pointe proposal—to conduct a regionwide, comprehensive study of its public lands and to take a critical look at proposals for residential development on TVA land. Several of the council members were new. Of the fifteen council members present, a minority of seven voted in favor of these recommendations, so they were not made to the TVA board.

33. R. Ferrar, "Agency to Keep Deciding on Land-Use Issues Singly," *Knoxville News-Sentinel,* 16 Nov. 2002.

34. M. Simmons, "Dairy Farm to be Safe from Homogenized Development," *Knoxville News-Sentinel,* 6 Jan. 2003, B1.

6. FOOD

1. RAND, Health Research Highlights, "The Health Risks of Obesity," RB-4549 (2002), accessible at http://www.rand.org/congress/health/0602/obesity/rb4549 (29 Apr. 2003).

2. J. J. Putnam and J. E. Allshouse, *Food Consumption, Prices and Expenditures, 1970-1997,* Statistical Bull. No. 965, 22-3. U.S. Dept. of Agriculture, Economic Research Service, Washington, D.C.. Apr. 1999.

3. RAND, "Health Risks of Obesity."

4. Putnam and Allshouse, *Food Consumption,* 17.

5. Ibid., 23.

6. Union of Concerned Scientists (UCS), "Myths and Realities about Antibiotic Resistance," *Food and Environment,* accessible at http://www.ucsusa.org/food_and_environment/antibiotic_resistance/page.cfm?pageID=248, webpage accessed 19 May 2004.

7. Consumers Union, "Consumer Reports Finds Nearly Half of Chickens Tested Contaminated: Analysis of Bacteria Shows Significant Resistance to Important Human Antibiotics" Food Safety Press Release, Consumers Union of the United States, Inc., 10 Dec. 2002. Accessible at http://www.consumersunion.org/food/chicken1202.htm (6 May 2003).

8. *Background Document for the Joint WHO/FAO/OIE Expert Workshop on Non-Human Antimicrobial Usage and Antimicrobial Resistance: Scientific Assessment,* Geneva, Switzerland, 1–5 Dec. 2003, 74.

9. UCS, "Myths and Realities."

10. Ibid.

11. M. Burros, "F.D.A. Announces Label Requirement for Artery Clogger," *New York Times,* 10 July 2003; C. Stark, "Trans Fatty Acids—Back in the News 2002," Cornell Cooperative Extension, Food and Nutrition, Timely Topics in Food, Diet and Health, Sept./Oct. 2002, Cornell University, Ithaca, N.Y. Accessible at http://www.cce.cornell.edu/food/fdharchives/091002/transfat.html (28 July 2003).

12. Burros, "F.D.A. Announces Label Requirement"; U.S. Food and Drug Administration, Center for Food Safety and Applied Nutrition, Office of Food Labeling, "Questions and Answers on Trans Fat Proposed Rule," Nov. 1999). Accessible at http://www.cfsan.fda.gov/~dms/qatrans.html, (28 July 2003).

13. Burros, "F.D.A. Announces Label Requirement."

14. Ibid.

15. Consumers Union, Public Services Projects Department, Technical Division, " Do You Know What You are Eating? An Analysis of U.S. Government Data on Pesticide Residues in Foods," Consumers Union of the United States, Inc., Feb. 1999. Accessible at http://www.consumersunion.org/food/do_you_know2.htm (28 July 2003).

16. Ibid.

17. U.S. Bureau of the Census, *Census of Agriculture* (1910–97).

18. S. A. Male, *The Tennessee Food System: Planning for Regeneration* (Emmaus, Pa.: Rodale Press, 1984), 4; personal observations of several East Tennessee vegetable growers.

19. Male, *Tennessee Food System,* 4–5.
20. U.S. Bureau of the Census, *1992 Census of Agriculture.*
21. *Knoxville News-Sentinel,* 6 Aug. 1996.
22. Richard Cartwright Austin, "The Spiritual Crisis in Modern Agriculture," *Christian Social Action* (Oct. 1994).
23. S. L. Fisher and M. Harnish, "Losing a Bit of Ourselves: The Decline of the Small Farmer," in *Proceedings of the 1980 Appalachian Studies Conference,* 72.
24. Ibid., 71.
25. J. Hightower, *Eat Your Heart Out: Food Profiteering in America* (New York: Vintage Books, 1976), 167.
26. N. Dorman-Hickson, "The Mood of Agriculture 1994," *Progressive Farmer* (May 1994), 16–19.
27. These recommendations are culled from Austin, "Spiritual Crisis," and Male, *Tennessee Food System.*
28. U.S. Department of Energy, *U.S. Agriculture: Potential Vulnerabilities* (Menlo Park, Calif.: Stanford Research Institute, 1969).
29. J. Hendrickson, "Energy Use in the U.S. Food System: A Summary of Existing Research and Analysis," *Sustainable Farming-REAP-Canada* 7 (Ste. Anne-de'Bellevue, Quebec, 1997).
30. M. Hora and J. Tick, *From Farm to Table: Making Connections in the Mid-Atlantic Food System,* Capital Area Food Bank, Washington, D.C., 2001, accessible at http://www.clagettfarm.org/Introduction.html (1 Apr. 2003).
31. R. Pirog, T. Van Pelt, K. Enshayan, and E. Cook, *Food, Fuel, and Freeways: An Iowa Perspective on How Far Food Travels, Fuel Usage, and Greenhouse Gas Emissions,* Leopold Center for Sustainable Agriculture, Iowa State University, Ames, Iowa, 2001, accessible at http://www.leopold.iastate.edu (22 Dec. 2002).
32. J. W. Hagen, D. Minami, B. Mason, and W. Dunton, *California's Produce Trucking Industry: Characteristics and Important Issues,* California State Univ., Center for Agricultural Business, California Agricultural Technology Institute, Fresno, Calif.: 1999.
33. According to the 2000 Census, there are approximately 3.1 million people living in our bioregion. American per capita consumption of bananas is 26.8 pounds (Putnam and Allshouse, *Food Consumption*).
34. T. Klotzbach, "U.S. Imports of Fresh Fruits, Vegetables and Cut Flowers Up," from *Tropical Produce Marketing News* (Spring 1995). Accessible at http://www.fintrac.com/gain/marketstats/tpm/spr95/usimport.html (1 May 2003).
35. U.S. Bureau of the Census, *Census of Agriculture,* 1997.
36. According to the 2000 Census, there are approximately 3.1 million people living in our bioregion. American per capita consumption of beef in the mid-nineties reached about 66 pounds (Putnam and Allshouse, *Food Consumption*), so the gross total for our bioregion is a little over 200 million pounds. A typical steer carcass yields about 600 pounds of red meat, though this varies according to cutting method and type of cattle being slaughtered. See Tyson Foods, Inc., *2001 Investor Fact Book* 2001, accessible at http://www.tysonfoodsinc.com/IR/publications/factbook/default.asp (28 July 2003).
37. J. MacDonald, M. Ollinger, K. Nelson, and C. Handy, *Consolidation in U.S. Meatpacking,* AER-785, U.S. Dept. of Agriculture, Economic Research Service, Food and Rural Economics Division, Washington, D.C., Feb. 2000.

38. Ibid., 14. The Great Plains includes Texas, Oklahoma, Kansas, Colorado, Nebraska, North Dakota, and South Dakota. The Corn Belt includes Minnesota, Iowa, Missouri, Illinois, Wisconsin, Michigan, Indiana, and Ohio. The West includes all states west of the Great Plains.
39. Ibid., 4; Tyson, Inc., *2001 Investor Fact Book,* 2.
40. U.S. Bureau of the Census, *Census of Agriculture,* 1997.
41. MacDonald et al., *Consolidation in U.S. Meatpacking,* 4.
42. U.S. Bureau of the Census, *Census of Agriculture,* 1997.
43. USDA, Foreign Agricultural Service, FAS Online, Dairy, Livestock and Poultry Division, U.S. Trade Data Collection, Cattle and Beef, Feb. 23, 2001, accessible at http://www.fas.usda.gov/dlp/beef/Beefpage.htm (3 Apr. 2003).
44. MacDonald et al., *Consolidation in U.S. Meatpacking,* 4.
45. Ibid.
46. Ibid., 5.
47. Consumers Union, "Feedlot Dust Side Bar." *Animal Factories: Pollution and Health Threats to Rural Texas* (Consumers Union of United States, Inc.: 2000), accessible at http://www.consumersunion.org/other/animal/feedlot.htm (19 May 2004)
48. E. Schlosser, *Fast Food Nation: The Dark Side of the All-American Meal* (New York: Houghton Mifflin, 2001), 169.
49. Ibid.
50. Ibid., 160.
51. Ibid.
52. Ibid., 176–77.
53. Ibid., 160–62.
54. Ibid., 162.
55. Ibid.; S. J. Hedges and D. Hawkins, "The New Jungle," *U.S. News and World Report,* 23 Sept. 1996, 34-45.
56. Schlosser, *Fast Food Nation,* 165.
57. MacDonald et al., *Consolidation in U.S. Meatpacking,* 5.
58. Tyson, Inc., *2001 Investor Hand Book,* 10.
59. The information in this paragraph comes from P. Singer, *Animal Liberation,* rev. ed. (New York: Avon Books, 1990), 119–23.
60. M. Simon, "Hog Factory Farming; Lagoons or Environmental Menace?" (WholeFoods.com, July 1999), Center for Informed Food Choices (Oakland, Calif.), accessible at http://www.informedeating.org/articles.html (28 July 2003).
61. J. Lee, "Neighbors of Vast Hog Farms Say Foul Air Endangers Their Health," *New York Times,* 11 May 2003.
62. M. Ollinger, J. MacDonald, and M. Madison, *Structural Changes in U.S. Chicken and Turkey Slaughter,* U.S. Department of Agriculture, Economic Research Service, Washington, D.C.: Sept. 2000, 15. The Southeast includes the states of Alabama, Arkansas, Georgia, Florida, Louisiana, Mississippi, North Carolina, South Carolina, and Tennessee.
63. Tyson, Inc., *2001 Investor Fact Book,* 6–9.
64. Most of the information in this section comes from Farm Sanctuary (Watkins Glen, N.Y.: 1 Jan. 2002), accessible at Poultry.org (28 July 2003). Information taken from other sources is noted in the text.
65. P. S. Goodman, "An Unsavory Byproduct: Runoff and Pollution," *Washington Post,* 1 Aug. 1999, sec. A1.

66. In developing this account of the basic structure of U.S. produce supply chains, we relied on the following: R. L. Cook, "The U.S. Fresh Produce Industry: An Industry in Transition," in *Postharvest Technology of Horticultural Crops,* 3d ed., Publication 3311, Univ. of California Division of Agriculture and Natural Resources, 2002, 5–30; L. Calvin and R. Cook, coordinators, M. Denbaly, C. Dimitri, L. Glaser, C. Handy, M. Jekanowski, P. Kaufman, B. Krissoff, G. Thompson, and S. Thornsbury, *U.S. Fresh Fruit and Vegetable Marketing: Emerging Trade Practices, Trends and Issues,* AER-795, U.S. Dept. of Agriculture, Economic Research Service, Market and Trade Economics Division, Washington, D.C., Jan. 2001; P. R. Kaufman, C. R. Handy, E. W. McLaughlin, K. Park, and G. M. Green, *Understanding the Dynamics of Produce Markets: Consumption and Consolidation Grow,* AIB-758, U.S. Dept. of Agriculture, Economic Research Service, Washington, D.C., Aug. 2000; Hora and Tick, *From Farm to Table.*

67. S. Moore, "Midnight in the Garden of Green," *Los Angeles Times,* 27 Dec. 2000, sec. 2(B).

68. Hora and Tick, *From Farm to Table.*

69. Kaufman et al., *Dynamics of Produce Markets,* 5.

70. Ibid., 10.

71. C. Dimitri and C. Greene, *Recent Growth Patterns in the U.S. Organic Foods Market,* AIB-777, U.S. Dept. of Agriculture, Economic Research Service, Market Trade Economics Division and Resource Economics Division, Washington, D.C., Sept. 2002.

72. U.S. Dept. of Agriculture, Agricultural Research Service, Alternative Farming Systems Information Center, "Organic Food Production" (24 Oct. 2002), accessible at http://www.nal.usda.gov/afsic/ofp (24 July 2003).

73. Dimitri and Greene, *Recent Growth Patterns,* 8, 10, 12, 16, 18–19.

74. J. Reganold, "Farming's Organic Future," *New Scientist,* 10 June 1989, 49–52.

75. J. Reganold, J. D. Glover, P. K. Andrews, and H. R. Hinman, "Sustainability of Three Apple Production Systems," *Nature* 410 (19 Apr. 2001): 926–30.

76. J. Jeavons, *How to Grow More Vegetables* (Berkeley, Calif.: Ten Speed Press, 1991), 70–97, 165–71.

77. C. Whitehead, personal communication.

78. Soil Association, *The Biodiversity Benefits of Organic Farming* (May 2000), accessible at http://www.soilassociation.org (25 July 2003).

79. Dimitri and Greene, *Recent Growth Patterns,* 3–4.

80. Ibid., 1.

81. Ibid., 3–4.

82. Ibid., 1.

83. Ibid., 2.

84. F. C. White, J. R. Hairston, W. N. Musser, H. F. Perkins, and J. F. Reed, "Relationship between Increased Crop Acreage and Nonpoint Source Pollution," *Journal of Soil and Water Conservation* 36 (1980): 175.

85. Tennessee Association of Conservation Districts and Tennessee Department of Agriculture, *A Matter of Natural Resources,* undated pamphlet.

86. SAMAB, "Southern Appalachian Assessment Aquatic Technical Report," (1996), 89.

87. TVA, *The First Fifty Years,* 111.

88. E. J. Plaster, *Soil Science and Management,* 3d ed. (New York: Delmar Publishers, 1997), 314.

89. U.S. Dept. of Agriculture, *Summary Report: 1997 Natural Resources Inventory,* Natural Resources Conservation Service, Washington, D.C., and Statistical Laboratory, Iowa State Univ., Ames, Iowa, rev. Dec. 2002, pp. 49–50. Accessible at http://www.nrcs.usda.gov/technical/NRI/1997/summary_report/ (19 May 2004). See Table 10, "Estimated average annual sheet and rill erosion on nonfederal land, by state and year," http://www.nrcs.usda.gov/technical/nri/1997/summary_report/table10.html for the North Carolina and Tennessee data.

90. USDA, *1997 Natural Resources Inventory.* See especially NRCS's map titled "Average Annual Soil Erosion by Water on Cropland and CRP Land, 1997," accessible at http://www.nrcs.usda.gov/technical/land/meta/m5058.html (28 July 2003).

91. Plaster, *Soil Science and Management,* 25.

92. Ibid., 20.

93. Ibid., 24.

94. J. Rissler and M. Mellon, *The Ecological Risks of Engineered Crops* (Cambridge, Mass.: MIT Press, 1996), 11.

95. Union of Concerned Scientists (UCS), Food and Environment, "What is Genetic Engineering?" accessible at http://www.ucsusa.org/food_and_environment (2 Feb. 2003); "What is Biotechnology?" accessible at http://www.ucsusa.org/food_and_environment (2 Feb. 2003).

96. See, e.g., G. G. Khachatouris, A. McHughen, R. Scorza, W. Nip, and Y. H. Hui, *Transgenic Plants and Crops* (New York: Marcel Dekker, Inc., 2002).

97. Rissler and Mellon, *Ecological Risks,* 17.

98. UCS, "Pharm and Industrial Crops: The Next Wave of Agricultural Biotechnology" (2002), accessible at http://www.ucsusa.org (3 Feb. 2003).

99. Rissler and Mellon, *Ecological Risks,* 20–22.

100. K. Hart, *Eating in the Dark: America's Experiment with Genetically Engineered Food* (New York: Pantheon, 2002), 4.

101. Ibid.

102. U.S. Environmental Protection Agency, "What are Biopesticides?" (18 June 2003), accessible at http://www.epa.gov/pesticides/biopesticides/whatarebiopesticides.htm (25 July 2003).

103. Hart, *Eating in the Dark,* 156.

104. C. Whitehead, *The Perpetual Garden Calendar* (Strawberry Plains, Tenn.: Planted Earth Press, 1995).

7. Energy

1. TVA, direct correspondence, Oct. 2002.

2. Dr. J. Scurlock, Univ. of Tennessee, personal communication, May 2003.

3. Working on-demand water heaters and passive solar water heaters may be seen at Narrow Ridge Earth Literacy Center in Grainger County, Tenn.

4. TVA, direct correspondence, Oct. 2002.

5. Energy Information Administration, *2001 Residential Energy Consumption Survey*. Table HC4-11b, accessible at http://ftp.eia.doe.gov/pub/consumption/residential/2001hc_tables/ac_percent.pdf (19 March 2003).

6. Energy Information Administration, "Table E1c World Per Capita Primary Energy Consumption, 1980–2001," accessible at http://www.eia.doe.gov/pub/international/iealf/tablee1c.xls (19 Mar. 2003).

7. TVA, *Energy Vision 2020*, vol. 2: T1.117.

8. TVA, *2001 Annual Environmental Report*, 6.

9. *Powell Valley News*, Oct. 18, 1994. See also TVA, *Energy Vision 2020*, vol. 2: T2.47.

10. Information supplied by Friends of the Clinch and Powell Rivers.

11. TVA, *2002 Annual Environmental Report*, 3.

12. TVA, *Energy Vision 2020*, vol. 1: 1.3.

13. U.S. General Accounting Office (GAO), *Tennessee Valley Authority: Financial Problems Raise Questions About Long-term Viability*, GAO/AIMD/RCED-95-134, Washington, D.C.: GPO, 1995, 20.

14. TVA, *The First Fifty Years*, 46.

15. M. Williams, "Watts Bar Plant May Get NRC Go-Ahead Soon," *Chattanooga Free Press*, 22 Jan. 1996, sec. B2.

16. GAO, *Tennessee Valley Authority*, 42.

17. TVA, *1996 Annual Report*, 32, 39.

18. TVA, *1997 Annual Report*, 4.

19. Roberto Grimac, "TVA Gets 'Lemon Award' for Worst Energy Ad in U.S.," *Tennessee Green* (Spring 1996), 7.

20. GAO, *Air Quality: TVA Plans to Reduce Air Emissions Further, but Could Do More to Reduce Power*, GAO-02-301, Washington, D.C.: GPO, 2002, 12.

21. Ibid., 13.

22. Ibid., 2.

23. Ibid., 19.

24. Ibid., 1.

25. TVA, *The First Fifty Years*, 70.

26. TVA, *Energy Vision 2020*, vol. 2: T1.117, T1.121.

27. Southern Alliance for Clean Energy, "Martin County, KY, Coal Waste Disaster," accessible at http://www.cleanenergy.org/energy/coal/KYdisaster/index.html#info (15 July 2003).

28. A. Gabbard, "Coal Combustion: Nuclear Resource or Danger?" *Oak Ridge National Laboratory Review* 26 (3 & 4) (1993).

29. M. Squillace, *Strip Mining Handbook*, n.p. Environmental Policy Institute and Friends of the Earth, 1990), 28.

30. Ibid., 19–20.

31. Save Our Cumberland Mountains (SOCM), "Does Reclamation Work?: A Study of Water Quality at Reclaimed Stripmine Sites in the Sewanee Coal Seam of Tennessee," Lake City, Tenn., 1992, 1.

32. C. Doering, "US Aims to Streamline Mountaintop Mining Permits," *Planet Ark*, Reuters News Service, 6 Feb. 2003; K. Ward Jr., "Mountaintop Removal Damage Proved: Bush Administration Proposes No Concrete Limits on New Mining Permits," *Charleston [West Virginia] Gazette*, 30 May 2003.

33. Squillace, *Strip Mining Handbook*, 28–34.

34. SOCM, "Does Reclamation Work?" 19–30.

35. M. Hopps, "Reforesting Appalachia's Coal Lands," *American Forests* (Nov./Dec. 1994), 40–50.

36. Duffy and Meier, "Do Appalachian Herbacious Understories Ever Recover from Clearcutting?" 196–201.

37. T. Ledford, assistant director of Univ. of Tennessee's Physical Plant, personal communication, Mar. 2002.

38. Ibid.

39. Scurlock, personal communication, May 2003.

40. TVA, "Sequoyah Nuclear Plant," accessible at http://www.tva.gov/sites/sequoyah.htm (19 Mar. 2003).

41. TVA, "Watts Bar Nuclear Plant," accessible at http://www.tva.gov/sites/wattsbarnuc.htm (19 Mar. 2003).

42. McKinney and M. Schoch, *Environmental Science*, 212–13; Raven, Berg, and Johnson, *Environment*, 215–18.

43. B. Martocci, TVA, personal communication, Mar. 2002.

44. P. Matthiessen, *In the Spirit of Crazy Horse* (New York: Viking, 1983).

45. TVA, direct correspondence, Oct. 2002.

46. TVA, *Energy Vision 2020*, vol. 2: T1.121–T1.122.

47. Ibid., T1.122.

48. TVA, *First Quarter 2000 Operational Report to Congress*, (Oct.–Dec. 1999), 1.

49. TVA, *2001 Annual Report*, 32.

50. TVA, direct correspondence, Oct. 2002.

51. U.S. Dept. of Energy (USDOE), "Yucca Mountain Future Repository?" accessible at http://tis.eh.doe.gov/whs/directory/eh51_templates/eh51_1d/news/ymrepository.html (20 Mar. 2003).

52. USDOE, "Spent Nuclear Fuel Transportation," accessible at http://www.ocrwm.doe.gov/wat/pdf/snf_trans.pdf (16 July 2003), 1.

53. State of Nevada, "Nucler Waste Transportation Routes," Jan. 1995, accessible at http://www.state.nv.us/nucwaste/states/us.htm (16 July 2003).

54. State of Nevada, "Earthquakes in the Vicinity of Yucca Mountain," accessible at http://www.state.nv.us/nucwaste/yucca/seismo01.htm (10 July 2003).

55. TVA, *2000 Annual Report*, 23.

56. Associated Press, "TVA Plans for Nuke Storage Ripped," Mar. 14, 2001. Accessible at http://www.cleanenergy.org/pressroom/storage.html (19 Mar. 2003).

57. TVA, *2001 Annual Report*, 49.

58. D. A. Blecker, "Deadly Dollars: The Economic Fallout from TVA's Watts Bar Unit 1," MSB Energy Associates (1995).

59. TVA, *2001 Annual Report*, 49.

60. TVA, "TVA Board Approves Browns Ferry Unit 1 Recovery, Extended Operation," 16 May 2002, accessible at http://www.tva.gov/news/releases/0502bferry.htm (16 July 2003).

61. M. Goldberg, "Federal Energy Subsidies: Not All Technologies Are Created Equal," *Renewable Energy Policy Project Research Report*, 11 (July 2000): 7.

62. TVA, *Energy Vision 2020*, vol. 2: T6.6–T6.7 and T6.34.

63. Scurlock, personal communication, May 2003.

64. McKinney and Schoch, *Environmental Science*, 507.

65. TVA, "Methane Gas Energy," accessible at http://www.tva.gov/greenpowerswitch/landfill.htm (16 July 2003).
66. Ibid.
67. TVA, *Green Power Switch News* 2 (1) (Winter 2002): 3.
68. USDOE, "Biopower," accessible at http://www.eere.energy.gov/biopower/feedstocks/index.htm (16 July 2003).
69. TVA, *Green Power Switch News* 2 (2) (Spring 2002): 2.
70. TVA, *Green Power Switch News* 2 (1) (Winter 2002): 3.
71. TVA, *Energy Vision 2020,* vol. 2: T6.31.
72. A. R. Hamlin, program administrator, Green Power Switch, TVA, personal communication, July 2002.
73. TVA, direct correspondence, Oct. 2002.
74. TVA, "TVA's Solar Power Generating Sites," accessible at http://www.tva.gov/greenpowerswitch/solar_sites.htm (16 July 2003).

8. consumption and waste

1. Public Broadcasting System (PBS), "Affluenza," accessible at http://www.pbs.org/kcts/affluenza (11 Aug. 2003).
2. Proverbs 22:7.
3. From a United Nations International Labor Organization report, quoted in Deborah Campbell, *Adbusters* 47 (May/June 2003).
4. C. Yeldell, "Grocer pains: With Wal-Mart Seeking a Piece of Their Pie, Other Food Retailers Try to Bag Their Own Niche," *Knoxville News-Sentinel,* 29 June 2003, C-1.
5. Ibid.
6. U.S. Department of the Interior, "U.S. Bureau of Mines," accessible at http://www.doi.gov/pfm/ar4bom.html (15 July 2003).
7. D. G. Rogich, "Material Use, Economic Growth and the Environment," *Capital Metals and Materials Forum* 1 (2) (1992).
8. J. E. Young and A. Sachs, "Creating a Sustainable Materials Economy," in *State of the World 1995,* ed. Lester Brown et al. (New York, Norton, 1995), 80.
9. Ibid., 81.
10. World Resources Institute, "Pollution and Waste Increasing in Five Countries Despite More Efficient Use of Resources," accessible at http://wri.igc.org/press/weightofnations.html (17 July 2003).
11. WorldWatch Institute, State of the World 2003, "'Impossible' Environmental Revolution Is Already Happening," accessible at http://www.worldwatch.org/ State of the World 2003 Worldwatch Institute News Release.htm (17 July 2003).
12. M. Lee, "State of the Planet," *Ecologist* 32 (7) (Sept. 2002): 10.
13. EPA, "Municipal Solid Waste—2000 Total Waste Generation," accessible at http://www.epa.gov/epaoswer/non-hw/muncpl/facts-text.htm (5 Sept. 2002).

14. EPA, "Municipal Solid Waste—Basic Facts," accessible at http://www.epa.gov/epaoswer/non-hw/muncpl/facts.htm (5 Sept. 2002).

15. J. Nolt et al., *What Have We Done? The Foundation for Global Sustainability's State of the Bioregion Report for the Upper Tennessee Valley and the Southern Appalachian Mountains* (Washburn, Tenn.: Earth Knows Publications, 1997), 159–60.

16. TDEC, "Annual Report to the Governor and General Assembly on the Solid Waste Management Act of 1991, Fiscal Year 2000–2001," Jan. 2002.

17. E. LeQuire, "Pay-as-You-Throw Pricing Relieves Burden on Landfills," *InSITES,* Univ. of Tennessee Waste Management Research and Education Insititute, Knoxville, 1996, 2.

18. GAO, "Fact Sheet for the Ranking Minority Member, Subcommittee on Interior and Related Agencies, Committee on Appropriations, House of Representatives, on the Forest Service Distribution of Timber Sales Receipts, Fiscal Years 1992–1994," GAO/FCED-95-237FS, Washington, D.C., 1995.

19. EPA, "Federal Disincentives: A Study of Federal Tax Subsidies and Other Programs Affecting Virgin Industries and Recycling," EPA 230-R-94-005, Office of Policy Analysis, Washington, D.C., 1994.

20. A. T. Durning, "Saving the Forests: What Will It Take?" World Watch Paper 117, Worldwatch Institure, Washington, D.C., 1993, 25.

21. EPA, "Municipal Solid Waste in the United States: 2000 Facts and Figures, Executive Summary," EPA530-S-02-001, 7.

22. Ibid.

23. City of Kingsport, "Recycle Program," accessible at http://www.kingsportpublicworks.com/recycle.html (5 July 2003).

24. Lee, "State of the Planet."

25. Reuters News Service, World Environment News, "US aluminum can recycling declined in 2001," accessible at http://www.planetark.org/avantgo/dailynewsstory.cfm?newsid=15730 (5 Sept. 2002).

26. E. LeQuire, "'Cookin' Up Compost: Knox, Blount Programs Reduce Organic Wastes Differently," *Tennessee Green* (Summer 1995): 7.

27. This information is from SOCM (Save Our Cumberland Mountains), which helped organize resistance at Shoat Lick Hollow.

28. National Resources Defense Council, "Too Good to Throw Away," accessible at www.nrdc.org/cities/recycling/recyc/chap2.asp (5 Sept. 2002).

29. SOCM, *1995 SOCM Annual Report,* Lake City, Tenn., 1995, 4 and *SOCM Sentinel* (Jan./Feb. 1995).

30. City of Bristol, Virginia, Public Works Department, "Waste Disposal," accessible at http://www.bristolva.org/WasteDisposal.htm (17 July 2003).

31. "Cigarette Butts Highway Litter Leader," *Knoxville News-Sentinel,* 11 Aug. 1996, B7.

32. EPA, *The National Biennial RCRA Hazardous Waste Report,* Executive Summary, EPA-530-S-01-001, June 2001, ES-3–ES-4.

33. Ibid., 91.

34. S. Barker, "EPA May Take on Coster Dump Site," *Knoxville News Sentinel,* 21 Aug. 2002, accessible at www.knoxnews.com/kns/politics/article/0,1406,KNS_356_1338114,00.html (28 Aug. 2002).

35. S. Barker, "Toxic Earth, Tainted Lives, Going to the Source: Differing Opinions Unearthed in Search for Pollution Remedy," *Knoxville News-Sentinel,* 22 Dec. 2002, A4.

36. S. Barker, "Three Parties Deny Role in Dumping," *Knoxville News-Sentinel*, accessible at www.knoxnews.com/kns/local_news/article/ 0,1406,KNS_347_1310643,00.html (7 Aug. 2002).

37. Material for most of this section was supplied in correspondence with J. L. Bonds, environmental scientist, Ground Water Protection Branch, EPA Region 4, and C. Tenut, Mining Section, TDEC. We are grateful for their help.

38. Unless otherwise noted, all previous material in this section is drawn from OREP, "Citizen's Guide to Oak Ridge."

39. R. Beck, "Woman's Cancer May Be Linked to Work in Radioactive Junkyard," *Knoxville Journal,* 13 Sept. 1985, A1.

40. The name of the sender is illegible on copies provided to the Foundation for Global Sustainability, but the memo is on Nuclear Division stationery and references a letter by Alvin M. Weinberg dated 24 Mar. 1969, which apparently also warns of the problem of sending plutonium to Witherspoon.

41. This memo, on Oak Ridge National Laboratory stationery, is from Julian R. Gissell to H. E. Seagren.

42. Beck, "Woman's Cancer."

43. Ibid.

44. DOE, *Remedial Investigation/Feasibility Study for the David Witherspoon, Inc. 901 Site, Kxoxville, Tennessee,* vol. I, DOE/OR/02-1503/V1&D1, Office of Environmental Restoration and Waste Management, Oak Ridge, Tennessee, 1996, 3–16.

45. These facts are drawn from the files on Witherspoon at TDEC, Division of Superfund's Knoxville office, and from DOE, *Remedial Investigation/Feasibility Study for the David Witherspoon, Inc.*

46. DOE, *Remedial Investigation/Feasibility Study for the David Witherspoon, Inc.,* 4–27.

47. Letter dated 11 July 2002 from Jessie Hill Robertson, assistant secretary for Environmental Management, U.S. Dept. of Energy, to Congressman John J. Duncan Jr.

48. M. Klesius, "The State of the Planet," *National Geographic* 202 (3) (Sept. 2002): 109.

49. J. E. Hilsenrath, "What those GDP Numbers Really Mean," *Knoxville News Sentinel,* 29 June 2003, WSJ 3.

9. Transportation

1. Tennessee Dept. of Transportation (TDOT), *State Transportation Plan* (Nashville: Tennessee Department of Transportation, 1994), I-3.

2. Center for Business and Economic Research (CBER), *Tennessee Statistical Abstract,* Univ. of Tennessee, Knoxville, 2003, Fig. 9.16.

3. Ibid., 6.

4. Tennessee State Government, "Tennessee Department of Safety/Department of Motor Vehicles Home Page," accessible at http://www.state.tn.us/safety (5 June 2003).

5. CBER, *Tennessee Statistical Abstract,* 3.

6. S. Allen, TDOT, personal communication, 17 July 2003.

7. T. Wilner, "How Green are We?" *Metro Pulse* 19 (49) 5 Dec. 2002.

8. R. Powelson, "Study Says Too Many Cars Boost Traffic Deaths, Pollution," *Knoxville News-Sentinel*, 18 Oct. 2002, 1; see also Ewing, Pendall, and Chen, *Measuring Sprawl and Its Impacts.*

9. Knoxville was eighth worst in 2002 and ninth worst in 2003; American Lung Association, "State of the Air 2003," accessible at http://www.lungusa.org/air2001/table3_02.html (7 Aug. 2003).

10. GAO, *Surface and Maritime Transportation—Developing Strategies for Enhancing Mobility: A National Challenge,* USGAO, Washington, D.C., 2002, 59.

11. Energy Information Administration, "Annual Energy Outlook 2003 With Projections to 2025," Energy Information Administration Home Page, 18 July 2003, accessible at http://www.eia.doe.gov/oiaf/aeo/index.html (25 July 2003).

12. CBER, *Tennessee Statistical Abstract,* table 7.

13. D. Gordon, *Steering a New Course: Transportation, Energy, and the Environment* (Washington, D.C.: Island Press, 1991), 35.

14. TVA, *Energy Vision 2020,* vol. 2, T1.90.

15. EPA, "National Air Quality: Emissions Trends Report 1995," Oct. 1996.

16. The Gas Guzzler Campaign, "Getting There: Strategic Facts for the Transportation Advocate," pamphlet, 1998.

17. M. Jaffe, "Study Shows 10 Percent of Cars Cause 50 Percent of Auto Pollution," *Philadelphia Inquirer,* 18 May 1995.

18. American Lung Association, "State of the Air 2003."

19. C. J. Burbank, "ITS and the Environment," *Public Roads* 58 (Spring 1995): 9.

20. S. Nadis and J. J. MacKenzie, *Car Trouble* (Boston, Mass.: Beacon Press, 1993), 52.

21. Ibid., 53.

22. Cheremisinoff, King, and Boyko, *Dangerous Properties,* 56–58.

23. Ibid., 67–69.

24. C. S. Papacostas and P. D. Prevedouros, *Transportation Engineering and Planning,* 2d ed. (Englewood Cliffs, N.J.: Prentice Hall, 1996), 480.

25. J. Satterfield, "Tightening Competition Urges Big Rigs to Speed," *Knoxville News-Sentinel,* 18 Nov. 1996, A1.

26. McKinney and Schoch, *Environmental Science,* 152.

27. EPA, "Health Assessment Document for Diesel Engine Exhaust," EPA/600/8-90/057F, Office of Research and Development, National Center for Environmental Assessment, Washington, D.C., 2002.

28. "Children's Exposure to Diesel Exhaust on School Buses," Environment & Human Health, Inc., 2002.

29. EPA, "Emission Standards for New Nonroad Engines," EPA420-F-02-037, Ann Arbor, Mich., 2002.

30. Sustainable Energy Institute, "Fact Sheet #1 The Problem With Paving," accessible at http://www.culturechange.org/factsheet1.html (17 July 2003).

31. Southern Appalachian Man and the Biosphere Cooperative (SAMAB), "Southern Appalachian Assessment Social/Cultural/Economic Technical Report" (1996), 177.

32. Ibid., 177–82.

33. Oak Ridge National Laboratory (ORNL), "Environmental Report (ER) for Section 8B of the Foothills Parkway in the Great Smoky Mountains National Park

(GSMNP), Volume 1 Summary," accessible at http://www.ornl.gov/ceea/ Foothills_Pkwy/summary.html (8 Aug. 2003)

34. H. H. Hahn and R. Pfeifer, "The Contribution of Parked Vehicle Emissions to the Pollution of Urban Run-Off," *Science Total Environment* (23 May 1994); Nadis and MacKenzie, *Car Trouble,* 17.

35. C. C. Lutes, R. J. Thomas, and R. Burnette, "Evaluation of Emissions from Paving Asphalts," EPA Office of Research and Development, 1994, 1–2, 72–75; see Cheremisinoff, King, and Boyko, *Dangerous Properties,* concerning carcinogenicity of the compounds mentioned.

36. TVA, "Navigation on the Tennessee River," accessible at http://www.tva.com/river/navigation/index.htm (18 July 2003).

37. TDOT, *State Transportation Plan,* VIII-7.

38. East Tennessee Development District (ETDD), *Evaluation and Update of the Transportation Plan,* Knoxville, Tenn., May 1995, 7.

39. Tennessee Valley Authority and U.S. Dept. of the Army, Corps of Engineers, direct correspondence.

40. G. Welty, "Railroads and the Environment," *Railway Age* 194 (2) (1993): 27.

41. TDOT, *Tennessee System Rail Plan—Task 4: Potential Intercity Passenger Rail Corridors,* Nashville, Tenn., 2002, 1.

42. Ibid., 9.

43. Ibid., 12.

44. Ibid.

45. Ibid., 5.

46. ETDD, *Evaluation and Update,* 12.

47. TDOT, "Tennessee Interstates Carrying Increasing Traffic Burden," accessible at http://www.tdot.state.tn.us/information-office/2001pr/Aug2001.htm (5 June 2003).

48. University of Tennessee Center for Business and Economic Research, "Profile of General Demographic Characteristics: 2000/Geographic Area: Knox County, Tennessee," accessible at http://bus.utk.edu/cber/census/sf3cnty.htm (14 Apr. 2003).

49. See also J. Nolt, *Down to Earth: Toward a Philosophy of Nonviolent Living* (Washburn, Tenn.: Earth Knows Publications, 1995).

10. Future prospects

1. B. Babbitt, foreword in J. Peine ed., *Ecosystem Management for Sustainability: Principles and Practices Illustrated by a Regional Biosphere Reserve Cooperative* (New York: Lewis Publishers, 1999).

2. B. Sterling, *Tomorrow Now: Envisioning the Next 50 Years* (New York: Random House, 2002).

3. K. Campbell, "Envisioning the Future, Today," *Christian Science Monitor* (26 Dec. 2002).

4. T. Wagner and P. Obermiller, *Valuing Our Past Creating Our Future: The Founding of the Urban Appalachian Council* (Berea, Ky.: Berea College Press, 1999).

5. Hudson and Tesser, *The Forgotten Centuries.*

6. P. White, Botany Dept., Univ. of North Carolina at Chapel Hill, personal communication.

7. A. W. Crosby, *The Columbian Exchange: The Biological Expansion of Europe, 900–1900* (Cambridge, Eng.: Cambridge Univ. Press, 1986).

8. E. Buckner, Dept. of Forestry, Wildlife and Fisheries, Univ. of Tennessee, Knoxville, personal communication.

9. Buckner and Turrill, "Fire Management," *Ecosystem Management for Sustainability,* ed. Peine, 329–47.

10. White, personal communication.

11. P. A. Delcourt and H. R. Delcourt, "Conservation of Biodiversity in Light of the Quatery Paleo-Ecological Record: Should the Focus Be on Species, Ecosystems, or Landscapes?" in W. J. Platt and R. K. Peet, eds., Use of Ecological Concepts in Conservation Biology: Lessons from the Southeastern Ecosystems, *Ecological Applications* (special edition, 2000).

12. N. Nicolas, C. Eagar, and J. Peine, "Threatened Ecosystem: High Elevation Spruce-Fir Forest," in *Ecosystem Management for Sustainability,* ed. Peine, 431–54.

13. Earth Share, "News and Resources," accessible at http://www.earthshare.org/ (30 Jan. 2003).

14. National Ocean and Atmospheric Agency (NOAA), accessible at http://www.noaa.gov/ (31 Jan. 2003).

15. J. D. Peine and C. Berish, "Climate Change: Effects in the Southern Appalachians," in *Ecosystem Management for Sustainability,* ed. Peine, 397–416.

16. S. Moore, fisheries biologist, Great Smoky Mountains National Park, personal communication.

17. J. D. Peine, "Nuisance Bears in Communities: Strategies to Reduce Conflict," *Human Dimensions of Wildlife* 6 (2001): 223–37; D. Whittaker and R. L. Knight, "Understanding Wildlife Responses to Humans," *Wildlife Society Bulletin* 26 (2) (1998): 312–17.

18. Intergovernment Panel on Climate Change (IPCC), accessible at www.ipcc.ch (28 Jan. 2003).

19. J. Lowy, "Scientists: Humans the Main Culprit in Global Warming," Common Dreams New Center, accessible at http://www.commondreams.org/headlines02/ 1217-05.htm (28 Jan. 2003).

20. J. Viegas, "Global Warming—Nature's Payback: The Temperature is Climbing, with Potentially Hazardous Results for the Earth," ABC News, formerly accessible at www.abcnews.go.com/ABC2000/abc2000science/globalwarming991108.html (30 Jan. 2003). Hardcopy in author's possession.

21. Peine and Berish, "Climate Change."

22. Buckner and Turrill, "Fire Management."

23. M. Grubb, J. C. Hourcade, O. Sebastian, "Keeping Kyoto: A Study of Approaches to Maintaining the Kyoto Protocol on Climate Change," International Network for Climate Policy Analysis, accessible at http://www.climate-strategies.org/ keepingkyoto.pdf (30 Jan. 2003).

24. United Nations Framework Convention on Climate Change (UNFCCC), "Status of Ratification," accessible at http://unfccc.int/ (12 May 2003).

25. U.S. Department of Energy (DOE), "Your Transportation: Alternative Fuel Vehicles," formerly accessible at http://www.energy.gov/transportation/sub/altfuel.html (30 Jan. 2003). Hardcopy in the author's possession.

26. Peine and Berish, "Climate Change."

27. Viegas, "Global Warming—Nature's Payback."

28. T. Friedman, "Repairing the World," editorial, *New York Times,* 16 Mar. 2002.

29. J. D. Peine and J. R. Renfro, "Visitor Use Patterns at Great Smoky Mountains National Park," USDI, National Park Service Southeast Region, Atlanta.

30. U.S. Dept. of Commerce, Census Bureau, accessible at http://www.census.gov/main/www/cen2000.html (30 Jan. 2002).

31. Conner and Hartsel, "Forest Area and Conditions," 357–402.

32. D. Wear and P. Bolstad, "Land-Use Changes in Southern Appalachian Landscapes: Spatial Analysis and Forest Evaluation," *Ecosystems* 1 (1998): 575–94.

33. K. Newton, "Bear/People Conflicts in Gatlinburg, Tennessee: An Analysis of the Social, Political, and Ecological Elements," Master's thesis, Univ. of Tennessee, Knoxville, 2000.

34. Wilbur Smith Associates, *Appalachian Development Highways: Economic Impact Studies* (Washington, D.C.: Appalachian Regional Commission, 1998).

35. F. Wegman, Dept. of Civil and Environmental Engineering, Univ.of Tennessee, Knoxville, personal communication.

36. K. Christen, *Sightline* (Fall–Winter 2002) Univ. of Tennessee, Energy, Environment, and Resource Center, Knoxville.

37. Little River Watershed Association (LRWA), accessible at http://www.blountweb.com/lrwa/ (31 Jan. 2003).

38. Little Tennessee River Watershed Association (LTRWA), accessible at http://www.littletennesseewatershed.org/ (31 Jan. 2003).

39. Nine Counties. One Vision. Accessible at http://ninecountiesonevision.org/ (18 June 2003).

40. J. Lamb, director of planning, Blount County, Tennessee, personal communication.

41. Balsam Mountain Preserve, accessible at http:www.balsammountainpreserve.com (23 June 2003).

42. S. Schlarbaum, R. Anderson, and F. Tompson-Campell, "Control of Pests and Pathogens," in *Ecosystem Management for Sustainability,* ed. Peine, 291–306.

43. W. Q. R. Burkman, R. Chavez, S. Cooke, S. DeLost, T. Luther, M. Mielke, M. Miller-Weeks, F. Peterson, M. Roberts, P. Seve, and D. Trawdus, *Northeastern Area Forest Health Report,* NA-TP-03-93, U.S. Dept. of Agriculture, Forest Service Northeastern Area, State and Private Forestry, Radnor, Pa., 1993.

44. Schlarbaum, Anderson, and Tompson-Campell, "Control of Pests and Pathogens."

45. J. Kotinsky, "The European Fir Trunk Louse," *Proceedings of the Entomolgical Society of Washington* 18 (1916): 14–16.

46. Schlarbaum, Anderson, and Tompson-Campell, "Control of Pests and Pathogens."

47. P. N. Annand, "A New Species of *Adelges,*" *The Pan-Pacific Entomologist* 1 (1924): 79–82.

48. Schlarbaum, Anderson, and Tompson-Campbell, "Control of Pests and Pathogens."

49. Buckner, personal communication.

50. White, personal communication.

51. Stegeman, "The European Wild Boar."

52. J. D. Peine and R. Lancia, "Control of Exotic Species: European Wild Boar," in *Ecosystem Management for Sustainability,* ed. J. Peine, 267–90.

53. S. P. Bratton, "The Effect of European Wild Boar on High-Elevation Vernal Flora in Great Smoky Mountains National Park," *Bulletin of the Torrey Botany Club* 101(4) (1974): 198–206.

54. L. B. Gary, "Big Hogs, Big Problems: Park Control of Wild Hogs Protects Fragile Ecosystems," *Sightline* 2(1) (2001): 14–16.

55. Schlarbaum, Anderson, and Tompson-Campell, "Control of Pests and Pathogens."

56. C. Johnson, biologist, Great Smoky Mountains National Park, personal communication.

57. R. Weiss, "West Nile Widening Toll Impact on North American Wildlife Far Worse Than on Humans," *Washington Post,* Dec. 28, 2002, accessible at http://www.washingtonpost.com/ac2/wp-dyn?pagename=article&node= &contentId=A45800-2002Dec27&notFound=true (28 Jan. 2003).

58. U.S. Forest Service (USFS), *Southern Forest Research Station News Bulletin,* Asheville, N.C., 22 Nov. 2002.

59. J. G. O'Brien, M. E. Mielke, S. Oak and B. Moltzan, "Pest Alert—Sudden Oak Death, Eastern United States. Oak Mortality Is Caused by a New Pathogen, Phytophthora Ramorum," NA-PR-02-02, USDA, Forest Service, State and Private Forestry, Northeastern Area, Jan. 2002.

60. American Chestnut Foundation, accessible at http://www.acf.org/ (15 Jan. 2003).

61. Southern Appalachian Mountain Initiative (SAMI), "Final Report," National Parks Conservation Association (NPCA), Asheville, N.C., 2002. Accessible at http://www.npca.org/flash.html (18 June 2003).

62. Tennessee Department of Environment and Conservation (TDEC), "TDEC Hot List for Fall, 2002."

63. SAMI, "Final Report."

64. North Carolina General Assembly, Senate Bill 1078, Section 10.

65. See TVA's homepage at http://www.tva.gov/ (18 June 2003).

66. J. Nolt and J. D. Peine, "The Evolution of Land Use Ethics and Resources Management: Coming Full Circle," in *Ecosystem Management for Sustainability,* ed. Peine, 41–59.

67. "Mountain Top Restoration: East Kentucky Will Benefit if Federal Law Is Enforced," *Lexington Herald-Leader,* 26 Sept. 1999.

68. C. Bolgisno, *Living in the Appalachian Forest* (Mechanicsburg, Pa.: Stackpole, 2002).

69. D. Wear and J. G. Greis, "The Southern Forest Assessment: Summary Report" in *Southern Forest Resource Assessment,* ed. Wear and Greis. 1-104.

70. USFS, "U.S. Forrest Resources," accessible at http://biology.usgs.gov/s+t/noframe/m1103.htm (1 Dec. 2002).

71. Ibid.

72. H. Hinote, "Framework for Integrated Ecosystem Management: The Southern Appalachian Man and the Biosphere Cooperative," in *Ecosystem Management for Sustainability,* ed. Peine. 81-98.

73. T. Watson, "Mountaintop Mining Halted," *USA Today* 8 May 2002.

74. Missouri Governor Committee on Chip Mills (MGCCM), "Governor's Advisory Committee on Chip Mills: Meeting Minutes, Department of Natural Resources," accessible at http://216.239.51.104/search?q=cache:tc6712ck4qwC:www.dnr.state.mo.us/ chipmills/min9905.htm+Solutions+to+chip+mills&hl=en&ie=UTF-8 (12 May 2003).

75. See Balsam Mountain Preserve at http:www.balsammountainpreserve.com (23 June 2003).

76. See Nine Counties. One Vision http://ninecountiesonevision.org/ (18 June 2003).

77. C. Lucash, B. Crawford, and J. Clark, "Species Repatriation: Red Wolf," in *Ecosystem Management for Sustainability,* ed. Peine, 225–46.

78. See American Chestnut Foundation at http://www.acf.org/ (15 Jan. 2003).

79. N. Ross, personal communication on Champion Paper Mill and its effect on the Pigeon River, 2003.

80. J. Burr, Tennessee Wildlife Resources Agency, personal communication.

81. TDEC, "TDEC Hot List for Fall, 2002."

82. G. Armas, "Knoxville Rated as One of the Most Sprawling Cities," *Maryville-Alcoa Daily Times,* accessible at http://www.korrnet.org/cappe/ DailyTimes2002-10-18.html (31 Jan. 2003).

83. U.S. Geological Survey (USGS), "Status and Trends of the Nation's Biological Resources," accessible at http://biology.usgs.gov/s+t/SNT/index.htm (30 Apr. 2003).

84. S. Ahn, J. DeSteiguer, R. Palmquist and T. Holmes, "Economic Analysis of the Potential Impact of Climate Change on Recreational Trout Fishing in the Southern Appalachian Mountains: An Application of a Nested Mulinomial Logit Model," *Climate Change* 45 (2002): 493–509.

85. J. Talley, "Demographic Shifts and Green Values in Rural America: A Southern Appalachian Case," Ph.D. diss., Univ. of Tennessee, Knoxville, 2001.

86. J. Peine, "Moving to an Operational Level: A Call for Leadership from the Southern Appalachian Man and Biosphere Cooperative," in *Ecosystem Management for Sustainability,* ed. Peine, 475–82.

11. Models of Sustainability

1. This section was written by Thomas Fraser.

2. U.S. Dept. of Agriculture press release, 29 Jan. 1938.

3. Tennessee Wildlife Resources Agency (TWRA), *Big Game Harvest Report,* 2001.

4. N. Gray, Great Smoky Mountains National Park public information officer, personal communication.

5. B. Minser, Dept. of Forestry, Wildlife, and Fisheries, Univ. of Tennessee, personal communication.

6. *Hellbender Press* 4 (4).

7. *[Maryville] Daily Times,* 11 Nov. 2002.

8. R. Brown, executive director of the Foothills Land Conservancy, personal communication.

9. Ibid.

10. B. Minser, personal communication.

11. K. Tarkenton, Tennessee Wildlife Resources Agency financial director, personal communication.
12. B. Minser, personal communication.
13. This and the remaining sections, except for the last, were written by Jonathan Scherch.
14. Intermediate Technology Development Group, "Frequently Asked Questions: What is 'appropriate' or 'intermediate' technology?" accessible at http://www.itdg.org/html/about_us/faq.htm (19 Dec. 2002).
15. D. Scanlin, "Appropriate Technology," accessible at http://www.acs.appstate.edu/dept/tech/AT/AT.htm (19 Dec. 2002).
16. Ibid.
17. Gaia Trust, "What is an ecovillage?" formerly at http://www.gaia.org/evis/whatisecovillage.html (18 Dec. 2002).
18. D. Orr, "Transformation or Irrelevance: The Challenge of Academic Planning for Environmental Education in the 21st Century," in *Academic Planning in College and University Environmental Programs: Proceedings of the 1998 Sanibel Symposium,* ed. P. B. Corcoran, J. L. Elder, and R. Tchen (Rock Spring, Ga., North American Association for Environmental Education, 1998), 34.
19. Ibid., 17.
20. See http://www.berea.edu/PR/Release/061102.html (18 Dec. 2002).
21. Narrow Ridge Earth Literacy Center, "Mission and Vision," accessible at http://www.narrowridge.org/mission.html (18 Dec. 2002).
22. Ibid.
23. United States Catholic Conference, *Renewing the Earth: A Resource for Parishes* (Washington, D.C.: United States Catholic Conference, Inc., 1994).
24. See Sequatchie Valley Institute, http://www.svionline.org/ (18 Dec. 2002).
25. Ocean Arks International, "Restorer Ecological Water Treatment Technology" (brochure), accessible at http://www.oceanarks.org/restorer/pdf/ocean_arks_restorer_brochure_page4.pdf (18 Dec. 2002).
26. Ibid.
27. *Little Sequatichie Cove Farm Newsletter,* (2002), 1.
28. This section was written by Athena Lee Bradley.
29. See Mountain Microenterprise Fund http://www.mtnmicro.org (21 July 2003).
30. See Mountainmade.com Home Page at http://www.mountainmade.com (21 July 2003).
31. See Appalachian Sustainable Development http://www.appsusdev.org (21 July 2003).
32. Ibid.
33. Ibid.
34. Appalmade, "Handcrafted in Appalachia," accessible at http://www.appalmade.bigstep.com (21 July 2003).
35. See Appalachian Spring Cooperative http://www.apspringcoop.com (21 July 2003).
36. This list of resources was compiled by Jonathan Scherch.

Index

A Land Imperiled was designed and typeset on a Macintosh computer system using QuarkXPress software. The body text is set in 11/14 Berkeley Book and display type is set in Cholla Unicase. This book was designed and typeset by Barbara Karwhite and manufactured by Thomson-Shore, Inc.